Bio-IT and AI

Series Editors

As the fields of Information Technology (IT) and Artificial Intelligence (AI) rapidly evolve, their potential to revolutionize biomedice, personalized healthcare, and drug discovery becomes increasingly evident. This series will provide comprehensive insights into how modern computational and IT methods, AI, generative AI, and machine learning (ML) techniques are being integrated into biomedicine and healthcare to advance research, diagnostics, and treatments.

The scope of the series includes, but is not limited to, the following:

- Fundamentals of Bio-IT and AI: Basic concepts, tools, theoretical and mathematical principles, methods, technical foundations, and software technologies and solutions;
- Latest developments in bioinformatics and systems biology research, including topics such as multi-omics data integration, applications of large language models, and network biology;
- AI in genomics and precision medicine: Exploring the use of AI and generative AI in analyzing genomic data, understanding genetic diseases, and the development of personalized treatment strategies;
- Machine learning, AI, and generative AI for drug discovery, focusing on how ML algorithms accelerate the process of drug discovery, from target identification to predicting drug responses;
- Biomedical and health informatics and data analytics, covering the use of big data and analytics, including patient data management, electronic health records (EHR) analysis, and predictive modeling for patient care.

Aim

The series is targeted at a broad audience that is keen on exploring both the fundamental theories of IT, AI and their practical uses in healthcare and biotechnology. It caters to various groups including academic researchers and students in the fields of biomedicine and computer science, specialists in IT and data science, AI and generative AI, HPC and clouds, healthcare professionals, and industry experts.

Azadeh Zamanifar · Amir Taherkordi ·
Amirfarhad Farhadi
Editors

Explainable Large Language Models in Healthcare Applications

Springer

Editors
Azadeh Zamanifar
Science and Research Branch, Department of Computer Engineering
Islamic Azad University
Tehran, Iran

Amir Taherkordi
University of Oslo
Oslo, Norway

Amirfarhad Farhadi
Iran University of Science and Technology
Tehran, Iran

ISSN 3059-2895 ISSN 3059-2909 (electronic)
Bio-IT and AI
ISBN 978-3-032-15087-5 ISBN 978-3-032-15088-2 (eBook)
https://doi.org/10.1007/978-3-032-15088-2

This Springer imprint is published by the registered company Springer Nature Switzerland AG
The registered company address is: Gewerbestrasse 11, 6330 Cham, Switzerland

Preface

The rapid evolution of artificial intelligence has led to big changes to healthcare, particularly through the progress of large language models (LLMs). However, the "black-box" nature of these models has significant challenges in clinical settings, where transparency and trust are important. Explainable AI (XAI) has emerged as an important field. It aims to make AI-driven decisions interpretable, trustworthy, and actionable for healthcare professionals and patients. The integration of XAI with LLMs has the potential to cover the gap between powerful computational models and the practical needs of clinicians. It enables more accurate diagnoses and treatment recommendations with clear, understandable reasoning.

In healthcare, explainable AI systems can analyze large and complex datasets. It ranges from electronic health records to genetic information. Moreover, it provides transparent understanding into their decision-making processes. This transparency is important for creating confidence between clinicians, patients, and regulators and for ensuring the ethical deployment of AI in medical contexts. LLMs, with advanced natural language processing capabilities, can generate human-readable explanations, and it makes complex model outputs more accessible and actionable. Implementing explainable LLMs in healthcare is not without challenges. Key issues are mitigating bias, preventing hallucinations, ensuring regulatory compliance, and managing computational complexity. Techniques such as post hoc explanation methods (e.g., SHAP and LIME), hybrid human-AI frameworks, and transparent architectures are used to address these concerns.

Explainable AI improves the quality of healthcare delivery and also empowers patients to take an active role in their care by making medical information more understandable. As XAI technologies advance, they will be integrated with wearable devices and remote monitoring systems. They personalize and improve patient outcomes. The goal is to improve a healthcare ecosystem where AI-driven recommendations are accurate, transparent, fair, and compatible with human values. This book aims to provide a comprehensive overview of the current direction, methodologies, challenges, and future directions in this field. By looking in to recent research and real-world applications, it helps clinicians, researchers, and developers in using the power of explainable LLMs to transform healthcare for the better life. In each chapter, we will include real case studies or comparative studies from scholars where applicable.

Chapter "Foundations of LLMs in Healthcare" explores how Large language models (LLMs) are rapidly transforming healthcare systems by enabling the analysis and synthesis of vast amounts of medical data. They include patient records, treatment plans, and clinical notes. These models support clinicians in making more accurate diagnoses and developing personalized treatment plans by integrating information from diverse sources such as electronic health records (EHRs), medical literature, and imaging data. LLMs are also used to automate and improve clinical documentation, reducing administrative burdens and allowing healthcare professionals to spend more time with patients. In practice, LLM-powered chatbots and virtual assistants help treat patients, answer routine medical questions, and facilitate communication between patients and providers. Fine-tuning LLMs on domain-specific medical data further improves their ability to interpret complex clinical information and support tasks like early disease detection and risk prediction.

Chapter "Role of Explainable AI" covers roles of Explainable AI (XAI) in healthcare systems. It also covers real case studies in applying explainable AI in healthcare systems. These are technologies that are transforming healthcare by making advanced data-driven understanding more accessible and trustworthy. XAI focuses on ensuring that AI systems provide clear, interpretable explanations for their predictions, which is important for building trust between clinicians and patients in medical environments. The core principles of XAI is transparency, interpretability, and accountability, that are essential for the safe adoption of AI in clinical practice.

Chapter "Explainability and Reliability of Large Language Models in Health Systems" covers the challenges of black-box LLMs in clinical environments The difficulty of understanding large language models (LLMs) in clinical environments creates significant challenges, particularly due to their lack of transparency in decision-making processes, that decreases clinicians' ability to validate recommendations. Explainability increases trust as healthcare professionals cannot recognize how inputs like patient data diagnose or treat outputs.

Chapter "Core Techniques for Explaining LLMs in Healthcare" covers core techniques for explaining large language models (LLMs) in healthcare. It focuses on making complex AI decisions transparent and interpretable to clinicians and patients. One fundamental method is SHapley Additive exPlanations (SHAP), that quantifies the contribution of each input feature to a model's prediction. It helps clinicians understand the reasoning behind diagnostic suggestions. Another widely used approach is Local Interpretable Model-agnostic Explanations (LIME), which creates simpler, interpretable models around individual predictions to explain LLM outputs. Today, patient engagement has become one of the key concepts in improving the quality of healthcare services.

Chapter "Designing Trustworthy and Explainable Clinical Decision Support Systems" covers designing trustworthy and explainable clinical decision support systems (CDSS). It is essential for improving healthcare quality and clinician adoption. Trustworthiness in CDSS means providing the right information, to the right person, in the right format, by the right channel, and at the right time. It ensures that recommendations are relevant and suitable within clinical workflows. Explainability improves trust by making the system's reasoning transparent.

Transfer learning is an important technique for enhancing the performance of large language models (LLMs) in clinical settings, particularly for explainable AI applications. It includes using pre-trained LLMs on extensive general datasets and then fine-tuning them on smaller, domain-specific clinical data to capture the nuanced medical knowledge necessary for healthcare tasks. Chapter "Transfer Learning for Explainable AI in Clinical LLMs", covers this.

Chapter "Explainable NLP in Healthcare: Enhancing Clinical Documentation and Information Extraction" covers explainable natural language processing (NLP) in healthcare systems. It plays an important role in improving clinical documentation and extracting valuable information from unstructured medical data. Healthcare generates large amounts of unstructured text, such as physician notes, lab reports, and discharge summaries, which are difficult to analyze without advanced tools.

Chapter "Case Studies of Explainable LLMs in Diagnosis, Treatment Planning, and Patient Interaction" covers case studies of explainable LLMS in diagnosis, treatment planning and patient interaction. It also contains synthetic framework and comparative analysis in real case studies too. Explainable large language models (LLMs) are increasingly applied in healthcare to enhance transparency in diagnosis, treatment planning, and patient interactions. These models combine advanced reasoning with interpretable outputs, and they enable clinicians to trust and validate AI-generated recommendations.

Reinforcement Learning (RL) improves Large Language Models (LLMs) by training them to generate human-understandable explanations for decisions, and it bridges transparency gaps in complex AI systems. RL frameworks like Policy Gradient and Actor-Critic algorithms optimize LLMs to produce step-by-step reasoning traces and enable models to justify outputs through natural language descriptions. Chapter "Reinforcement Learning in Healthcare: From Treatment Optimization to the Challenge of Explainability with Large Language Models" covers this.

Chapter "Evaluating Explainability: Metrics, Benchmarks, and Human-Centered Evaluation Methods" covers evaluating explainability in AI particularly in healthcare. We also cover comparative analysis and benchmarking. It involves assessing how well AI-generated explanations help users understand model decisions. Key evaluation metrics focus on performance, interpretability, and the quality of explanations. It ensures they are accurate, useful, and actionable.

Legal and regulatory considerations in explainable large language models (LLMs) within healthcare systems are fundamental to ensuring patient safety, data privacy, and accountability. As healthcare AI tools affect diagnostic and therapeutic decisions, regulations require transparent and interpretable AI models so clinicians and patients can understand and trust the underlying reasoning behind these tools. Chapter "Legal and Regulatory Considerations" covers this.

Chapter "Future Directions—Human-AI Collaboration, Adaptive Explanations, and Regulatory Readiness" covers the future of explainable large language models (LLMs) in healthcare. It focuses on enhanced human-AI collaboration, adaptive explanations, and regulatory readiness. Human-AI collaboration will evolve into a seamless partnership that clinicians and AI agents work interactively.

We would like to thank the contributors of this book for their timely and positive response to the received review comments. Moreover, we acknowledge individuals who have helped us in this project. Last but not least, we thank Springer Publishing Team for their continuous supporting since the beginning stage to the production of this edited volume. We sincerely hope that the chapters of this book can help future research studies not only in academics or engineering domain but also in the healthcare context.

Tehran, Iran — Azadeh Zamanifar
Oslo, Norway — Amir Taherkordi
Tehran, Iran — Amirfarhad Farhadi

Contents

Foundations of LLMs in Healthcare

Alireza Taheri, Azadeh Zamanifar, and Amirfarhad Farhadi

Abbreviation

ADMET	Absorption, Distribution, Metabolism, Excretion, and Toxicity
AI	Artificial Intelligence
ASR	Automatic Speech Recognition
BERT	Bidirectional Encoder Representations from Transformers
CAD	Computer-Aided Diagnosis
CDS	Clinical Decision Support
CoT	Chain of Thought
CPG	Clinical Practice Guideline
DPO	Direct Preference Optimization
DRP	Drug-Related Problem
EHR	Electronic Health Record
EMR	Electronic Medical Record
ICL	In-context learning
ICT	Integrated Care Team
LLM	Large Language Model
LoRA	Low-Rank Adaptation
MDT	Multi-disciplinary Team
MLLM	Multimodal Large Language Model
MQA	Medical Question-Answering

A. Taheri · A. Zamanifar (✉)
Department of Computer Engineering, SR.C Islamic Azad University, Tehran, Iran
e-mail: azamanifar@iau.ac.ir

A. Taheri
e-mail: alirezataheri@srbiau.ac.ir

A. Farhadi
School of Computer Engineering, Iran University of Science and Technology, Tehran, Iran
e-mail: am_farhadi@mail.iust.ac.ir

A. Zamanifar et al. (eds.), *Explainable Large Language Models in Healthcare Applications*, Bio-IT and AI, https://doi.org/10.1007/978-3-032-15088-2_1

NLP	Natural Language Processing
PCP	Primary Care Clinician
PEFT	Parameter-Efficient Fine-Tuning
PPO	Proximal Policy Optimization
QLoRA	Quantized Low-Rank Adaptation
RAG	Retrieval Augmented Generation
RNN	Recurrent Neural Network
SFT	Supervised Fine-Tuning
SOAP	Subjective, Objective, Assessment, and Plan
STD	Sexually Transmitted Disease
USMLE	United States Medical Licensing Exam

1 Introduction

The emergence of large language models (LLMs) has marked a truly remarkable turning point in the field of artificial intelligence (AI), captivating widespread interest across numerous domains due to their extraordinary abilities in language comprehension and content generation. These sophisticated models, exemplified by prominent advancements such as OpenAI's GPT-3, GPT-4, and Google's Bard, have demonstrated an unparalleled capacity to understand and produce text that closely resembles human language [1–3]. Their fundamental prowess lies in their capability to swiftly and precisely process and analyze immense volumes of textual data, a feat that far exceeds human capabilities in terms of speed and efficiency [4–6]. A particularly notable aspect of their development is their proficiency as few-shot or zero-shot learners, enabling them to adapt and perform novel tasks with minimal or no prior specialized training after being exposed to vast datasets. This impressive technical aptitude signifies a profound shift in how machines interact with and process human language, laying the groundwork for revolutionary changes across various sectors [5].

The healthcare sector, with its inherent complexity, heterogeneity, and overwhelming scale of data, stands at the precipice of a paradigm shift brought about by the integration of LLMs. While the deployment of these powerful AI tools has been met with considerable enthusiasm for their potential to provide highly proficient responses to free-text queries and demonstrate a nuanced understanding of professional medical knowledge, it has also sparked apprehension [1]. LLMs are increasingly viewed as a formidable force multiplier, capable of assisting healthcare professionals grappling with information overload by automating the analysis of medical texts, extracting crucial insights, and applying that knowledge [7]. This transformative potential ignites hope for a new era characterized by enhanced efficiency, expedited diagnostic processes, and an elevated quality of patient care. The aim of this chapter is to delve into the foundational aspects of LLMs within the healthcare and medical domains, exploring their evolution, diverse applications, and the critical challenges that must be addressed for their responsible and effective implementation.

The trajectory of natural language processing (NLP), which underpins LLM development, has been characterized by a continuous series of significant advancements, with each innovation building upon the strengths and limitations of its predecessors. Early recurrent neural networks (RNNs) established the groundwork for retaining contextual information in NLP tasks, but their limitations in capturing long-range dependencies became apparent [8]. A pivotal moment arrived with the introduction of transformers, a groundbreaking architecture that effectively addressed the challenge of modeling distant word relationships. This innovation provided the impetus for the emergence of sophisticated language models like Llama 2 and GPT-4, which, bolstered by extensive training data, have elevated NLP to a level of understanding and text generation that closely approximates human-like language [1, 6]. The core aspects of LLMs encompass their substantial scale, typically involving hundreds of millions to billions of parameters, their pre-training on vast textual corpora, and subsequent fine-tuning for specific tasks. This multi-stage process enables LLMs to acquire comprehensive linguistic knowledge and adapt effectively to diverse NLP tasks [9].

Recognizing the unique challenges posed by medical text, which often features complex terminology, lexical ambiguity, and variable usage, tailored adaptations of general models have been introduced within the healthcare domain. Models such as BioBERT [10] and ClinicalBERT [11], built upon the BERT architecture, were developed using large biomedical and clinical corpora like PubMed abstracts and MIMIC-III data. These specialized biomedical language models have consistently demonstrated superior performance in domain-specific NLP tasks compared to their general-purpose counterparts [1]. GatorTron [12], notably, stands out as one of the largest clinical language models, trained on over 90 billion words from deidentified clinical notes, PubMed articles, and Wikipedia [2]. Furthermore, the evolution has led to multimodal large language models (MLLMs), representing a groundbreaking advancement that allows for processing and generating content across multiple modalities including text, images, audio, and video. This capability is critical for healthcare, enabling systems like Visual ChatGPT [13] or SkinGPT-4 to interpret and integrate information from various sensory inputs, bridging the gap between human communication and machine understanding in complex scenarios like medical imaging analysis [6].

The applications of LLMs in healthcare are rapidly expanding and are poised to profoundly impact various facets of medical practice and patient experience. They can revolutionize patient journeys from pre-consultation to diagnosis and management. In clinical decision support and diagnosis, LLMs contribute by analyzing patient symptoms, medical records, and pertinent clinical data to aid in identifying potential illnesses or conditions with a certain degree of accuracy [1, 2, 6, 8]. For instance, transformer models have been used to predict COVID-19 diagnosis, and LLMs have been explored for diagnosing Alzheimer's disease and dementia [7]. In oncology, LLMs like ChatGPT have been investigated for assisting breast tumor boards and guiding therapy planning in primary breast cancer cases, showing promising, albeit sometimes imprecise, alignment with expert decisions and guidelines. Their integration with computer-aided diagnosis (CAD) systems in radiology and imaging facilitates early identification of abnormalities and generates more precise diagnostic interpretations. Studies have shown that models like GPT-4 can effectively assess clinical acuity in emergency departments, performing comparably with human physicians in triage [5, 6]. Similarly, SkinGPT-4 has been introduced as an interactive dermatology diagnostic system, utilizing visual LLMs to interpret skin disease images and provide treatment recommendations, demonstrating high diagnostic accuracy [14]. LLMs are also showing utility in neurodegenerative disorders by predicting conditions like dementia based on speech data and aiding in diagnostic predictions [7].

Beyond diagnosis, LLMs are transformative tools for patient care and engagement. They offer the promise of revolutionizing healthcare delivery by providing personalized recommendations, customized treatment strategies, and continuous monitoring of patient progress. Virtual medical assistants and health chatbots, powered by LLMs, are revolutionizing healthcare by offering continuous and personalized health-related support, including medical advice, health condition monitoring, and mental health support. These tools can offer explanations in plain language, making complex medical terminology understandable for patients, as seen in efforts to simplify radiology reports or medication counseling. The multilingual capabilities of LLMs are also vital for enhancing communication in diverse populations, bridging language barriers for patients with varying linguistic needs [2, 6, 7].

In medical record analysis and administrative tasks, LLMs exhibit remarkable efficiency in reviewing and summarizing extensive volumes of medical literature, patient records, and EHRs [5–7]. This capability aids researchers and clinicians in staying current with cutting-edge developments and evidence-based methodologies, fostering informed practices [1]. LLMs streamline administrative procedures by automating documentation, medical coding, billing, and generating clinical referrals or discharge summaries, thereby reducing operational costs and allowing medical staff to focus more on direct patient care [15]. While these tasks are often labor-intensive, current research indicates a significant gap in systematically evaluating LLMs for these high-value, immediately impactful administrative applications, with a predominant focus on medical knowledge assessment [16].

For medical research and education, LLMs are poised to reshape the landscape. In drug discovery, they significantly impact by scrutinizing intricate molecular structures, discerning promising therapeutic compounds, and forecasting their efficacy and safety profiles [1]. Chemical language models have demonstrated notable achievements in de novo drug design. In medical education, LLMs have shown the ability to pass demanding examinations such as the USMLE and generate realistic clinical vignettes, serving as valuable self-learning tools and complementing existing student-directed learning methods [17]. They can also improve the efficiency of research article writing by automating tasks like literature review, text generation, and guiding manuscript style. Beyond these areas, LLMs hold nascent but promising roles in specialized medical fields such as nephrology, gastroenterology, and allergy and immunology, where their full potential is still being explored [6, 17].

Despite the immense potential, the integration of LLMs into healthcare presents complex and significant challenges that demand careful scrutiny and resolution to ensure their safe and effective use. A primary concern is model explainability and transparency. LLMs often function as black boxes, making it difficult to discern the underlying processes leading to their conclusions [18]. In healthcare, where decisions have profound repercussions, understanding the logic behind AI-generated outputs is imperative for practitioners to build trust and validate recommendations. Inadequate transparency can undermine trustworthiness and hinder the validation of clinical recommendations, emphasizing the continuous pursuit of interpretable frameworks [5, 7, 18].

2 What Are LLMs?

LLMs are essentially advanced computer programs built using deep neural networks. They learn by being pre-trained on massive amounts of text data from the internet. Their fundamental skill is to predict the next word or token in a sequence, given the words that came before it. This ability allows them to understand and generate human-like text, producing remarkably coherent and often clever responses to questions or instructions you give them [19].

To delve deeper, LLMs are rooted in transformer-based neural network architectures, initially trained through unsupervised multitask learning by performing next-word prediction on vast datasets. This initial phase allows them to develop instrumental knowledge. This means they gain knowledge through their ability to successfully execute tasks by using next-word generation as their primary tool.

While LLMs excel at formal linguistic competence, which is understanding the rules and statistical patterns of language, there's an ongoing discussion about whether they truly possess worldly knowledge. Worldly knowledge, as humans exhibit, relies on world models. These are internal, structure-preserving, and behaviorally efficacious representations of real-world entities and processes, such as objects, scenes, and agents. While LLMs are trained purely on text, some research suggests that their compression mechanisms during next-word prediction *could* allow them to approximate underlying data-generating processes, akin to forming aspects of these world models, especially in highly constrained, non-language domains. However, this ability is less consistently demonstrated and quantitatively weaker when dealing with the vast complexity of natural language and real-world scenarios [19].

A significant challenge LLMs face, especially with long inputs like lengthy documents or multi-turn conversations, is the management of their memory. This memory, crucial for the attention mechanism, grows linearly with input length, leading to increased consumption and slower decoding during inference. To address this, innovations like SnapKV [20] have been developed. SnapKV is a fine-tuning-free method that dramatically reduces the KV cache size without significantly impacting accuracy. It works by identifying and selecting only the most crucial attention features from the prompt, primarily from an observation window at the end, and combining them with important features from the prompt. This intelligent compression ensures that the LLM retains vital information for generation. This optimization allows LLMs to achieve constant decoding speeds regardless of input length and process exceptionally long contexts, demonstrating substantial improvements. While such methods significantly enhance the efficiency of LLMs in handling long contexts, they primarily optimize the generative aspect and KV cache, meaning they cannot inherently extend a model's long-context capability if the model fundamentally struggles with complex long contexts or lacks strong underlying performance. These advancements underscore the continuous effort to bridge the gap between LLMs' impressive linguistic abilities and their capacity to robustly process and understand information over extended contexts in a manner that approaches human-like knowledge, which can be necessary for the healthcare industry.

3 LLMs in Healthcare

LLMs represent a significant advancement in AI, adept at predicting and assembling word sequences to generate coherent text, answer complex inquiries, and execute intricate instructions. While their general capabilities allow them to summarize vast amounts of information, their application to the highly specialized and critical domain of healthcare necessitates a tailored approach. The field of healthcare demands precision and domain-specific understanding, making it essential to adapt these powerful but general-purpose models for nuanced medical contexts [9, 21]. Figure 1 illustrates the overall process of fine-tunning LLMs using medical data.

Fig. 1 Fine-tunning LLMs on medical data

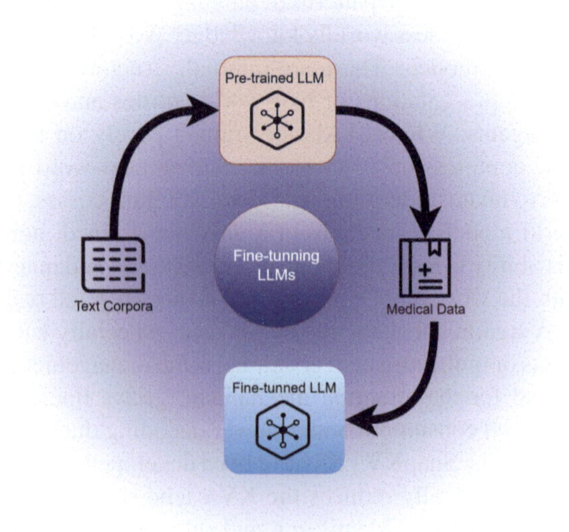

The primary method for equipping LLMs for healthcare applications is fine-tuning, a process where a pre-trained model undergoes further training on a custom, domain-specific dataset to adapt it for specialized tasks. This is crucial because, once initially trained, an LLM's knowledge is static and may not be sufficiently detailed or current for a specific field like medicine, often relying on general information that falls short of clinical needs. Fine-tuning effectively bridges the gap between the broad knowledge base of foundational LLMs and the specific expertise required in medical scenarios. It allows for the acquisition of domain-specific knowledge and optimization for specialized tasks such as health report summarization or disease detection from patient records. This adaptation leads to better performance in relevant tasks while preserving broader, out-of-domain knowledge, and it is significantly more resource-efficient than training an LLM from scratch [9, 22]. Furthermore, fine-tuning helps mitigate critical limitations of LLMs, such as the generation of hallucinations that could be harmful in a medical context, and it aids in addressing biases inherited from original training data, which could otherwise exacerbate health inequities. The sheer computational cost associated with globally fine-tuning models with billions of parameters also makes specialized, efficient techniques necessary [23].

The fine-tuning process typically follows a workflow that includes data preparation, model selection, the fine-tuning itself, and rigorous validation. Data preparation for medical LLMs involves meticulous cleaning, normalization, and formatting of task-specific datasets, which can combine extensive open medical question-answering data or leverage synthetic data generation to overcome data scarcity [24, 25]. Model selection involves choosing an appropriate pre-trained LLM, such as those from the Llama-2 or Mistral architectures, considering its compliance with the target task and available computational resources. Several methodologic approaches and techniques are utilized to fine-tune LLMs for healthcare:

One foundational approach is Supervised Fine-Tuning (SFT), where the model is trained on task-specific labeled data, learning to predict labels as accurately as possible. Within SFT, transfer learning initializes a model with weights from a large, general dataset and then trains it with limited, task-specific data. Multitask learning involves fine-tuning models on multiple related tasks simultaneously to leverage their similarities. Instruction-tuning specifically trains LLMs to follow explicit instructions for tasks like summarization, translation, or question answering, allowing them to generalize to new instances of those tasks. This can also extend to multi-task fine-tuning, where all task data is integrated into a unified optimization process to learn shared medical knowledge [9, 23].

Another significant method is Reinforcement Learning from Human Feedback (RLHF), which integrates human evaluators' knowledge to allow the model to adjust and develop in response to real-world input. Techniques within RLHF include reward modeling, where humans assign rankings or ratings to model outputs to guide behavior; proximal policy optimization (PPO), which modifies the model's strategy to maximize expected rewards; comparative ranking, where humans rank multiple outputs to provide relative feedback; and preference feedback, where human experts select preferred outputs, especially useful when numeric reward assignment is difficult [9, 26].

Given the massive scale of modern LLMs, Parameter-Efficient Fine-Tuning (PEFT) techniques are critical for reducing computational and storage requirements while maintaining performance comparable to full fine-tuning. Key PEFT methods include Low-Rank Adaptation (LoRA), which freezes pre-trained weights and injects small, trainable rank decomposition matrices into the transformer architecture, drastically reducing the number of trainable parameters [9, 21, 23]. A highly popular and memory-efficient variant is Quantized Low-Rank Adaptation (QLoRA), which uses various techniques to achieve significant memory savings. Other PEFT approaches include Prefix tuning and Prompt tuning, which involve adding trainable, continuous prompts or embeddings to the input sequence [27, 28] More advanced PEFT frameworks have also emerged, such as MOELoRA, which combines the Mixture-of-Experts (MOE) architecture with LoRA, using multiple low-rank experts and a task-motivated gate function to learn distinct parameters for various multi-task medical applications efficiently [23]. Studies indicate that LoRA-Tuning and Prefix-Tuning show strong efficacy in medical visual language models, achieving performance on par with global fine-tuning with reduced training costs [24].

Beyond traditional fine-tuning, several other strategies enhance LLMs for healthcare. In-context learning (ICL) allows a pre-trained LLM to perform new tasks by providing a few examples directly in the prompt, guiding its predictions without gradient-based training. Retrieval Augmented Generation (RAG) is a powerful technique that integrates LLMs with information retrieval systems. When a query is received, the model first fetches relevant documents from a large corpus and then uses this retrieved information, alongside its pre-trained knowledge, to produce accurate and contextually rich responses with citations. This is especially valuable for accessing up-to-date and source-specific data that may not have been part of the original model's training [9, 22].

For increased robustness and ethical performance, advanced techniques like Model Merging and a dedicated Alignment phase are employed. Model merging combines similar models to leverage their individual knowledge and produce a more robust outcome. The alignment phase, often using Direct Preference Optimization (DPO), adjusts models based on human preferences to mitigate risks and ensure safer interactions. Finally, sophisticated Prompt Engineering strategies further boost performance during inference. These include Chain of Thought (CoT) prompting, which encourages LLMs to generate intermediate reasoning steps to break down complex problems and achieve more accurate answers, and Medprompt, a strategy that combines CoT with in-context learning, K-nearest neighbor few-shots, and self-consistency methods (aggregating outputs from multiple model runs) to achieve state-of-the-art results [22].

These diverse fine-tuning techniques equip LLMs to address a wide array of specialized medical tasks, from aiding diagnosis and treatment planning to automating patient interactions, generating clinical documentation, and assisting in medical research and education. Examples include models like EchoGPT [29], an Llama-2 model fine-tuned with QLoRA for echocardiography report generation, and

CohortGPT [30], built on ChatGPT/GPT-4 for clinical trial patient recruitment using Chain of Thought prompting. However, constant verification and human review of model outputs remain crucial to ensure clinical accuracy and safety in real-world applications.

4 Applications and Use Cases

LLMs represent a transformative advancement in AI that has rapidly begun to reshape the landscape of numerous fields, with a particularly profound impact on healthcare. These sophisticated models, characterized by their immense number of parameters and training on colossal volumes of textual data, possess the remarkable ability to comprehend, process, and generate human-like text with exceptional accuracy and speed. In the healthcare domain, LLMs are proving to be a powerful force multiplier, aiding healthcare professionals, providers, and patients alike in navigating the ever-expanding complexities of medical knowledge and clinical practice. Their core value lies in their capacity to synthesize vast quantities of medical literature, patient records, and diverse clinical research, extracting crucial insights and automating tasks that traditionally consume significant time and resources.

The deployment of LLMs is ushering in a new era characterized by intelligent systems designed to bolster decision-making, expedite diagnostic processes, and elevate the quality of patient care. From streamlining administrative workflows to enhancing direct patient interactions, LLMs offer innovative solutions across a broad spectrum of healthcare applications [31]. They can facilitate tasks such as summarizing complex clinical notes, interpreting lab results, aiding in medical diagnosis, and generating personalized patient education content. Beyond direct clinical support, LLMs are also poised to revolutionize medical education and research, assisting with literature reviews, guiding manuscript writing, and even generating new research hypotheses [2, 32]. Figure 2 presents the applications of LLMs in healthcare.

4.1 Diagnosis and Decision Support

Medical decision support (CDS) is a critical function within healthcare, dedicated to enhancing clinical decision-making by providing timely and evidence-based guidance at the point of care. This field encompasses the use of various tools and systems to integrate targeted clinical knowledge with patient-specific information and other health data, ultimately aiming to optimize clinical workflows and improve patient outcomes. Traditionally, CDS systems, often rules-based, have encountered limitations such as alert fatigue, where an overload of irrelevant alerts leads to them being ignored by clinicians [33, 34]. The exponential growth of medical data, including electronic medical records (EMRs), physiological signals, and imaging data, has created a complex landscape where traditional methods of data management and

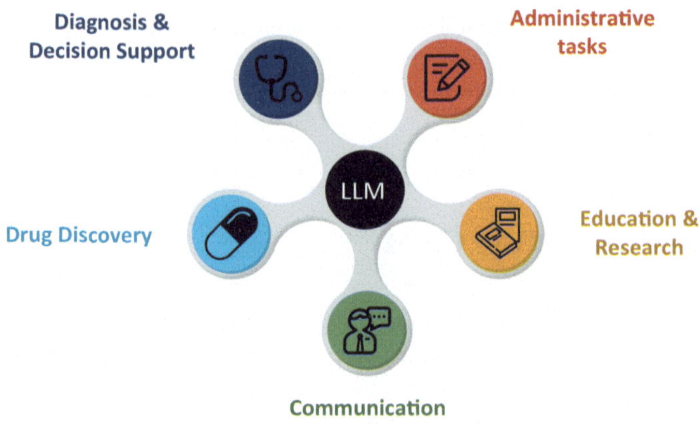

Fig. 2 Applications of LLMs in healthcare

retrieval are often insufficient, making it challenging for healthcare professionals to stay updated and extract actionable insights. LLMs are emerging as transformative tools in this domain, offering advanced capabilities to process, analyze, and synthesize vast amounts of diverse medical information, thus providing a promising solution to these challenges [35].

The primary goals of integrating LLMs into medical decision support are extensive and aim to revolutionize various facets of healthcare delivery. Fundamentally, LLMs seek to enhance the accuracy and efficiency of clinical decision-making by delivering reliable, evidence-based support promptly. A significant objective is to alleviate the considerable administrative workload on physicians and other healthcare professionals, streamlining diagnostic processes, optimizing documentation, and reducing the time spent on non-patient-facing tasks. LLMs are also geared towards providing personalized treatment recommendations, tailoring suggestions to individual patient conditions, medical histories, and specific needs, moving beyond generalized guidelines. Furthermore, they aim to improve information retrieval and synthesis, enabling healthcare providers to navigate complex unstructured data, extract critical insights, and integrate new medical research into clinical practice in real-time. By offering more personalized, intuitive, and less disruptive alerts, LLMs are designed to address the inherent limitations of conventional rules-based CDS, thereby mitigating alert fatigue and improving system adoption. Ultimately, these goals converge on supporting complex medical scenarios that require multi-modal data interpretation and collaborative reasoning, while consistently ensuring patient safety by identifying potential medication errors and adverse events before they occur [34–36]. Significant achievements have already been realized through the integration of LLMs in medical decision support, demonstrating their potential across various clinical applications.

One key area of achievement lies in diagnostic enhancement and multi-modal reasoning. LLMs have shown remarkable capabilities in streamlining diagnostic processes, optimizing radiology reports, and generating personalized treatment recommendations. They can transcribe physician observations and diagnostic findings through voice-to-text technology, automatically generating standardized diagnostic reports and supporting radiologists in high-pressure environments [35, 36]. LLMs also contribute to quality control in radiology by identifying redundant content and logical errors, assessing the completeness of reports in real-time, and enhancing their accuracy and consistency. Furthermore, when integrated with computer vision technology, LLMs can rapidly analyze imaging data to identify abnormal regions and provide preliminary diagnostic and treatment suggestions, which is crucial in emergency care settings. In the context of COVID-19 outpatient treatment, LLMs enhanced with Clinical Practice Guidelines (CPGs) have demonstrated improved performance in providing accurate recommendations [33]. LLMs are also being explored for their utility in specialized medical fields such as personalized oncology, cardiology, and orthopedics. Some LLMs, like Med-PaLM-2 and GPT-4, have achieved expert-level competence in Medical Question-Answering (MQA) benchmarks, with Med-PaLM-2 even surpassing the passing score on the MedQA dataset and its answers being preferred over those provided by physicians in various metrics including factuality and reasoning [33, 36].

Another significant achievement is in medication safety and management. LLM-based CDS utilizing RAG frameworks have been developed to support safe medication prescription, addressing the critical issue of patient safety. These systems aim to overcome the limitations of traditional rules-based CDS by providing prescribing error alerts that are relevant to the patient's context and institutional guidelines, thereby reducing alert fatigue. Studies have shown that when employed in a co-pilot mode the accuracy, recall, and F1 scores for identifying Drug-Related Problems (DRPs) are significantly optimized, particularly for moderate to severe DRPs. This co-pilot approach has been shown to double the relative accuracy of DRP identification compared to using LLM alone, signifying a promising hybrid solution that merges human expertise with AI's analytical capabilities to enhance patient safety in medication management. LLMs can also provide detailed medication information, including potential drug interactions and side effects, directly to patients or clinicians [34].

The development of adaptive and collaborative decision-making frameworks represents another remarkable achievement. The MDAgents framework [37], for instance, introduces a novel multi-agent system that emulates real-world hierarchical diagnosis procedures in clinical practice. This framework dynamically assigns LLMs to operate independently or collaboratively within groups based on the assessed complexity of a medical task, categorizing them as low, moderate, or high complexity. Low complexity cases are handled by a single Primary Care Clinician (PCP) agent, while moderate complexity cases involve a Multi-disciplinary Team (MDT) for collaborative discussion and consensus-building. For high complexity scenarios, an Integrated Care Team (ICT) is recruited, requiring extensive coordination and report generation from multiple specialists. MDAgents has demonstrated superior accuracy compared to previous solo and static group methods in 7 out of 10 medical benchmarks, highlighting the effectiveness of adaptive strategies. This framework also exhibits efficiency by intelligently calibrating the number of collaborating agents, leading to fewer computational API calls while maintaining high accuracy, which is crucial for scalable and cost-effective medical AI solutions. Furthermore, integrating external medical knowledge and incorporating moderator reviews into MDAgents significantly boosts accuracy, showcasing a synergistic effect between structured collaboration and domain-specific knowledge retrieval. The framework's robustness is evident in its ability to perform well even with varying temperatures, a parameter affecting creativity, and it shows a clear trend towards consensus among agents across different data modalities, demonstrating its capacity to integrate and reconcile diverse information.

4.2 Drug Discovery and Development

Drug discovery is a fundamentally complex and time-consuming initial phase within the broader drug research and development (R&D) pipeline, traditionally characterized by substantial resource investment and a high rate of failure. Historically, the journey to bring a single new small-molecule drug to market could span an average of 10–15 years and incur costs exceeding US $2 billion [38]. This demanding process necessitated a vast interdisciplinary understanding, drawing from basic science research, medicinal chemistry, intricate biological assays, and advanced pharmaceutical technologies [39]. The inherent challenges, including the difficulty in accurately predicting drug candidate efficacy and safety, the labor-intensive nature of synthesizing and testing numerous compounds, and the significant costs of equipment and skilled personnel, have historically presented formidable barriers to innovation [40].

However, the emergence of AI and, more specifically, LLMs, has marked a transformative shift in this landscape, introducing innovative methodologies designed to accelerate drug discovery, significantly reduce costs, and enhance success rates. With their expansive datasets, numerous parameters, and robust multitasking capabilities, LLMs are now revolutionizing various stages of the drug R&D pipeline. The overarching goal of integrating LLMs is to achieve a more highly automated, precise, and cost-efficient drug discovery process, ultimately enabling pharmaceutical companies to innovate more effectively and bring patient-centric therapies to market faster [41].

The achievements and applications of LLMs span across critical aspects of drug discovery. In drug target identification and validation, LLMs are adept at mining extensive biomedical literature, genomics data, and patent analyses to uncover crucial connections between targets, genes, and diseases [42]. They can analyze gene-related literature, genomics, and omics data to prioritize candidate genes and interpret protein structures, identifying structural domains, drug-binding pockets, and active sites of protein-based drug targets. Notable models like Geneformer have successfully identified therapeutic targets for conditions such as cardiomyopathy, and Quote-Target is used to identify potentially druggable protein targets using sequence-based transformer protein language models.

For molecular design and de novo drug design, LLMs facilitate the creation of novel drug-like molecules or the optimization of existing ones with desired properties [43]. They assist in structure-based drug design by generating SMILES strings from molecular graphs, and in de novo design, they create new drug candidates without a template, aiming for high affinity to specific targets or improved characteristics. Specific models such as MolGPT, cMolGPT, DrugLLM, and FSM-DDTR have been developed to accelerate the design of innovative drug molecules [39].

In the realm of ADMET (Absorption, Distribution, Metabolism, Excretion, and Toxicity) prediction, LLMs are vital for understanding these crucial properties, which are essential for filtering out compounds with undesirable characteristics early in. LLMs, including general-purpose ones like ChatGPT, can help comprehend and predict ADMET properties through question-answering models. Domain-specific models like PharmaBench have been developed for ADMET prediction, and LLM4SD leverages LLMs to extract meaningful hypotheses from ADMET data, often outperforming traditional methods [38].

LLMs are also transforming chemistry experiments and in-silico simulations. In chemistry robotics, LLMs can generate code to program robotic manipulators for complex chemistry and synthetic reactions, translating user requirements into experimental protocols and robot instructions. For retrosynthetic planning and reaction prediction, they recommend feasible synthetic routes and predict chemical reaction outcomes, which is crucial for preparing high-throughput screening experiments, with models such as Molecular Transformer and Chemformer contributing significantly. Furthermore, in protein–ligand interaction prediction, LLMs help forecast how drugs interact with protein targets, a fundamental step for rational drug design and virtual screening, with AlphaFold-Multimer and PSICHIC leveraging LLM capabilities for these predictions [40, 41].

In lead optimization, LLMs assist in enhancing a lead compound's efficacy, stability, and safety through chemical modifications or protein sequence adjustments. They achieve this by predicting how structural changes will affect properties, thereby guide chemists and minimize trial-and-error [38]. Models like DrugAssist [44], REINVENT [45], MERMAID [46], LigGPT [47], and ProteinDT [48] are specifically developed for molecular and protein optimization. For drug repurposing, LLMs efficiently identify new therapeutic uses for existing drugs, which can save significant time and costs compared to traditional discovery methods. They analyze existing literature and clinical data to pinpoint potential drug candidates, with examples including DrugReAlign [49] and DrugAgent [50]. Lastly, LLMs serve as powerful assistants for knowledge management, offering fast and accurate information retrieval from vast scientific and medical literature, compound databases, and patent documents. They simplify complex scientific and medical concepts for better communication and understanding and can generate code for various computational tasks throughout the drug discovery process. The synergistic integration of LLMs with other AI and machine learning techniques, such as deep generative models like VAEs, GANs, and transformers, is not only expediting traditional processes but also fostering innovative approaches to the complex challenges of drug discovery [39, 41].

Zheng et al. [38] made a deep investigation into how these advanced computational models could revolutionize the process, including uncovering target-disease linkages, interpreting complex biomedical data, enhancing drug molecule design, predicting drug efficacy and safety profiles, and optimizing clinical trial processes. They explored two main paradigms of LLMs, specialized and general-purpose, and detailed their applications across the three critical stages of drug development: understanding disease mechanisms, drug discovery, and clinical trials. This work achieved a comprehensive overview for researchers and practitioners, highlighting the expansive and revolutionary role LLMs can play in streamlining numerous aspects of drug development, from automating detailed literature reviews and identifying therapeutic targets to facilitating intricate clinical trial operations.

Chakraborty et al. [43] introduced a significant innovation by comprehensively highlighting AI-enabled ChatGPT and LLMs as a transformative force in drug discovery and development, signaling a new paradigm for pharmaceutical innovation. Their approach, serving as their own methodology, involved providing an in-depth editorial discussion that systematically explored how these advanced computational tools are being applied across the entire drug research and development pipeline. They detailed how LLMs contribute to initial insights for drug target discovery, such as understanding protein structures, structural domains, and drug-binding pockets. Furthermore, their discussion encompassed how LLMs accelerate the drug discovery process by aiding in the blueprinting of new compounds, predicting features like pharmacodynamics, pharmacokinetics, and toxicity, and assisting in early-stage tasks such as computing compound assortment and generating input files. They also elaborated on LLMs' utility in predicting and explaining drug-drug interactions and identifying potential candidates for drug repurposing. Qualitatively, this work achieved a profound understanding that LLMs are reducing the time and cost associated with early drug discovery, enhancing the comprehension of intricate molecular properties, and ultimately accelerating the development of new therapeutic molecules, which is poised to bring significant benefits to patients.

4.3 Administrative Tasks and Documentation

In healthcare, administrative tasks and documentation, particularly the creation of medical notes like discharge summaries and SOAP (Subjective, Objective, Assessment, and Plan) notes, are fundamental for patient care continuity, communication among providers, and effective management. However, these processes are traditionally time-consuming, prone to inconsistencies and errors, and represent a significant administrative burden that detracts from direct patient care and contributes to physician burnout.

To address these critical challenges, researchers have utilized LLMs as an innovative solution to automate and enhance medical documentation. One approach involves deploying specialized LLMs, such as Mistral-7B, that are meticulously refined through SFT with labeled, real-world patient data to accurately generate comprehensive discharge notes for cardiac patients [51]. This methodology also employs PEFT techniques, including QLoRA, to optimize computational resource usage while maintaining high performance. The MediNotes [52] framework further advances this by integrating LLMs with RAG and Automatic Speech Recognition (ASR) to process both text and real-time voice inputs from medical conversations, automatically transcribing them into structured SOAP notes and facilitating quick retrieval of relevant information. Similarly, the HEAL model [53], a continuously trained LLaMA2-based LLM, was purpose-built for medical conversations and automated scribing, utilizing diverse data including proprietary medical datasets and explanation tuning to interpret complex medical dialogues and generate physician-approved SOAP notes without human intervention.

These innovative approaches have led to substantial achievements, qualitatively transforming healthcare documentation. The generated notes have been recognized for significantly improving documentation efficiency and the continuity of care, receiving high marks from medical experts for clinical relevance, completeness, readability, and utility in health-related planning. The HEAL model has shown to surpass human scribes and other advanced models in identifying and summarizing critical medical information, demonstrating a superior ability to achieve high completeness and correctness in medical note generation while operating almost instantly, thereby vastly improving efficiency. The MediNotes framework has also proven highly effective, with 75% of its generated notes deemed clinically usable without manual corrections and contributing to a substantial reduction in administrative burden on healthcare providers, enhancing operational efficiency and physician well-being. These advancements underscore the profound potential of LLMs to streamline operations, overcome traditional bottlenecks, and drive significant progress in medical documentation.

Jung et al. [51] introduced a significant innovation by directly addressing the inefficiencies and inaccuracies inherent in the manual creation of medical discharge notes for cardiac patients through the application of AI. Their approach involved deploying specialized LLMs, specifically the Mistral-7B model, which they meticulously refined using SFT on a substantial dataset of actual patient records from a cardiology department. This methodology leveraged detailed patient progress notes as input to generate comprehensive discharge summaries. To optimize resource use, they also incorporated PEFT techniques like QLoRA. This groundbreaking work achieved a considerable improvement in documentation efficiency and continuity of care. Qualitatively, the discharge notes generated by Mistral-7B received high praise from medical experts for their clinical relevance, completeness, readability, and utility in supporting informed decision-making and care planning. The model demonstrated a high degree of accuracy in capturing essential patient information, including medical conditions, treatments, and follow-up recommendations, while maintaining consistent quality and comprehensibility, thereby streamlining healthcare documentation practices and reducing administrative burdens.

Goyal et al. [53] introduced a significant innovation by developing a cost-effective, custom, domain-specific LLM named HealAI, designed to surpass generic models like GPT-4 in medical note-writing tasks. Their approach involved an end-to-end development cycle for designing this specialized LLM. This began with instruction fine-tuning on an existing open-source LLM to enhance its ability to follow diverse medical documentation instructions, followed by SFT using actual medical note-creation data derived from GPT-4 prompting and human scribes. To address challenges like long patient-doctor interactions and improve accuracy, they designed a note-creation pipeline incorporating RAG. They also focused on refining the model to produce doctor-preferred medical writing styles and enable doctors to instruct the LLM to make edits, significantly saving time. Qualitatively, this work achieved superior performance in medical note-writing compared to larger, generic LLMs, with results demonstrating enhanced precision, reduced biases, and prevention of hallucinations, which are critical for accuracy in healthcare. The custom LLM also produced a more acceptable and consistent note-writing style, significantly streamlining the creation of SOAP notes from doctor-patient conversations and contributing to the alleviation of administrative burdens for healthcare professionals. Ultimately, their model offers a cost-effective and customizable solution with greater transparency and opportunities for continuous improvement, directly impacting clinician well-being and patient care.

Yuan et al. [54] introduced a significant innovation by developing HEAL, the first continuously trained 13B LLaMA2-based LLM purpose-built for medical conversations and automated scribing, specifically designed to generate comprehensive medical notes without requiring human intervention. Their approach involved systematically training this LLM on a diverse range of data, including non-medical public datasets to maintain general generative capabilities, medical public datasets to instill a deep understanding of medical concepts, and proprietary medical datasets comprising real-world doctor-patient conversations, EHRs, and SOAP notes. They further enhanced HEAL through explanation tuning, which allowed the model to better comprehend complex medical instructions and the reasoning behind medical documentation. Qualitatively, this work achieved remarkable success; HEAL surpassed human scribes and other comparable models in terms of correctness and completeness in medical note generation and matched the performance of high-end models like GPT-4. The model demonstrated a superior ability to identify and summarize critical medical information, leading to a substantial reduction in overlooked details compared to other systems. This innovation drastically improved efficiency and time-saving capabilities in medical transcription, generating physician-approved SOAP notes almost instantly and significantly alleviating the administrative burden on healthcare professionals.

Leong et al. [52] proposed a significant innovation with MediNotes, a generative AI framework designed to alleviate the substantial administrative burden of medical documentation, particularly the creation of SOAP notes. Their approach integrates LLMs, RAG, ASR to process both text and voice inputs, either in real-time from ambient medical conversations or from recorded audio. This framework employs fine-tuned LLM models, such as LLaMA3-8B, alongside PEFT and QLoRA to ensure computational efficiency, and Instruction Tuning to guide the model in generating structured and coherent notes. Furthermore, MediNotes includes a query-based retrieval system, allowing rapid access to relevant medical information. Qualitatively, this work achieved substantial improvements, demonstrating that MediNotes significantly enhances the accuracy, efficiency, and usability of automated medical documentation, offering a robust solution that reduces the administrative burden on healthcare professionals and improves the overall quality of clinical workflows. The generated notes were frequently deemed clinically usable without the need for manual corrections and achieved satisfactory levels of completeness, leading to a positive reception from both physicians and patients and a strong belief that the system could enhance operational efficiency and physician well-being. The MediNotes model consistently outperformed other competitive models, including GPT4o, in generating medical reports that were accurate, semantically relevant, and possessed human-like quality and fluency.

Medical records management is a foundational aspect of healthcare, involving the comprehensive handling of patient information from creation to storage, retrieval, and eventual disposition. Traditionally, this has been a complex, time-consuming, and error-prone process, even with the advent of EMRs and EHRs. EHRs, serving as digital repositories of patients' medical history, encompass a vast array of information, from demographics and medications to clinical notes and lab results, but their sheer volume, complexity, and unstructured nature, particularly in clinical notes, pose significant challenges for interpretation and analysis. LLMs are emerging as transformative tools in this domain, offering advanced capabilities to overcome these hurdles and revolutionize how medical records are managed [55, 56].

The primary goals of integrating LLMs into medical records management are multifaceted: to significantly enhance the accuracy and efficiency of documentation, reduce the considerable administrative workload faced by healthcare professionals, and improve the accessibility and utility of patient information for better clinical reasoning and decision-making. By leveraging their ability to understand context, capture semantic relations, and generate human-like text, LLMs aim to streamline processes that were previously manual, resource-intensive, or prone to human error. This also includes addressing the inherent limitations of conventional NLP techniques, which often struggle with the intricate nuances, jargon, abbreviations, and context-specific terminologies prevalent in medical language.

Significant achievements have been demonstrated across several key areas. One prominent application is the automatic generation and summarization of clinical notes. Systems like AscleAI have been designed to automatically generate clinical notes directly from dialogues between patients and clinicians, dramatically reducing the cognitive load on healthcare providers who previously had to write notes while simultaneously examining patients. This capability not only increases efficiency and accuracy but also allows clinicians to concentrate more fully on their patients. The system also streamlines note summarization, providing concise overviews of lengthy dialogues or multiple historical clinical notes. Feedback from clinicians indicates that such summarization is useful for quickly understanding treatment history and confirming information explained to patients. Beyond internal clinician use, LLMs can transform complex discharge summaries into patient-friendly language, thereby improving patient comprehension and engagement with their healthcare journey.

Studies have shown that LLM-generated summaries can be perceived as equivalent or even superior to those created by medical experts in terms of completeness, correctness, and conciseness, pointing to their potential to reduce documentation burdens [15].

Another crucial area of achievement is efficient information retrieval and conversational querying of medical records. LLMs enable clinicians to quickly extract desired information from a multitude of past clinical notes through conversational interfaces, mitigating the difficulties associated with reviewing unorganized notes, unfamiliar abbreviations, or notes written by other clinicians with different formats and subjective interpretations [55]. This ability to query medical records and obtain satisfactory answers in an easy-to-understand language, even when dealing with notes from various departments, greatly improves the review process and supports diagnostic assistance.

Furthermore, LLMs are making significant strides in privacy protection and anonymization of sensitive patient data within medical records, a critical concern given the highly sensitive nature of health information. Researchers have developed novel, no-code privacy protection frameworks that leverage LLMs and natural language interaction for intelligent de-identification of medical data. This involves automatically identifying and tagging sensitive information, such as names, addresses, contact details, national ID numbers, hospital names, and even highly sensitive medical information like sexually transmitted disease (STD) or pregnancy-related data. The use of techniques like Few-shot Prompting for entities with fixed patterns and RAG for those requiring external knowledge bases (e.g., hospital directories or disease descriptions) has led to high precision (97%) and recall (95%) in extracting these sensitive entities. Beyond mere extraction, LLMs can also classify sensitive data into different risk tiers, allowing for dynamically adapted privacy strategies that balance data security with the utility needed for research or operational purposes. This advancement democratizes access to sophisticated anonymization tools for clinicians, reduces the reliance on IT professionals, and maintains the statistical analysis value of the original data [57].

4.4 Communication

LLMs are rapidly transforming the landscape of patient interaction and communication in healthcare by leveraging their ability to process extensive data and generate human-like responses. This domain encompasses all touchpoints where patients engage with medical information, healthcare providers, and administrative systems, aiming to democratize information access, foster better understanding, and streamline the overall healthcare experience. The core objective of integrating LLMs in this area is to enhance patient satisfaction, improve communication efficiency, and alleviate the significant administrative burden that often complicates patient care journeys [58].

The primary goal of LLMs in patient interaction and communication is to make complex medical knowledge more accessible and comprehensible to non-experts, thereby empowering individuals to make informed decisions about their health. This involves tackling specialized medical terminology, technical jargon, and syntactic complexity found in biomedical literature and patient records. By offering personalized and dynamic approaches to information delivery, LLMs aim to reduce educational disparities and shift the focus from passive information reception to active, critical engagement with health content. Furthermore, LLMs strive to optimize operational efficiency within healthcare systems, indirectly benefiting patients by reducing waiting times and improving the quality of care delivery [59].

Significant achievements have been realized in several key areas. In patient education and information dissemination, LLMs excel at simplifying complex medical texts and providing clear, understandable answers to health-related inquiries. Studies have shown that patients often perceive LLM-generated information as useful, complete, and user-friendly, with a majority even preferring LLMs over traditional search engines for answering medical questions. This capability to translate complex medical language into accessible formats, often incorporating real-time information, is crucial for improving health literacy and enabling patients to navigate their conditions better [58, 60]. For instance, a randomized study involving almost 5,000 participants demonstrated a statistically significant 3.9% absolute increase in comprehension when reading simplified texts compared to original versions, with the most notable gains observed in complex PubMed articles (14.6% improvement). Participants also reported a substantial increase in perceived ease and confidence when using simplified texts. This suggests that LLM-driven simplification can not only improve understanding but also reduce cognitive load and enhance user confidence when engaging with challenging medical content [60].

Beyond direct information provision, LLMs are improving direct patient-provider communication and teleconsultations. They offer informational support, diagnostic assistance, and aid in communication, enhancing efficiency and potentially leading to cost and time savings. LLMs can also support personalized patient care and provide multilingual support, which is particularly beneficial for diverse patient populations. While human urologists were still rated higher for the understandability of medical advice, LLMs were found to make a competent impression and were broadly understood. LLMs also contribute to streamlining administrative processes such as appointment scheduling, medical invoicing, and insurance claim processing [58]. By automating these tasks, LLMs reduce errors and free up healthcare professionals, allowing them to focus on more complex patient inquiries and direct care, ultimately alleviating burnout and improving the overall quality of clinical workflows. Researchers are developing robust interactive evaluation frameworks, such as MMD-Eval, to assess the proactive communication and diagnostic capabilities of medical LLMs in multi-turn consultation scenarios, moving beyond simple question-answering tasks to better simulate real-world patient interactions. These frameworks utilize structured medical records to ensure accuracy and avoid issues like hallucinations, demonstrating that LLMs can accurately identify doctor's intent and understand dialogue context [59].

4.5 Research and Education

Medical education, research, and training represent a crucial domain within healthcare where LLMs are poised to revolutionize traditional teaching methodologies and streamline various academic and clinical processes. This field encompasses the entire learning journey for aspiring physicians and healthcare professionals, from preclinical education and clinical clerkships to postgraduate training and research endeavors [61]. The primary goal of integrating LLMs here is to address significant challenges such as information overload, time constraints for educators, and the demand for personalized learning experiences, ultimately enhancing students' knowledge, skills, and overall competence. These models aim to make medical knowledge more accessible, reduce educational disparities, and shift the focus of learning from rote memorization to deeper understanding, critical thinking, problem-solving, and practical application of concepts in clinical settings [62].

In medical education, LLMs like ChatGPT and GPT-4 are deployed in diverse ways to achieve these objectives. They serve as powerful tools for curriculum development, assisting educators in needs assessments, identifying content gaps, and creating measurable learning objectives and customized curricula tailored to individual student needs. For teaching methodologies, LLMs augment traditional lectures by providing real-time clarifications and additional resources, facilitate small-group discussions by generating thought-provoking questions, and create realistic virtual patient scenarios for students to practice clinical reasoning, decision-making, and communication skills in a safe environment. They also act as virtual mentors, offering instant feedback and personalized guidance during clinical rotations. Achievements in this area include the generation of interactive medical case studies and personalized study plans, complete with concise summaries, flashcards, and practice questions, significantly optimizing study time and knowledge retention. Furthermore, LLMs play a vital role in assessment and evaluation, helping design comprehensive plans, align methods with learning objectives, and provide prompt feedback and rubrics to students [61, 63].

Beyond foundational learning, LLMs have demonstrated remarkable aptitudes in medical training and skill development. They have shown competence in standardized clinical exams, with models like ChatGPT achieving passing scores on all written steps of the USMLE and performing strongly in specialized board exams across various medical disciplines, including ophthalmology, otolaryngology, general surgery, and cardiology. This indicates their potential as a cost-effective alternative for exam preparation materials and as a valuable reference for medical students and educators [62]. In clinical training, LLMs contribute to simulation-based learning by generating patient encounters that allow students to practice history-taking and clinical reasoning, providing a greater range of training materials than traditional methods. For medical note-writing support, LLMs can transcribe patient interactions into clinical notes, which can serve as templates for educators to evaluate student history-taking and can extract relevant details from EHRs, streamlining documentation and potentially elevating the quality of care by reducing overlooked details.

Specialized domain-specific LLMs, known as BioLLMs such as BioBERT, ClinicalBERT, and ChatCAD, have been specifically fine-tuned on vast medical texts and EHRs to enhance natural language processing tasks, improve research accessibility, provide targeted question answering, and support computer-aided diagnosis outputs, thereby offering more precise clinical decision support and information management [61, 62].

In the realm of medical research, LLMs are invaluable for streamlining the entire research process, from initial strategy drafting to manuscript preparation. They can retrieve pertinent references, fetch abstracts, extract key study details, suggest suitable statistical tests, and even produce sample codes for software use. Furthermore, LLMs assist in academic writing, improving the flow, organization, style, formatting, and translation of manuscripts, which is particularly beneficial for non-native English speakers. Qualitatively, the outputs of LLMs in these areas, when properly guided, have been praised for their clinical relevance, completeness, readability, and utility in supporting informed decision-making and care planning. The continuous advancements in LLM technology, including the integration of plugins for updated information and image generation, further enhance their capabilities to bridge knowledge gaps and provide diverse learning modalities. These collective achievements signify that LLMs are not merely supplementary tools but are becoming integral to creating a more efficient, effective, and learner-centered medical education [61–63].

5 Challenges and Concerns

The integration of LLMs into healthcare settings, while promising, faces a complex array of concerns and challenges that must be meticulously addressed for their responsible and effective deployment. These issues span from fundamental operational pitfalls to intricate ethical and societal implications, necessitating a comprehensive understanding for anyone exploring the foundations of LLMs in this critical domain. Figure 3 presents the challenges and risks associated with utilizing LLMs in healthcare.

A fundamental and immediate concern revolves around the accuracy and reliability of information generated by LLMs, particularly given their general-purpose nature and training on vast public internet datasets, rather than exclusively verified medical sources. This inherent design means LLMs prioritize generating plausible-sounding content rather than ensuring absolute factual correctness. Consequently, they can produce inaccurate, incomplete, or misleading information, which, in a healthcare context, carries significant risks for patient well-being [31, 64].

A more specific manifestation of this inaccuracy is hallucination, where LLMs create entirely fictitious or false information that has no basis in reality or the provided input. This can manifest in several ways, such as the fabrication of non-existent references to substantiate false claims, or providing incorrect medical instructions. For instance, an LLM might suggest that a patient needs to fast for several hours before a bone scan, a procedure that typically does not require fasting, potentially causing

Fig. 3 Challenges and risks associated with using LLMs in healthcare

unnecessary inconvenience or even harm. The danger is amplified by the LLMs' ability to present such misinformation with persuasive and confident prose, making it difficult for users, including healthcare professionals, to discern its credibility, and potentially leading to inappropriate changes in treatment or patient adherence.

Beyond outright fabrication, LLMs also struggle with inconsistency, a lack of reproducibility in their responses. The same question might elicit different answers depending on minor variations in the prompts or even when asked by different users, primarily due to the probabilistic nature of LLM output generation. This non-deterministic behavior is a significant barrier in medical applications where consistent and predictable results are paramount for patient safety, reliable diagnosis, and sound clinical decision-making. Furthermore, many LLMs suffer from an outdated knowledge base, as their training data has a specific cutoff date. This means they may not possess the most current medical information, recent advancements in treatment protocols, or updated clinical guidelines, potentially leading to the dissemination of obsolete or non-standard-of-care advice [31, 32].

A significant challenge lies in the lack of interpretability, or the black box nature, of LLM algorithms. Unlike traditional rule-based systems, LLMs typically do not provide transparent reasoning or a clear audit trail for their generated responses. This opacity makes it exceedingly difficult for healthcare professionals to understand the rationale behind an LLM's suggestions or to validate the basis of its recommendations, which is crucial in critical clinical scenarios, where understanding the *why* is as important as the *what*. This fundamental lack of transparency can severely impede trust and acceptance among physicians and patients, especially when the potential for inaccurate medical advice could have detrimental consequences [2].

Data-related limitations form a substantial and intricate set of concerns, with bias being a prominent issue. LLMs are trained on enormous datasets, often sourced from the public internet, which can inherently contain and reflect existing societal biases related to factors such as ethnicity, gender, socioeconomic status, or geographical location. If not carefully managed, LLMs risk perpetuating and even amplifying these biases and disparities within healthcare, potentially leading to unequal treatment, misdiagnosis, or inappropriate recommendations for marginalized populations. For example, a model trained predominantly on data from fair-skinned individuals might be less effective at identifying subtle signs of skin cancer in darker-skinned individuals, resulting in delayed diagnosis and treatment for a specific demographic. Similarly, if the training data heavily features urban patient populations, the LLM might fail to accurately represent or address health conditions prevalent in rural communities. These embedded biases, often subtle, are challenging to detect and correct, requiring continuous monitoring and evaluation [64].

The privacy and security of sensitive patient data are paramount ethical and legal considerations when integrating LLMs into healthcare. There is a tangible risk that identifiable patient information could be inadvertently leaked directly from training examples embedded within the model's responses. Moreover, LLMs may be susceptible to inferential disclosure, where sensitive attributes about a patient could be inferred from the model's general responses, even if direct identifiers are not explicitly provided. Healthcare organizations are justifiably cautious about providing raw, sensitive patient data to widely accessible LLMs, such as those that rely on cloud services or public APIs, due to these privacy concerns and strict legal compliance requirements like HIPAA. Additionally, complex questions arise regarding the ethical management of patient consent and data ownership when LLMs are used to analyze or generate insights from sensitive health information, especially when data crosses state or national boundaries with differing privacy regulations [2, 31].

From an operational and human-centric perspective, there is a significant risk of over-reliance and dependency on LLMs by both healthcare professionals and patients. This dependency could diminish critical thinking, independent clinical judgment, and potentially lead to the deskilling of practitioners who might increasingly defer to algorithmic assessments rather than applying their own expertise. The intent is for LLMs to augment, not replace, human expertise, but this crucial distinction may become blurred with increasing integration, potentially undermining the crucial human element in patient care, which encompasses empathy and understanding [65].

Further adding to the complexity, LLM outputs often suffer from non-comprehensiveness, meaning they might provide incomplete, generic, or overly cautious advice. For instance, an LLM might suggest a partial therapy without all necessary components, rendering the entire treatment plan insufficient and potentially leading to suboptimal patient outcomes or incorrect diagnoses if not subjected to thorough expert human supervision. The quality of an LLM's output is also highly dependent on the quality and specificity of the user's input, making it prone to errors when faced with vague or complex information. Moreover, the environment in which an LLM is used significantly influences the appropriateness of its output. Models might recommend treatments or protocols that are inappropriate for specific patient populations (e.g., adult care protocols being suggested for pediatric patients), particular regions with different healthcare standards, or even current local standards of care, if not adequately localized or trained for specific contexts [32].

Broader ethical and systemic challenges include the potential for LLMs to exacerbate existing health disparities. Factors such as the financial costs associated with accessing more advanced, often subscription-based, LLM models like GPT-4, the varying readability levels of their generated responses, and general accessibility barriers (e.g., digital literacy or internet access) can create an unequal playing field. This inequity could potentially widen existing gaps in healthcare access and outcomes, benefiting those with resources while disadvantaging marginalized or underserved populations [31].

The lack of clear definitions for LLMs' roles and responsibilities in healthcare and research is another complex issue that needs urgent clarification. Fundamental questions persist regarding accountability for AI-generated content in clinical practice: who is liable if an LLM provides incorrect advice that harms a patient? How can healthcare systems effectively distinguish between AI-generated and human-authored text, particularly in documentation? And how should differing views between physicians and AI recommendations be reconciled in shared decision-making processes? LLMs are intended to serve complementary roles rather than acting as full substitutes for physicians, but the exact boundaries and implications of this remain to be fully articulated and regulated.

Finally, the increasing ease of deployment of some LLMs, especially open-source models that can be fine-tuned and run on resource-constrained devices, raises significant concerns about their potential misuse or malicious training. There is a risk that these models could be deliberately trained to disseminate deeply falsified medical information or detrimental content, posing serious public health threats. Such malicious outcomes could severely undermine public trust in AI technologies within healthcare, making the development and implementation of robust safeguards, stringent terms of use, and clear deployment specifications absolutely crucial to prevent such misuse. These extensive and interconnected challenges underscore the critical need for rigorous validation, the development of clear regulatory guidelines, and continuous, collaborative efforts among AI developers, healthcare professionals, patients, and policymakers to ensure LLMs are integrated safely, ethically, and effectively into clinical practice [2]. Table 2 provides a concise overview of the key challenges and risks associated with using LLMs in healthcare, detailing their specific implications.

6 On the Horizon

The future of LLMs in healthcare is anticipated to be profoundly transformative and ubiquitous, promising to revolutionize patient care, medical research, and operational efficiencies across the sector. This evolution will necessitate continuous advancements in LLM capabilities, strategic applications, and robust frameworks for responsible integration.

A key aspect of future development involves significantly enhancing the inherent capabilities of LLMs, moving beyond text-only processing to encompass MLLMs.

Table 2 Key Challenges and Risks of LLM Integration in Healthcare

Challenge	Description
Hallucination	LLMs create entirely fictitious or false information that has no basis in reality or the provided input, which can manifest as fabricated references or incorrect medical instructions
Accuracy and reliability	LLMs, prioritizing plausible content over factual correctness, can produce inaccurate, incomplete, or misleading information, posing significant risks to patient well-being
Inconsistency	LLMs struggle with a lack of reproducibility in their responses, where the same question might elicit different answers due to the probabilistic nature of output generation, hindering consistent and predictable medical results
Outdated knowledge	Many LLMs suffer from an outdated knowledge base with a specific data cutoff date, meaning they may not possess the most current medical information, treatment advancements, or clinical guidelines
Lack of interpretability	LLMs often function as black boxes, not providing transparent reasoning or clear audit trails for their responses, making it difficult for healthcare professionals to understand or validate their suggestions
Bias	LLMs trained on vast internet datasets can inherently contain and reflect existing societal biases, potentially perpetuating disparities in healthcare through unequal treatment or misdiagnosis for marginalized populations
Privacy	There is a tangible risk of inadvertent leakage of identifiable patient information directly from training examples, or inferential disclosure of sensitive attributes from general responses
Over-reliance	Both healthcare professionals and patients face a significant risk of over-reliance and dependency on LLMs, which could diminish critical thinking, independent clinical judgment, and potentially lead to the deskilling of practitioners

These advanced models are expected to integrate diverse data sources, including textual, image, and tabular data, potentially leading to performance levels beyond human capability in certain tasks. For instance, newer models like GPT-4o and Claude 3.5-sonnet already demonstrate multimodal processing, integrating online research and private knowledge bases to support complex decision-making. Future LLMs are envisioned to analyze not just images in radiology and pathology, but also sound and video formats, expanding their versatile diagnostic and analytical potential. Furthermore, there is a promising avenue in enhancing LLMs to interpret and generate biomolecular data, which could significantly advance genomics and personalized medicine by predicting individual treatment responses based on genetic profiles. The integration of adaptive learning capabilities in real-time is also on the horizon, allowing LLMs to act as dynamic aids during critical procedures like surgery or emergencies, analyzing live data from medical devices to offer instant decision support.

In terms of applications, LLMs are poised to redefine CDS and diagnosis. They are expected to become indispensable tools for physicians, enhancing diagnostic accuracy, personalizing treatment plans, and optimizing operational workflows. This includes improving predictions by considering patients' comprehensive medical backgrounds, such as comorbidities, risk factors, and medication lists. The integration of LLMs into EHR systems will enable more accurate diagnoses through sophisticated NLP techniques applied to clinical notes and patient histories. The future of CDS envisions a co-pilot model, merging human expertise with the analytical prowess of AI to deliver higher-quality healthcare and improve patient outcomes.

Patient care and engagement will also see significant advancements. LLMs will serve as personalized virtual assistants for patients, offering health guidance, support, and simplifying complex medical information into understandable language, including multilingual translations. This could extend to automated post-procedure counseling, medication counseling, and lifestyle modifications. For administrative tasks, LLMs will further streamline processes by automating documentation, coding, and billing, thereby reducing the workload on medical staff and allowing them to focus more on direct patient care. In areas like mental health, LLMs hold potential to address the critical need for psychiatric care services, supplementing shortages of professionals and enhancing patient compliance. The use of LLMs in telehealth and telemedicine is also expected to expand, providing accessible care to patients, particularly in rural areas. Automated note-taking and summarization will continue to improve, with LLM-generated summaries potentially outperforming human experts in completeness and conciseness, significantly reducing documentation burdens and improving continuity of care.

The realm of drug discovery and development is another area ripe for revolution, with LLMs accelerating processes from target identification to molecular design and clinical trial analysis. Future LLMs are expected to automate chemistry experiments, assist in interactive de novo molecule generation, design novel therapeutic modalities, and uncover new biochemical and pharmacological principles [43].

In medical education and research, LLMs hold immense potential to transform learning experiences and accelerate scientific discovery. They can offer innovative solutions for personalized learning, intelligent tutoring, and content generation. LLMs will be capable of producing clinical case studies, acting as virtual patients or tutors, developing course plans, and providing personalized feedback and assistance. For researchers, LLMs can improve the efficiency of article writing by automating tasks like literature reviews, generating text, and guiding manuscript style and formatting. They can also help match patients to clinical trial opportunities, thereby facilitating recruitment and access to breakthrough treatments. Academic institutions are encouraged to develop guidelines for ethical LLM use, including proper disclosure and citation, and provide training for both students and educators [63].

7 Conclusion

This chapter has illuminated the profound and rapidly evolving role of LLMs within AI and their burgeoning impact across the healthcare sector. From their fundamental capacity to comprehend and generate human-like text, LLMs have undergone a crucial evolution, necessitating specialized adaptations like BioLLMs to meet the unique demands of medical terminology, complex data, and clinical precision. We have explored how these advanced models are poised to instigate a paradigm shift in virtually every facet of medicine, promising enhanced efficiency, improved diagnostic processes, and an elevated quality of patient care.

We have seen the diverse and impactful applications of LLMs, ranging from revolutionizing clinical decision support and diagnosis by aiding in the identification of illnesses, optimizing radiology reports, and enhancing medication safety through co-pilot models, to transforming patient care and engagement by offering personalized recommendations, virtual assistance, and simplifying complex medical information. Beyond direct patient interaction, LLMs are proving indispensable in streamlining medical record analysis and administrative tasks, automating documentation, summarizing vast quantities of clinical data, and improving information retrieval efficiency, while also pioneering methods for privacy protection and data anonymization. Furthermore, their influence extends significantly into drug discovery and development, where they accelerate target identification, molecular design, ADMET prediction, and even automate chemistry experiments, as well as into medical research and education, by facilitating curriculum development, offering personalized learning experiences, aiding exam preparation, and enhancing academic writing. These advancements collectively underscore the immense potential for LLMs to become integral tools that augment human capabilities and drive innovation throughout healthcare.

However, this transformative potential is intrinsically linked to significant and complex challenges that demand unwavering attention and responsible mitigation strategies. Paramount among these are concerns regarding the accuracy and reliability of LLM-generated information, particularly the risk of hallucinations that can produce misleading or entirely false content in critical clinical scenarios. The black box nature of many LLMs also presents a lack of interpretability, making it difficult for healthcare professionals to understand the reasoning behind their suggestions and thus hindering trust and validation. Furthermore, LLMs carry inherent data biases from their training, which could perpetuate or even exacerbate existing health disparities, and raise critical questions about privacy and security of sensitive patient data. Issues of over-reliance, non-comprehensiveness, and the potential for malicious misuse further underscore the necessity for caution and robust safeguards.

In conclusion, while LLMs represent a truly remarkable leap forward in AI with the capacity to reshape healthcare in profound ways, their responsible and effective integration is not merely an option, but an imperative. Realizing their full promise will require continuous efforts in rigorous validation, transparent explainability, meticulous bias mitigation, and robust privacy protection. Ultimately, LLMs should be viewed as powerful augmentative tools for human expertise, not replacements, necessitating a collaborative future where interdisciplinary teams of AI researchers, healthcare professionals, ethicists, and policymakers work together to develop and implement these technologies safely, ethically, and equitably, always with patient well-being at the core.

References

1. Nazi, Z.A. and W. Peng. *Large language models in healthcare and medical domain: A review*. in *Informatics*. 2024. MDPI.
2. Yang, R., et al., *Large language models in health care: Development, applications, and challenges*. Health Care Science, 2023. **2**(4): p. 255-263.
3. Cascella, M., et al., *Evaluating the feasibility of ChatGPT in healthcare: an analysis of multiple clinical and research scenarios*. Journal of medical systems, 2023. **47**(1): p. 33.
4. Hamid, R. and S. Brohi, *A review of large language models in healthcare: Taxonomy, threats, vulnerabilities, and framework*. Big Data and Cognitive Computing, 2024. **8**(11): p. 161.
5. Bedi, S., et al., *A systematic review of testing and evaluation of healthcare applications of large language models (LLMs)*. medRxiv, 2024: p. 2024.04. 15.24305869.
6. Lin, C. and C.-F. Kuo, *Roles and potential of large language models in healthcare: a comprehensive review*. Biomedical Journal, 2025: p. 100868.
7. Mumtaz, U., A. Ahmed, and S. Mumtaz, *LLMs-Healthcare: Current applications and challenges of large language models in various medical specialties*. arXiv preprint arXiv:2311. 12882, 2023.
8. Nassiri, K. and M.A. Akhloufi, *Recent Advances in Large Language Models for Healthcare*. BioMedInformatics, 2024. **4**(2): p. 1097-1143.
9. Anisuzzaman, D., et al., *Fine-tuning large language models for specialized use cases*. Mayo Clinic Proceedings: Digital Health, 2025. **3**(1): p. 100184.
10. Lee, J., et al., *BioBERT: a pre-trained biomedical language representation model for biomedical text mining*. Bioinformatics, 2020. **36**(4): p. 1234-1240.
11. Huang, K., J. Altosaar, and R. Ranganath, *Clinicalbert: Modeling clinical notes and predicting hospital readmission*. arXiv preprint arXiv:1904.05342, 2019.
12. Yang, X., et al., *Gatortron: A large clinical language model to unlock patient information from unstructured electronic health records*. arXiv preprint arXiv:2203.03540, 2022.
13. Wu, C., et al., *Visual chatgpt: Talking, drawing and editing with visual foundation models*. arXiv preprint arXiv:2303.04671, 2023.
14. Zhou, J., et al., *Pre-trained multimodal large language model enhances dermatological diagnosis using SkinGPT-4*. Nature Communications, 2024. **15**(1): p. 5649.
15. Taheri, A., A. Farhadi, and A. Zamanifar, *Application of GenAI in Clinical Administration Support*, in *Application of Generative AI in Healthcare Systems*. 2025, Springer. p. 91-117.
16. CHCP. *Medical Administrative Assistant Career Guide*. 2025 [cited 12 May 2025]; Available from: https://www.chcp.edu/blog/what-does-a-medical-administrative-assistant-do/.
17. Sallam, M. *ChatGPT utility in healthcare education, research, and practice: systematic review on the promising perspectives and valid concerns*. in *Healthcare*. 2023. MDPI.
18. Ahmad, M.A., I. Yaramis, and T.D. Roy, *Creating trustworthy llms: Dealing with hallucinations in healthcare ai*. arXiv preprint arXiv:2311.01463, 2023.
19. Yildirim, I. and L. Paul, *From task structures to world models: what do LLMs know?* Trends in Cognitive Sciences, 2024. **28**(5): p. 404-415.
20. Li, Y., et al., *Snapkv: Llm knows what you are looking for before generation*. Advances in Neural Information Processing Systems, 2024. **37**: p. 22947-22970.

21. Christophe, C., et al., *Med42--evaluating fine-tuning strategies for medical LLMs: full-parameter vs. parameter-efficient approaches.* arXiv preprint arXiv:2404.14779, 2024.
22. Gururajan, A.K., et al., *Aloe: A family of fine-tuned open healthcare llms.* arXiv preprint arXiv: 2405.01886, 2024.
23. Liu, Q., et al. *When moe meets llms: Parameter efficient fine-tuning for multi-task medical applications.* in *Proceedings of the 47th International ACM SIGIR Conference on Research and Development in Information Retrieval.* 2024.
24. Chen, J., et al. *Can llms' tuning methods work in medical multimodal domain?* in *International Conference on Medical Image Computing and Computer-Assisted Intervention.* 2024. Springer.
25. Farhadi, A. and A. Taheri, *Application of GenAI in Synthetic Data Generation in the Healthcare System*, in *Application of Generative AI in Healthcare Systems.* 2025, Springer. p. 67-89.
26. Bill, D. and T. Eriksson, *Fine-tuning a llm using reinforcement learning from human feedback for a therapy chatbot application.* 2023.
27. Ansari, M.S., et al., *Lightweight Clinical Decision Support System using QLoRA-Fine-Tuned LLMs and Retrieval-Augmented Generation.* arXiv preprint arXiv:2505.03406, 2025.
28. Hoque, F., *Enhancing Large Language Model Performance in Healthcare: Fine-tuning LLaMA 2 with QLoRA on MIMIC-IV Data.* 2025, Itä-Suomen yliopisto.
29. Chieh-Ju Chao, M., et al., *EchoGPT: A Large Language Model for Echocardiography Report Summarization.* Preprint, medRxiv. https://www.medrxiv. org/content/https://doi.org/10.1101/ 2024.01, 2024. **18**: p. v1.
30. Guan, Z., et al., *Cohortgpt: An enhanced gpt for participant recruitment in clinical study.* arXiv preprint arXiv:2307.11346, 2023.
31. Wang, L., et al., *Applications and concerns of ChatGPT and other conversational large language models in health care: systematic review.* Journal of Medical Internet Research, 2024. **26**: p. e22769.
32. Busch, F., et al., *Current applications and challenges in large language models for patient care: a systematic review.* Communications Medicine, 2025. **5**(1): p. 26.
33. Oniani, D., et al. *Enhancing large language models for clinical decision support by incorporating clinical practice guidelines.* in *2024 IEEE 12th International Conference on Healthcare Informatics (ICHI).* 2024. IEEE.
34. Ong, J.C.L., et al., *Development and testing of a novel large language model-based clinical decision support systems for medication safety in 12 clinical specialties.* arXiv preprint arXiv: 2402.01741, 2024.
35. Li, J., et al., *Large language models-powered clinical decision support: enhancing or replacing human expertise?* 2025, Elsevier. p. 1–4.
36. Karttunen, P., et al., *Large Language models in healthcare decision support: A review.* Available at SSRN 4892593, 2023.
37. Kim, Y., et al., *Mdagents: An adaptive collaboration of llms for medical decision-making.* Advances in Neural Information Processing Systems, 2024. **37**: p. 79410-79452.
38. Zheng, Y., et al., *Large language models in drug discovery and development: From disease mechanisms to clinical trials.* arXiv preprint arXiv:2409.04481, 2024.
39. Chakraborty, C., et al., *Ai-enabled language models (LMs) to large language models (LLMs) and multimodal large language models (MLLMs) in drug discovery and development.* Journal of Advanced Research, 2025.
40. Guo, H., et al., *A Survey of Large Language Model for Drug Research and Development.* IEEE Access, 2025.
41. Guan, S. and G. Wang, *Drug discovery and development in the era of artificial intelligence: From machine learning to large language models.* Artificial Intelligence Chemistry, 2024. **2**(1): p. 100070.
42. Lu, J., et al., *Large Language Models and Their Applications in Drug Discovery and Development: A Primer.* Clinical and Translational Science, 2025. **18**(4): p. e70205.
43. Chakraborty, C., M. Bhattacharya, and S.-S. Lee, *Artificial intelligence enabled ChatGPT and large language models in drug target discovery, drug discovery, and development.* Molecular therapy Nucleic acids, 2023. **33**: p. 866-868.

44. Ye, G., et al., *Drugassist: A large language model for molecule optimization.* Briefings in Bioinformatics, 2025. **26**(1): p. bbae693.
45. Loeffler, H.H., et al., *Reinvent 4: modern AI–driven generative molecule design.* Journal of Cheminformatics, 2024. **16**(1): p. 20.
46. Safdar, E.A. and N.A. Safdar, *Mermaid Syndrome-A Systematic Review.* Journal of American Medical Science and Research. https://doi.org/10.51470/AMSR, 2024. **2**.
47. Bagal, V., et al., *Liggpt: Molecular generation using a transformer-decoder model.* ChemRxiv.
48. Liu, S., et al., *A text-guided protein design framework.* Nature Machine Intelligence, 2025: p. 1–12.
49. Wei, J., et al., *DrugReAlign: a multisource prompt framework for drug repurposing based on large language models.* BMC biology, 2024. **22**(1): p. 226.
50. Inoue, Y., et al., *Drugagent: Multi-agent large language model-based reasoning for drug-target interaction prediction.* ArXiv, 2025: p. arXiv: 2408.13378 v4.
51. Jung, H., et al., *Enhancing clinical efficiency through llm: Discharge note generation for cardiac patients.* arXiv preprint arXiv:2404.05144, 2024.
52. Leong, H.Y., Y. Gao, and S. Ji. *A gen ai framework for medical note generation.* in *2024 6th international conference on artificial intelligence and computer applications (ICAICA).* 2024. IEEE.
53. Goyal, S., et al. *Healai: A healthcare llm for effective medical documentation.* in *Proceedings of the 17th ACM International Conference on Web Search and Data Mining.* 2024.
54. Yuan, D., et al., *A continued pretrained llm approach for automatic medical note generation.* arXiv preprint arXiv:2403.09057, 2024.
55. Han, J., et al. *AscleAI: A LLM-based clinical note management system for enhancing clinician productivity.* in *Extended Abstracts of the CHI Conference on Human Factors in Computing Systems.* 2024.
56. Li, L., et al., *A scoping review of using large language models (LLMs) to investigate electronic health records (EHRs).* arXiv preprint arXiv:2405.03066, 2024.
57. Gong, M., et al., *Privacy Protection for Chinese Electronic Medical Records Using Large Language Models: Effectiveness Evaluation and Application of LLM Models in Medical Data Tasks.* medRxiv, 2025: p. 2025.07. 27.25332177.
58. Vishwanath, A.B., V.K. Srinivasalu, and N. Subramaniam, *Role of large language models in improving provider–patient experience and interaction efficiency: A scoping review.* Artificial Intelligence in Health, 2024. **2**(2): p. 1-10.
59. Liu, R., et al. *Interactive evaluation for medical llms via task-oriented dialogue system.* in *Proceedings of the 31st International Conference on Computational Linguistics.* 2025.
60. Carl, N., et al., *Evaluating interactions of patients with large language models for medical information.* BJU international, 2025. **135**(6): p. 1010-1017.
61. Benítez, T.M., et al., *Harnessing the potential of large language models in medical education: promise and pitfalls.* Journal of the American Medical Informatics Association, 2024. **31**(3): p. 776-783.
62. Lucas, H.C., J.S. Upperman, and J.R. Robinson, *A systematic review of large language models and their implications in medical education.* Medical education, 2024. **58**(11): p. 1276-1285.
63. Abd-Alrazaq, A., et al., *Large language models in medical education: opportunities, challenges, and future directions.* JMIR medical education, 2023. **9**(1): p. e48291.
64. Mirzaei, T., L. Amini, and P. Esmaeilzadeh, *Clinician voices on ethics of LLM integration in healthcare: a thematic analysis of ethical concerns and implications.* BMC Medical Informatics and Decision Making, 2024. **24**(1): p. 250.
65. Ullah, E., et al., *Challenges and barriers of using large language models (LLM) such as ChatGPT for diagnostic medicine with a focus on digital pathology–a recent scoping review.* Diagnostic pathology, 2024. **19**(1): p. 43.

Role of Explainable AI

Amirfarhad Farhadi, Atoosa Hosseinian, Azadeh Zamanifar, and Amir Taherkordi

Abbreviation

AI	Artificial Intelligence
XAI	Explainable Artificial Intelligence
ML	Machine Learning
DL	Deep Learning
DNN	Deep Neural Networks
CNN	Convolutional Neural Network
EHR	Electronic Health Record
SHAP	Shapley Additive Explanations
LIME	Local Interpretable Model-agnostic Explanations
FDA	Food and Drug Administration
GDPR	General Data Protection Regulation
HIPAA	Health Insurance Portability and Accountability Act
GMLP	Good Machine Learning Practice

The original version of the chapter has been revised. A correction to this chapter can be found at https://doi.org/10.1007/978-3-032-15088-2_13

A. Farhadi (✉) · A. Zamanifar
School of Computer Engineering, Iran University of Science and Technology, Tehran, Iran
e-mail: am_farhadi@mail.iust.ac.ir

A. Zamanifar
e-mail: azamanifar@iau.ac.ir

A. Hosseinian
Department of Computer Engineering, SRC, Islamic Azad University, Tehran, Iran
e-mail: a.hosseinian@iau.ir

A. Taherkordi
Department of Informatics, University of Oslo, Oslo, Norway
e-mail: amirhost@ifi.uio.no

A. Zamanifar et al. (eds.), *Explainable Large Language Models in Healthcare Applications*, Bio-IT and AI, https://doi.org/10.1007/978-3-032-15088-2_2

PCCPs	Predetermined Change Control Plans
MCCV	Monte Carlo Cross-Validation
DSC	Dice Similarity Coefficient
JI	Jaccard Index
VOE	Volumetric Overlap Error
ASSD	Average Symmetric Surface Distance
HD	Hausdorff Distance
GANs	Generative Adversarial Networks
LLMs	Large Language Models
CT	Computed Tomography
TCD	Transcranial Doppler
CAM	Class Activation Mapping
Grad-CAM	Gradient-weighted Class Activation Mapping
RICE	Randomized Input Concatenation
GMP	Generative Model Predictions
CLIX-M	Clinician-Informed XAI Evaluation Checklist
CDSS	Clinical Decision Support Systems
EU AI Act	European Union Artificial Intelligence Act
MDR	Medical Device Regulation
IVDR	In Vitro Diagnostic Regulation
PHI	Protected Health Information
BPC	Bias Prevention Components
XAI Orchestrator	A framework for multimodal XAI methods
BenchXAI	A rigorous benchmarking framework for XAI
RISE	Randomized Input Sampling for Explanation
LRP	Layer-wise Relevance Propagation
SHAP	Shapley Additive Explanations
LIME	Local Interpretable Model-agnostic Explanations
MCCV	Monte Carlo Cross-Validation
MDR/IVDR	Medical Device Regulation/In Vitro Diagnostic Regulation

1 Introduction to AI in Healthcare

Over the past few decades, artificial intelligence (AI) has greatly changed the field of medicine [1, 2]. Early systems relied on simple, manually created rules for clear reasoning, but they didn't work very well [3]. In contrast, modern AI in healthcare uses machine learning (ML) and deep learning (DL) to handle huge amounts of medical data [1]. Today's DL models can match or even do better than human experts on difficult tasks. For example, a type of AI known as convolutional neural networks (CNNs) has proven to be very accurate in diagnosing problems from medical images [2, 4]. Similarly, ML is now a key part of genetics, patient monitoring, and predicting health outcomes, providing new ways to diagnose diseases, forecast patient futures, and offer personalized care [5]. Combining digital health data with AI is a major driver of precision medicine, allowing diagnoses and treatments to be customized based on a patient's genes, environment, and lifestyle [4]. This shift toward personalized care has a wider industry impact: AI is changing how we process and analyze patient data, making healthcare one of the biggest fields benefiting from the rise of AI [5]. Historically, AI's journey in healthcare began with symbol-based and rule-based approaches [3]. The roots of explainable AI (XAI) in healthcare trace back to the mid-twentieth century. Early expert systems like MYCIN translated clinical knowledge into specific "if–then" rules [6]. This made their reasoning clear, but they weren't effective at handling uncertainty [6, 7]. Researchers first used models like decision trees, which had "a set of clear rules" and features created by experts [3]. These hand-crafted models were easy to understand but lacked the accuracy needed for crucial diagnostic tasks [3]. Machine learning has certainly evolved from statistical techniques to more advanced deep neural networks in its quest for better results. However, machine learning owes its origins to earlier powerful automatic methods, such as kernel machines and graphical models, as well as neural networks. This creates a trade-off: deep learning models are the most accurate, but they're hard to understand, while simpler models, like decision trees or linear classifiers, are easier to interpret but not as accurate [8]. The complexity of DL and large language models (LLMs) makes them act as opaque "black boxes" [2, 8], and clinicians struggle to understand the reasoning behind a deep network's output [9]. Many healthcare AI models provide answers "without stating the logic behind them," making it "difficult to identify errors, biases, or inconsistencies" [10]. This lack of clarity has slowed down the widespread use of AI. Also, being able to explain how AI works is important for dealing with moral and technical issues like accountability and patient privacy [4]. Transparent AI builds trust, ensures fairness, and makes sure that the AI's results match what doctors expect. However, trust remains a big challenge. This concern has led to a stronger focus on Explainable AI (XAI) in healthcare research [10]. New laws are also addressing his demand for transparency. For example, the EU's 2024 Artificial Intelligence Act requires that high-risk medical AI systems be able to explain their decisions [11]. This means AI needs to be able to explain how it came to its conclusions, especially for important tasks like diagnosis. All this research and new regulation show that the future of AI in healthcare depends on two things: it must be understandable, and it must be built with human needs in mind [12]. Even with these advancements, there are still many challenges to overcome. Informal or temporary explanations can be misleading if not properly validated [13]. Different explanation methods may conflict, and relying on "wrong or inconsistent explanations" without question could jeopardize patient safety [13]. There is also an ongoing debate about the rules for explainability: experts are split on whether explanations should depend on the specific context or be the same in all situations [12]. AI applications still have problems with things like data privacy and bias, which are still issues.

1.1 Definition and Scope of Explainable AI (XAI)

Black-box, grey box, and white box AI

In healthcare, there is increasing skepticism toward the adoption of "black-box" AI models. Such systems are often regarded as opaque because the computational processes connecting inputs to outputs are too complex to trace, with deep neural networks representing a prime example—delivering high flexibility but at the expense of transparency [14]. Not knowing how these systems work can also lead to both unintentional and intentional biases that affect their recommendations, predictions, and decisions [15]. To tackle these concerns, XAI models are usually divided into three main categories: black-box, grey-box, and white-box approaches. Black-box models, like deep learning and ensemble methods, offer high predictive accuracy but lack transparency [10]. Grey-box models, such as statistical approaches, strike a moderate balance between performance and interpretability. On the other hand, white-box models—like graphical models, linear models, rule-based systems, and decision trees—are easier to understand, though they often sacrifice some accuracy. Ideally, healthcare systems should use models that strike a balance, offering enough transparency while still delivering good performance [10].

Interpretability versus Transparency

Explainability is a term with two main parts: interpretability and transparency [16]. An interpretable model is one whose decision process is easy for a person to understand, like a small decision tree or a simple linear model [3]. A transparent model lets you look inside it and see how it turns inputs into outputs. You can "trace the algorithm to generate the output," which isn't possible with "black-box" models [17]. It's important to know that explainability is more than just being transparent. It includes the model's ability to give clear, understandable reasons or justifications for its results, which should be suited to the person using it [16]. In practice, this usually involves using interpretability techniques, like highlighting the most important features, so users can understand why the model made a particular decision [18]. Understanding the difference between these terms is important. While explainability is sometimes incorrectly used interchangeably with interpretability or transparency, true explainability requires both an understandable reason and the clear communication of that reason to the user [19].

Explainable AI (XAI)

It is a part of artificial intelligence that makes AI models more understandable and reliable [4]. Unlike complex, opaque "black-box" systems, XAI methods help us see and understand why and how a model reached a certain conclusion [17]. XAI is defined as AI that "communicates reasoning to users" and makes its internal logic clear [20]. Figure 1 shows transforming complex models into Interpretable models using Interpretability methods.

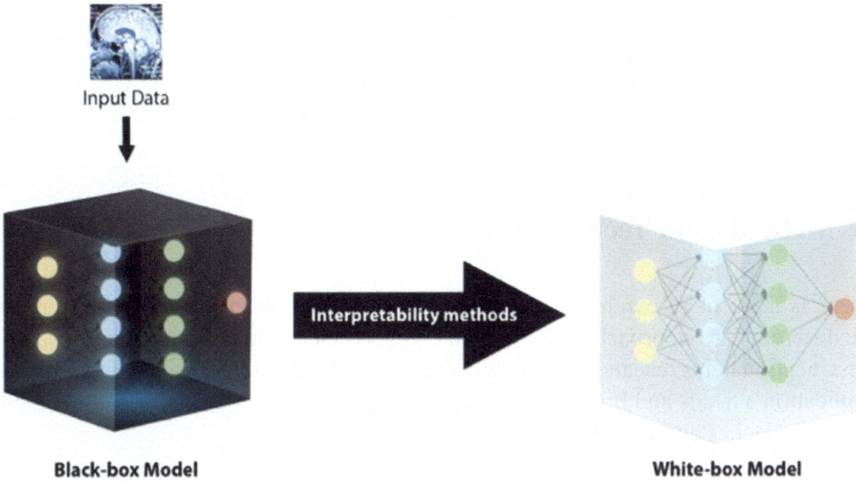

Fig. 1 Transforming complex models into interpretable models using interpretability methods

1.2 Trust, Responsibility, and Moral Aspects of (XAI)

Trust in healthcare AI goes beyond just technical transparency—it also includes ethical issues like patient autonomy, fairness, and accountability. From the patient's point of view, lack of clarity can affect informed consent and autonomy, shifting decision-making power from doctors and patients to algorithms, a situation known as "digital paternalism" [19]. For this reason, fairness and safety must always be at the heart of how AI systems are designed and used [9]. However, explainability alone cannot guarantee trust. Some XAI techniques might give explanations that are incomplete or even misleading [19]. Also, human medical decisions aren't always clear or logical; they often rely on gut feelings or unconscious biases. So, expecting AI systems to be completely open could lead to an unfair double standard [19]. Because of these issues, various regulations appeared, such as the EU AI Act (effective in 2024), which classifies most AI/ML-based medical software as high-risk. This means taking strict steps like managing risk, keeping good records, protecting data, and having people watch over things [21]. The FDA in the U.S. treats AI/ML software as medical devices. It is essential to make sure that the software goes through the right approval processes, including Good Machine Learning Practice (GMLP), Predetermined Change Control Plans (PCCPs), and ongoing monitoring throughout the software's lifecycle [22]. The GDPR controls health data in the EU by limiting the processing of sensitive data (Article 9) and automated decision-making (Article 22). It also often requires Data Protection Impact Assessments [23]. The WHO's guidance talks about ethical principles like benefit, safety, transparency, accountability, equity, and sustainability. It also advises on how to run things, including for big multimodal models [24]. HIPAA protects Protected Health Information (PHI) in the U.S. through rules for privacy, security, and breaches. The cybersecurity laws will tighten up in the year 2025 [25]. Hence, these frameworks assert that those who design and operate health AI must abide by the appropriate legal standards (on an EU level, e.g., the AI Act, the MDR/IVDR, the FDA, the GDPR, and HIPAA) and ethical rules (such as the WHO).

2 Methods and Techniques

There is a lot of research in this area, but it is important to organize and group the different XAI approaches so that you can get a clear and complete picture. This classification helps researchers and clinicians choose the best methods based on each application's needs and limitations. Below, we present a detailed review of the main XAI approaches and how they are categorized.

2.1 Attribution and Non-Attribution Explanation

Attribution explanation

In explainable AI, approaches based on attribution identify and highlight the input components that most significantly influence the model's prediction. These techniques typically produce localization maps that emphasize significant regions of an image or attributes of the data that affect the decision. This allows for a clear view of how the model's output connects to specific input elements, illustrating which factors the model considers in reaching its conclusions [26].

Non-Attribution Explanation

In contrast, non-attribution approaches go beyond just focusing on specific input areas and instead aim to reveal the internal reasoning process of the model. These methods look at factors like stability, sensitivity, and how the model behaves overall. Instead of just pointing out which features or pixels are most important, they offer a wider view of how and why a prediction is made. These explanations are handy for debugging and gaining a deeper understanding of how the model makes its decisions [26]. Figure 2 shows XAI methods by Interpretability Type: Attribution vs. Non-Attribution Explanation.

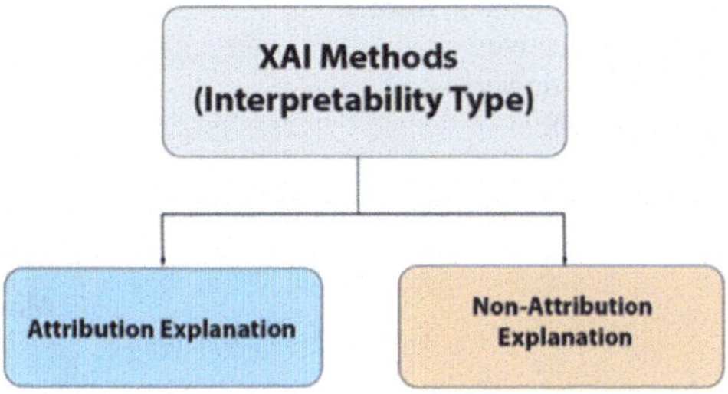

Fig. 2 XAI methods by interpretability type: attribution versus non-attribution explanation

2.2 *Intrinsic (Ante-hoc) and Post-hoc Methods*

Intrinsic (Ante-hoc)

The Latin word "ante," which means "before," is where the word "ante-hoc" comes from. This approach makes the model's transparency a natural part from the beginning. A few instances are decision trees, rule-based models, monotonic GBM, and Bayesian rule lists. These models are easy to understand on purpose, but they are often not as accurate as more complicated models [7].

Post-hoc

The term post-hoc is also derived from Latin, meaning "after" [27]. Explanations are generated after model training to interpret complex black-box systems. Methods such as LIME, SHAP, CAM, Grad-CAM, and surrogate models fall into this category. They preserve model accuracy but provide less transparency [26, 27]. From the reviewed studies, it is noted that Class Activation Mapping (CAM) and its extensions, such as Grad-CAM, Grad-CAM ++ , Guided Grad-CAM, Score-CAM, and others, are among the most frequently used explanatory tools in the field of medical imaging [28]. Figure 3 shows XAI methods in terms of Integration: Ante-hoc vs. post-hoc.

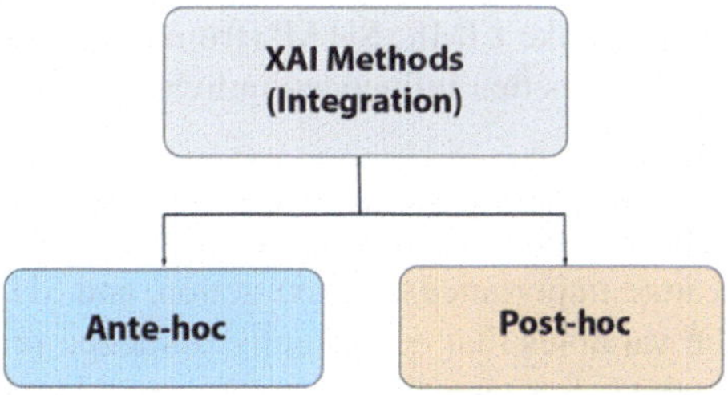

Fig. 3 XAI methods in terms of integration: ante-hoc versus post-hoc

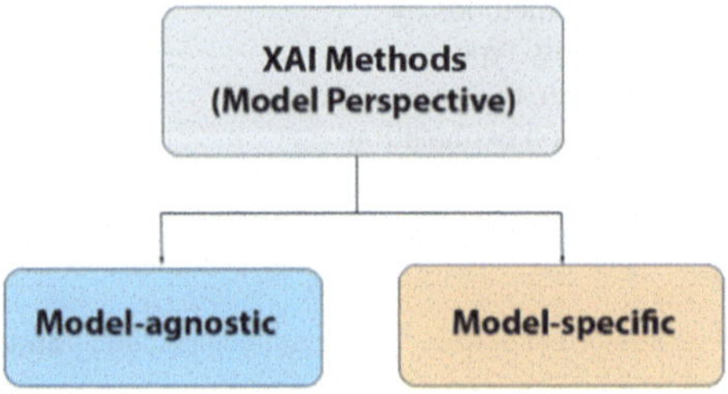

Fig. 4 XAI methods from a model perspective: model-agnostic versus model-specific

2.3 *Model-Specific and Model-Agnostic*

Model-specific

This refers to methods that rely on the internal structure of a particular machine learning (ML) or deep learning (DL) model. The model-specific approach uses the model's unique architecture and parameters to provide explanations that are tailored to that specific design [26].

Model-agnostic

A model-agnostic approach aims to explain how a model makes decisions, without depending on its internal structure [27]. The goal is to provide a general idea of how the model works and to point out the key elements that affect its predictions or decisions [26]. Figure 4 shows XAI methods from a model perspective.

2.4 *Local and Global*

Local

Focuses on explaining a single prediction and why a specific decision was made. Methods like LIME, SHAP (from a local perspective), and saliency maps highlight the features that influenced individual outcomes [26, 29].

Global

Tells you how the model works as a whole across all of the data. Techniques like feature importance, rule extraction, and SHAP (from a global perspective) highlight the variables that consistently influence predictions [27, 28]. Figure 5 shows XAI methods from a coverage perspective: local versus global.

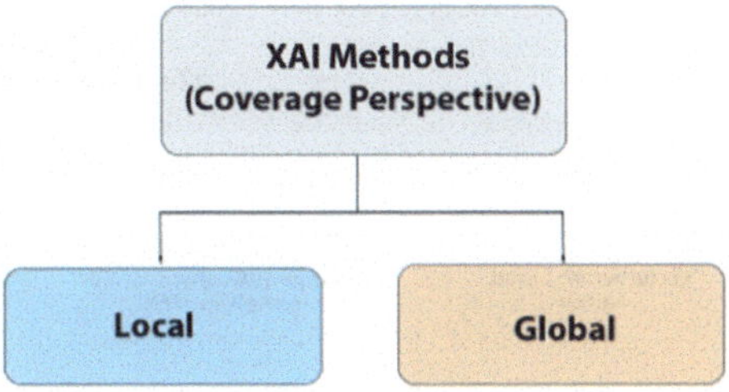

Fig. 5 XAI methods from a coverage perspective: local versus global

2.5 Other Important Methods

LIME

Is a well-known method that changes the input data and trains a simple model locally to mimic the black-box behavior, which helps explain individual predictions [26, 30].

SHAP

Is another popular method that uses Shapley values from game theory to determine the contribution of each feature, providing both local and global insights into the model's predictions [27, 30]. However, Perturbation-based approaches, such as LIME and SHAP, are especially resource-intensive since they require repeated model evaluations and the retraining of surrogate models, which leads to substantial overhead. Compared to perturbation-based techniques, gradient-based methods (e.g., Grad-CAM, LRP, RISE) are typically more efficient than perturbation-based techniques. Grad-CAM requires only a single backpropagation pass, enabling its use in real-time medical imaging [26].

Class Activation Mapping (CAM)

It is a widely used visualization approach that helps interpret and evaluate the behavior of machine learning or deep learning models in medical image analysis [27]. The core concept of CAM is to derive a weight matrix from the activations of the final convolutional layer in a CNN [27]. It enables clinicians and practitioners to see which regions or features of an image the model considers most significant for making a particular decision [26].

Gradient Class Activation Mapping (Grad-CAM)

It is a widely used XAI technique in image analysis. Over the years, CAM has improved, leading to the creation of many helpful extensions [27]. Grad-CAM is one of the most popular XAI methods for analyzing images. It is based on the original CAM but is more flexible and can work with more types of CNN architectures, even those that don't have a global average pooling layer [26]. It provides a localization map that shows the areas that are most important to the model's prediction via the gradient of a chosen target (like a class output) in relation to the last convolutional layer [26].

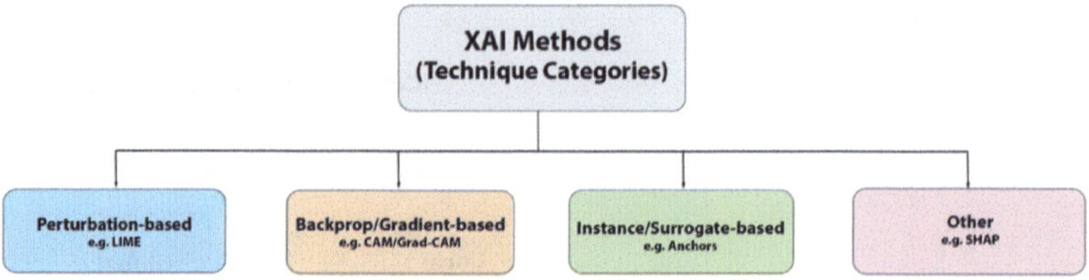

Fig. 6 Technique categories of XAI (Explainable AI) methods

Surrogate models

Another important type of model is the surrogate model. Surrogate XAI techniques entail the training of interpretable models, like linear regression, to approximate the behavior of neural networks [30]. These methods make the decision-making process clearer by using simpler models instead of more complicated black-box ones. For example, in the Anchors method, key features influencing the model's prediction are identified, and a specific explanation is provided for that prediction. These methods are especially helpful when you need to explain a particular decision made by the model [26]. On the other hand, combining XAI with multimodal data and utilizing multiple AI models (ensemble and hybrid) for holistic evaluation can significantly improve the trustworthiness and practical use of XAI methods in clinical practice [26]. Figure 6 shows technique categories of XAI (Explainable AI) methods.

Beyond taxonomy, we need to explain other dimensions that are different from one task to another:

Problem Type

AI can support a wide range of medical tasks, including detection, classification, diagnosis, segmentation, prediction, and improving image quality [28]. The choice of an explanatory approach often varies depending on the specific task to ensure the most suitable explanation is provided.

Input format

The specific type of input data you have will have a big effect on which method of interpretation you should use. In medical AI, inputs may include images (like X-rays, MRIs, and CT scans), text, numbers, or groups of numbers. Multi-modal approaches that integrate multiple data types demand more sophisticated interpretability frameworks [26, 27].

Output format

Their objective is to classify the different types of output formats, which include numerical, rule-based, textual, and visual, as well as combinations of these [28].

- Visual. includes saliency maps, heatmaps, attention maps, CAM/Grad-CAM, and t-SNE plots, making the model's reasoning visible [26–28].

- Textual. descriptions of rules or feature attributions in plain language. When used for prediction tasks, text-based explanations are especially useful [28].
- Numerical. explanations are used for many things, but they are most useful for prediction and classification. SHAP and LIME are the most common methods. They use plots and violin plots (for SHAP) and Pareto charts (for LIME) to show numerical information [28].
- Rule-based. methods were used to provide explanations for the diagnoses, predictions, and classifications generated by the algorithms. Relies on symbolic logic and extracted decision rules, such as if–then rules or Anchor [28].
- Example-based. Among the studies, example-based explanations were the least common. They included methods such as prototypes, counterfactuals, patch similarity, content-based image retrieval (CBIR), and TraCE, and were always paired with other output formats [28].

2.6 Evaluation Framework

A good explanation in health-AI must provide necessary and sufficient information to close users' knowledge gaps, be clear and understandable, explain predictive certainty, accurately match the model's input–output mapping, remain objective, and withstand small perturbations in input data without changing the output. These attributes directly inform evaluation criteria across XAI research [15]. The reviewed literature identifies several key metrics as essential: fidelity (the degree to which the explanation aligns with the model's decisions), robustness (the capacity to maintain stability amidst changes), comprehensibility (the ease of understanding), plausibility (the alignment with domain knowledge), fairness (the equity of explanations), and actionability (the ability to facilitate clinical decision-making) [13, 20, 31]. In imaging, quantitative metrics like DSC, JI, VOE, ASSD, and HD are used as ground truth comparisons, although they have limited ability to capture explainability [32]. Several methodologies for evaluating XAI have been reported across the literature. In this section, some of the evaluation methods, along with a summary of their performance, are presented:

BenchXAI

is a rigorous benchmarking framework for explainable AI in healthcare/biomedicine that evaluates 15 XAI methods. It uses Monte Carlo cross-validation to make things less uncertain and statistical tests, like the Wilcoxon signed-rank and Friedman tests, to compare performance. BenchXAI normalizes attribution outputs (including saliency maps) across different explainers based on their rank. This makes sure that comparisons are fair and direct, which is a big problem with previous evaluation methods [30].

CLIX-M

In a different review, Brankovic et al. (2025) present CLIX-M, a checklist consisting of 14 items developed with feedback from clinicians. It looks at XAI explanations in four areas: Purpose, Clinical attributes, Decision attributes, and Model attributes. Clinical attributes emphasize relevance to the discipline, coherence (rationality), and actionability; decision attributes encompass correctness, confidence, consistency, robustness, and causal validity; and model attributes address narrative reasoning, bias/fairness, troubleshooting, interpretation, and the limitations of XAI. CLIX-M is designed to embed evaluation both during system development and in clinical settings, supporting physicians in assessing explanations in situ [13].

XAI Orchestrator

Pahud de Mortanges et al. (2024) examine XAI for multimodal and longitudinal medical imaging data and suggest the development of an XAI Orchestrator. They outline six key evaluation metrics—faithfulness, robustness, localization, complexity, randomization, and axiomatic—based on Quantus. They also mention that these metrics would be necessary for the orchestrator, with additional considerations for evaluating large language models (LLMs). They discuss issues arising from variations in data (such as discrepancies in scanners and protocols) and emphasize the need for multicenter validation and methods to apply XAI effectively across various domains [31] (Table 1).

Conclusion

In summary, the purpose of this section is not to affirm the superiority of any single Explainable AI (XAI) technique. Rather, what is our guiding purpose is to elucidate the meaning of each family of techniques, classifying its advantages, disadvantages, and suitability with respect to its aim. Two general dimensions define the most essential distinctions: (1) integration timing-the so-called models that are inherently interpretable were created under integrate-then-predict (ante-hoc) premises, whereas explanatory add-ons are only applied after training (post-hoc) when approximation fidelity rather than accuracy is prioritized; and (2) scope of explanation-pertaining to specific local knowledge in justifying individual predictions as opposed to global deduction on model functioning. Another important distinction is based on whether the techniques put forward are model-specific (e.g., CAM and Grad-CAM for CNNs, etc.), offering very tight visual mappings, or those that are model agnostic (e.g., LIME, SHAP, surrogate models), allowing for more flexible operationalization across different architectures. The overriding message is that no single method can satisfy all needs, and the choice must be governed by the "right method for the right data and task" principle. From data-and task-oriented considerations, for images, gradient and activation-based methods such as Grad-CAM are particularly powerful as they are fast and emphasize spatially pertinent regions; for tabular, signal, or textual data, perturbation-based methods such as LIME and SHAP provide both local and global feature importance; and for multimodal scenarios, there is a need to combine several

Table 1 Comparative overview of XAI evaluation frameworks in healthcare (BenchXAI, CLIX-M, XAI Orchestrator)

Evaluation framework	Objective	Methods and approaches	Methods used
BenchXAI [30]	Comprehensive benchmarking of post-hoc XAI on multimodal biomedical data (clinical, medical image/signal, and molecular)	Compares 15 XAI methods across tasks and modalities; introduces a sample-wise normalization of attribution scores to enable fair, cross-explainer comparison	MCCV; statistical tests (Wilcoxon, Friedman); extensive visual analytics; rank-preserving normalization to $[-1,1]$ for attribution outputs
CLIX-M [13]	Clinician-informed, human-grounded evaluation and reporting checklist for XAI components in AI-powered CDSS, used during development and clinical evaluation phases	It consists of 14 items across four categories: Purpose, Clinical Attributes, Decision Attributes, and Model Attributes. It also provides suggested metrics and Likert-type rating scales to assess clinical relevance, coherence, and actionability	Checklist-based assessment covering correctness, confidence, consistency, robustness (incl. ensemble explainers), causal validity, narrative reasoning, bias/fairness, model troubleshooting, interpretation, and XAI limitations; guidance on where to report (Methods/Results/Discussion)
XAI orchestrator [31]	Concept of a virtual assistant that coordinates explanations for multimodal and longitudinal data in clinical imaging; proposes properties and workflow integration	Defines desired properties (adaptive, hierarchical, interactive, uncertainty-aware, time-effective, modular, privacy-preserving, drift-resilient) and RAG/LLM-based orchestration design; focuses on clinical usability	Highlights six classes of quantitative metrics (from Quantus style): faithfulness, robustness, localization, complexity, randomization, axiomatic; notes the need to extend metrics for LLM evaluation. Separately emphasizes challenges from data heterogeneity (e.g., scanner/protocol differences), underscoring the importance of multicenter validation and domain adaptation for XAI

methods objectively. All these families have their own pros and cons: gradient-based methods are fast but architecture-dependent and sensitive to layer choice; perturbation-based methods are flexible but computationally expensive and sometimes unstable; whereas surrogate or rule-based models are interpretable but can diverge from the logic of the original model. A clear recommendation that derives from this is to choose a small but complementary set of methods for every case and ensure that this addresses the "what" (key features or regions) and the "why" (general reasoning) while being assessed systematically regarding fidelity, stability, actionability, and fairness.

3 Applications in Healthcare

3.1 Applications in Clinical Practice

Medical Imaging (Radiology, Pathology, etc.)

XAI is frequently used in medical imaging to assist in interpreting intricate deep learning models for tasks such as disease identification, classification, and segmentation. Typical techniques offer visual interpretations (such as saliency maps or heatmaps) on images to emphasize the characteristics that affect the model's predictions. The literature shows a surge in XAI for imaging, primarily targeting MRI, CT, and X-ray applications for lung and brain pathologies. Most explanations are **local** (case-specific) rather than global, reflecting the dominance of tools like Class Activation Mapping (CAM) and its variants (Grad-CAM, etc.) in providing human-interpretable image highlights. We will now talk about some important studies:

Champendal et al. (2023) performed a scoping review of 228 studies regarding XAI in medical imaging. The majority of explanations were post-hoc (222/228 in total; 77 agnostic and 145 model-specific), and the dominant output was visual (186/228). CAM (Class Activation Mapping) and its variants (Grad-CAM) were utilized most frequently to generate heat maps (n = 96). The tasks most commonly explained were classification (89 studies), prediction (47), diagnosis (39), detection (29), and segmentation (13). With respect to scope, 78.1% of explanations were local, 5.7% were global-only, and 16.2% combined local and global views. XAI applications concentrated on MRI (32%; n = 73), radiography (20%; n = 47), and CT (20%; n = 46). The authors emphasize that future XAI work should be more user-centred—incorporating the needs and workflows of end users such as physicians and radiographers, rather than remaining developer-centric [28].

As cited in Sheu et al., Mohapatra said that an explainable AI (XAI) model was made to find pneumonia in chest X-rays. It used a flexible convolutional neural network (XAI-ICP) with a feedback mechanism that makes its classification accuracy better all the time. The model was first trained on the NIH Chest X-ray dataset and then fine-tuned with a dataset specific to pneumonia through transfer learning. It reached an accuracy of about 93%. Localized saliency maps show the XAI feature of the model by highlighting important areas in the X-ray images, such as effusion, infiltrates, and consolidation. These maps help us understand what the model is focusing on by showing how its reasoning matches clinical findings like consolidation or cardiomegaly. This method not only makes the AI's decisions more accurate, but it also makes them easier to understand, which means that they will be in line with what medical experts would expect [32].

Mohapatra (as cited in Darapaneni et al.) mentions that this study utilized explainable AI (XAI) strategies for diagnosing COVID-19 pneumonia from CT lung scans. A CNN is used for pneumonia detection, with class activation maps (Grad-CAM) and lesion segmentation. The XAI overlay indicated to the clinicians the areas the model considered, such as GGO and consolidation, thereby confirming that these were indeed areas of infection. The model compared the Class Activation Maps to the segmented lesions to calculate the percentage of lung involvement due to COVID-19, ensuring that the AI only focused on lung tissue, disregarding other structures. The explainable model averaged 0.87 for multi-label classification (discerning COVID-19 from other pneumonia types) on the F1-score and 0.8066 for lesion segmentation on the Dice coefficient. These outcomes demonstrate the robust performance of the model while enhancing the transparency of its diagnostic reasoning, i.e., ensuring that the decisions made by AI were in tandem with the clinician's expectations [32].

Mohapatra (as cited in Ullah et al.) states that this study created an ensemble classifier that combines deep CNN features with traditional machine learning methods to find COVID-19 in chest images in response to the COVID-19 pandemic. The researchers employed Grad-CAM to emphasize significant regions on X-rays/CT scans and utilized t-SNE plots to analyze feature clustering, thereby validating and interpreting the decisions made by the ensemble model. These XAI methods showed that the model was right to focus on the lungs' most important pathological areas. Clinicians were able to trust the model's almost perfect accuracy (~99% on two COVID datasets) because they could clearly see heatmap evidence showing the lesions that affected the prediction. This example shows how XAI was very important for fixing bugs in a model and building trust in an imaging-based diagnostic tool during the COVID-19 pandemic [32].

The study of Mohapatra (as cited in Nafisah and Muhammad) recalled all about the research accomplished by Mohapatra et al. (2025) that establishes an explainable deep learning (DL) method for identifying tuberculosis (TB) in chest radiographs (CXR). The first procedure splits the lung regions in CXR images into smaller parts, then customizes those regions for the model to focus on the right points. Subsequently, used XAI to indicate those lung parts that have a confirmed TB presence condition. Then, radiologists would evaluate why the AI gave a particular X-ray a TB-positive annotation by using visual explanations like highlighting TB's lesion characteristics instead. There seem to be quite accurate numbers in this model: its accuracy is about 99.1%, and its AUC percentage is 99.9%, at which it was performed on three public TB datasets. Clear heatmap reasons showed areas linked to pulmonary infiltrates, which probably encouraged doctors to trust this AI in terms of screening because the areas highlighted complied with aware TB patterns [32] (Table 2).

Risk Prediction and Prognosis

XAI helps find the things that affect a model's risk estimates in prognostic modeling and risk prediction. This makes it easier to check AI against clinical knowledge. These applications frequently utilize tabular clinical data or multimodal inputs, wherein techniques such as feature importance (SHAP, LIME) or rule-based explanations can improve transparency. XAI wants to help doctors trust models that predict outcomes

Table 2 Summary of case studies in medical imaging

Title of the study	Methods used	Tasks explained	Evaluation metrics	Key insights	Context/outcome
Champendal et al. (2023)—XAI for medical imaging [28]	Post-hoc XAI, Grad-CAM, Heatmaps	Classification, Prediction, Diagnosis, Detection, Segmentation	Accuracy (~ 99%), AUC (~ 99%)	Majority of explanations are post-hoc, XAI methods focused on visual outputs	Healthcare and medical imaging, need for more user-centered XAI
Mohapatra (as cited in Sheu et al.)—Pneumonia detection in hest X-rays [32]	Reconfigurable CNN, Transfer learning, Saliency maps	Pneumonia Classification	Accuracy (~ 93%)	Localized saliency maps provide interpretability and clinician alignment	Improved interpretability and accuracy in pneumonia detection
Mohapatra (as cited in Darapaneni et al.)—XAI for COVID-19 detection [32]	CNN, Grad-CAM, Lesion segmentation	COVID-19 Diagnosis	F1-score (~ 0.87), Dice coefficient (~ 0.8066)	Combining Class Activation Maps with segmentation to quantify lung involvement	Enhanced trust and diagnostic transparency for COVID-19 detection
Mohapatra (as cited in Ullah et al.)—Ensemble classifier for COVID-19 detection [32]	Ensemble classifier, CNN, Grad-CAM, t-SNE	COVID-19 detection	Accuracy (~99%), t-SNE clustering	Ensemble approach ensures high performance and clinician trust with visual validation	High clinician trust achieved with clear visual explanations during pandemic
Mohapatra (as cited in Nafisah and Muhammad)—Tuberculosis detection in CXR [32]	DL Approach, XAI, Lung segmentation, Heatmaps	Tuberculosis detection	Accuracy (~ 99.1%), AUC (~ 99.9%)	XAI used to help radiologists validate TB diagnosis with clear lesion heatmaps	Clinician confidence in AI for screening and high performance in TB detection

or assess patient risk by explaining predictions like the chance of death or getting sick. Some important examples are:

Metsch et al. (2025) presented BenchXAI, a framework intended to assess 15 post-hoc XAI methodologies across various biomedical tasks. One of the tasks involved utilizing clinical tabular data (from the Heart Failure dataset comprising 299 patients) to forecast mortality. The authors examined the potential of XAI methods to emphasize established prognostic factors, including serum creatinine and ejection fraction, which prior studies have recognized as robust indicators of survival. Gradient-based methods like Integrated Gradients, DeepLIFT, DeepLiftShap, and GradientSHAP consistently produced strong and reliable results across a range of tasks, including predicting heart failure, classifying images of diabetic retinopathy, and working with synthetic signal data. On the other hand, methods like Deconvolution, Guided Back-propagation, and LRP-α1-β0 didn't work as well on some tasks, which made the results less stable. Overall, BenchXAI stresses that choosing XAI methods carefully can help find clinically relevant features (like age, high blood pressure, and anemia) that match model predictions with clinical reasoning. But some methods could lead to wrong interpretations if they aren't checked properly [30].

As Mohapatra's review points out, Moreno-Sánchez (2023) created an explainable AI model for predicting Chronic Kidney Disease (CKD) early on. The model employed standard clinical variables and an Extreme Gradient Boosting (XGBoost) classifier, attaining 99.2% accuracy (SD = 0.8) on novel data and 97.5% accuracy through fivefold cross-validation. The explainability analysis revealed that hemoglobin was the primary predictor, followed by urine specific gravity and hypertension. These results are consistent with established medical knowledge, as anemia and hypertension are recognized risk factors for renal impairment. The study bolstered confidence in the model's capacity for early diagnosis and clinical decision support by demonstrating that its predictions were grounded in clinically pertinent features [32].

According to Joyce et al. (2023), who summarized explainable AI for mental health applications, it comprises a TIFU framework (i.e., Transparency and Interpretability for Understandability) to enhance how clinicians comprehend AI predictions in psychiatry. That speaks to where one could draw on a time of making predictions on mood worsens using an XGBoost model during COVID-19 lockdowns. It is through the use of SHAP (Shapley Additive Explanations) that key features influencing predictions on mood deterioration could be identified. Joyce and the team suggest that post-hoc explanations could drive simpler, clearer, and more understandable predictive models that emphasize those important features. They suggest that initially complicated black-box models be used for exploratory undertakings, but then simplified or better yet morph into transparent models (e.g., knowledge distillation or rule extraction) eventually. This way, predictive accuracy would be preserved while ensuring that risk factors align with what doctors would expect. This would make AI more acceptable in psychiatric decision-making [14].

Okada et al. (2023) underscore that in emergency medicine, explainability is essential for clinicians to have confidence in AI tools employed for triage and outcome prediction in high-pressure, time-critical scenarios. They talk about a test of an AI model that was supposed to pick up on cardiac arrest from emergency call audio. Because there was no explanation, people didn't trust the system's alerts as much. The authors use feature-attribution methods like SHAP (Shapley Additive Explanations) to show how explainability can help with this problem. For instance, a SHAP waterfall plot showed how the predicted chance of survival was affected by things like the time it took to get to the hospital, whether someone saw the event, their age, and their sex. These visual breakdowns clarify the model's reasoning behind its predictions of low or high survival rates for clinicians. Okada et al. say, though, that in emergencies, explanations need to be short and to the point, since doctors need quick, clear information to make decisions quickly. In general, they say that XAI can make risk predictions (like triage scores or survival estimates) easier to understand, but only if it is made for the needs of emergency medicine [17].

Hossain et al. (2023), as analyzed by Mohapatra et al. (2025), established an explainable AI framework for forecasting thyroid dysfunction, particularly differentiating between hypothyroidism and hyperthyroidism. The model used routine lab test results and feature attribution methods like LIME and SHAP to show how inputs like TSH and T3 levels affected its predictions. The explanations showed that very high or low TSH levels were strongly linked to predictions of hypothyroidism, which is what doctors already know. The study did not concentrate on overall accuracy; its primary contribution was illustrating how explainability facilitated clinicians' comprehension of the rationale behind the model's outputs, thereby fostering trust in its application for endocrine decision-making [32] (Table 3).

Clinical Decision Support Systems

AI is used by clinical decision support systems (CDSS) to help with diagnosis, treatment recommendations, or prescribing. In these situations, XAI is very important for building the right level of trust. Recent studies investigate the influence of explanations on user behavior and the design of Clinical Decision Support Systems (CDSS) incorporating Explainable Artificial Intelligence (XAI) that is congruent with clinical reasoning.

Nagendran et al. (2023). This research quantitatively evaluated the contribution of explanations toward an AI-based clinical decision support system (CDSS) for prescribing in critical care. In a study of 86 ICU physicians, Nagendran and others studied how the "AI Clinician" aided in sepsis in four different scenarios: none, AI, peer human suggestions, and AI suggestions with explanations. The explanation was that of a feature-importance display indicating which patient variables (blood pressure, lactate, etc.) were most influential to the AI's treatment suggestion. Findings revealed that AI recommendations had a large effect on the doctors' prescribing, yet providing explanations did not aid in garnering acceptance of those recommendations over AI advice alone. Interestingly, clinicians exposed to explanations reported greater confidence and perceived the AI as more useful; however, their decisions did not appear to align more closely with the AI's outputs relative to any of the other groups. The authors, therefore, express that more basic feature-oriented explanations may not suffice for such highly complex tasks as ICU dosing and warn that

Table 3 Summary of case studies in risk prediction and prognosis

Study title	Methods	Task explained	Evaluation metrics	Key insights	Context outcome
Moreno-Sánchez (2023) [32]	XGBoost + XAI (feature importance)	Early prediction of chronic kidney disease (CKD)	Accuracy: 99.2% (SD = 0.8), 97.5% (fivefold CV)	Hemoglobin most important; followed by urine specific gravity & hypertension	XAI revealed medically plausible features, improving clinicians' trust
Joyce et al. (2023) [14]	TIFU framework; SHAP; XGBoost (literature example)	Mental health prediction (mood deterioration during COVID-19)	Not focused on accuracy; emphasis on interpretability	SHAP identified key risk features; proposed distillation into transparent models	Transparent models align with clinicians' reasoning, improving AI acceptance in psychiatry
Okada et al. (2023) [17]	XAI in emergency medicine; SHAP attribution	Outcome prediction in emergency/critical care (e.g., cardiac arrest survival)	Illustrative examples (SHAP plots), not standard metrics	SHAP plots explained predictions (time-to-hospital, age, witnessed status)	Explanations must be concise and context-aware for emergency physicians
Hossain et al. (2023) [32]	XAI framework; LIME and SHAP	Predicting thyroid dysfunction (hypo- vs. hyperthyroidism)	No accuracy emphasized; focus on interpretability	Extreme TSH values were strong predictors, consistent with medical knowledge	Explainability improved clinician trust and supported endocrine decision-making

self-reported trust does not necessarily equate to real-world behavior change. Accordingly, they opine that different forms of explanation-more elaborate, more intuitive, or otherwise-will be needed to effect real change in actual clinical decision-making, other than simply making clinicians feel better about using the AI [33].

Bienefeld et al. (2023) undertook a year-long co-design study of explainable AI (XAI) for a clinical decision support system (CDSS) in the neuro-ICU, concentrating on the prediction of delayed cerebral ischemia (DCI) following aneurysmal subarachnoid hemorrhage. The study included 112 stakeholders, such as AI developers and clinicians, and found a major difference in how they thought about things: developers cared more about how easy it was to understand algorithms and how valid data was, while clinicians cared more about how relevant it was to patients and how patient-centered it was. The study also showed that there are real problems in acute care, since ICU staff need quick, useful explanations instead of long, technical ones. The authors propose the utilization of causal inference models to better correspond with clinicians' cause-and-effect reasoning, alongside tailored, patient-specific explanations appropriate to the clinical context. The study underscores that efficient XAI in critical care must reconcile data-centric and patient-centric perspectives, employing hybrid methodologies to satisfy the requirements of both developers and clinicians [11].

Rosenbacke et al. (2024) performed a systematic review of 10 empirical studies to analyze the impact of explainable AI (XAI) on clinicians' trust in AI-driven decision support. The results were inconsistent: in five studies, offering explanations alongside AI recommendations enhanced clinicians' trust relative to unaccompanied outputs. Three studies, however, found no effect, and two found both positive and negative effects, depending on the kind of explanation. The review stressed that explanations that are clear, short, and relevant to the clinic are more likely to make doctors more willing to follow AI advice. Conversely, explanations that are overly intricate, discordant with clinical reasoning, or difficult to interpret may diminish trust. The authors also stressed how important it is to calibrate trust: just increasing trust isn't always a good idea if it makes people too dependent on AI. The conclusion is that XAI can make people more trusting of clinical decision support systems if the explanations are well-made. However, if the explanations are bad, they may make people less trusting or cause them to trust the wrong thing [29].

Räz et al. (2025) present a forward-thinking paper that outlines a conceptual roadmap for integrating XAI into future clinical routines, using a hypothetical stroke management scenario as an example. The authors point out that current FDA-approved AI tools in stroke care (such as algorithms that detect intracerebral hemorrhage or large-vessel occlusion on CT scans) provide limited explainability—usually just highlighting the suspected area on the scan. While this is helpful, this one-dimensional explanation (a heatmap on the brain image) doesn't capture the full scope of the clinical decision-making process. Räz et al. argue that future XAI-based systems should function as **collaborative agents** in multi-step care pathways. In their stroke scenario, an ideal XAI assistant would: (Phase 1) start gathering and pre-processing patient data (NIH stroke scale, history) and share it with the team in real-time; (Phase 2) not only pinpoint the blocked artery on imaging, but also proactively scan for *all* abnormalities (to save radiologists from missing small infarcts elsewhere); (Phase 3) join clinical discussions by **highlighting key patient information** that clinicians haven't brought up (e.g. a prior stroke on record that might explain current deficits) and even conduct on-the-fly literature searches for relevant treatment guidelines; and (Phase 4) provide an ongoing *dashboard* in the rehabilitation phase to track the patient's progress on various metrics, with explanations of any changes. Achieving this requires XAI to be **context-aware and interactive**. Explanations must adapt to the user's role (nurse vs. neurologist) and context (ER vs. rehab ward). The paper identifies three escalating challenges: (1) context- and user-dependent explanations (tailoring the detail and format to what's needed at that moment); (2) a genuine **AI–clinician dialogue**, where the AI can clarify or justify when questioned, rather than just outputting static explanations; (3) developing *socially intelligent* AI that understands team dynamics and trust. While largely conceptual, Räz et al. underscore that current XAI techniques (saliency maps, simple feature scores) are **too inflexible** for such purposes. To integrate into complex care, XAI needs to evolve to provide the right information at the right time, perhaps through hierarchical explanations (brief overview with option to drill down) and continuously learning to communicate in clinically meaningful ways. This vision pushes beyond today's one-off explanations towards XAI as a cooperative decision support *partner* in medicine [34].

Baig et al. (2024): In a review of Generative AI for personalized care plans, Baig and colleagues emphasize explainability as an essential criterion for any AI-driven clinical decision framework. They say that generative models, like GPT or variational autoencoders, could suggest personalized treatment plans, open up new possibilities for personalized medicine. However, they are worried about their "black-box" nature. The authors highlight that model transparency and explainability are major barriers to adoption. In fact, making sure clinicians understand the reasoning behind an AI-recommended care plan is considered just as important as the accuracy of the plan itself. One of their listed action points is promoting algorithmic transparency to enhance trust among healthcare professionals. They stress that when an AI recommends a certain therapy sequence, the CDSS should also be able to explain its decision process understandably (for example, by pointing to patient data and clinical guidelines that support the recommendation). The review concluded that successful implementation of generative artificial intelligence in clinical decision support must integrate strong explainability and oversight mechanisms, which may include bridging the generative suggestions and traditional rule-based or XAI approaches, so that the clinician can trust that "the decision-making process is understandable and explainable.". In summary, although generative AI can enhance personalized care, Baig et al. argue that without solving the XAI challenge—providing clear, *human-understandable justifications* for AI outputs—wider clinical adoption will be stunted [5] (Table 4).

Table 4 Summary of case studies in clinical decision support systems

Study title	Methods	Task explained	Evaluation metrics	Key insights	Context outcome
Nagendran et al. (2023) [33]	Experimental study with 86 ICU physicians; AI Clinician (sepsis CDSS); Feature-importance XAI	Evaluating AI recommendations (with/without explanations) for prescribing in critical care	Quantitative adoption analysis; self-reported trust	AI advice strongly influenced prescribing; explanations did not increase adoption; gap between perceived trust and behavior	Feature-based explanations insufficient for ICU dosing; need more intuitive strategies
Bienefeld et al. (2023) [11]	Year-long co-design study with 112 stakeholders (clinicians & developers)	XAI for predicting delayed cerebral ischemia (DCI) in neuro-ICU	Qualitative (focus groups, co-design feedback)	Mismatch between developers' data-driven focus and clinicians' need for clinical plausibility; importance of rapid, actionable explanations	Recommend causal inference & personalized explanations; bridging patient- vs. data-centric reasoning
Rosenbacke et al. (2024) [29]	Systematic review of 10 empirical studies on XAI & trust	How XAI influences clinicians' trust in AI-based CDSS	Comparative synthesis of trust outcomes	5 studies: ↑ trust; 3: no effect; 2: mixed effects; clear, relevant explanations ↑ trust, complex/confusing ones ↓ trust	Trust calibration critical—avoid both overtrust and undertrust; explanation quality decisive

(continued)

Table 4 (continued)

Study title	Methods	Task explained	Evaluation metrics	Key insights	Context outcome
Räz et al. (2025) [34]	Conceptual framework; stroke management scenario (imaginary)	Roadmap for integrating XAI into multi-step clinical routines	Conceptual (no performance metrics)	Three challenges: (1) context-/user-dependent explanations, (2) genuine AI–clinician dialogue, (3) socially intelligent AI	XAI must evolve from static outputs (heatmaps) to interactive, adaptive partners in care
Baig et al. (2024) [5]	Review of generative AI in personalized care plans; focus on explainability & transparency	Generative AI for individualized treatment planning	Conceptual (no performance metrics)	Black-box nature is barrier; transparency as important as accuracy; robust oversight essential	Adoption requires combining GenAI with XAI/rule-based systems to ensure understandable, trusted outputs

3.2 Applications Beyond Clinical Settings

Personalized Medicine

Personalized (precision) medicine is the application of patient-specific information, including that on their genetics, environment, and lifestyle, in customizing medical treatments. This has the potential to make care much more efficient and effective. AI has shown promise in precision medicine in identifying complex patterns that relate an individual's patient characteristics to the most suitable therapies, for instance, matching patients to the most effective drug. With this use of AI, however, trust formed by the clinicians in these therapeutic recommender systems is dependent on the degree of transparency. Explainable AI (XAI) techniques are therefore very important in transforming AI insights into understandable human rationale, consequently building clinician confidence in precision medicine applications [4]. XAI methods would serve to narrow the gap between complex models and the expert knowledge of medicine and ultimately determine whether AI-guided personalized treatments will find acceptance in practice [4]. By way of example, explainable deep learning models would highlight patient-specific factors influencing a prediction, by which healthcare providers may validate whether or not the factors correlate with clinical reasoning [4]. Development of XAI approaches specific to a particular patient's data and needs for personalized healthcare has thus been emphasized by researchers [10]. Overall, XAI is thought to be a catalyst for trustworthy precision medicine, in that it helps justify individualized AI-recommended interventions and enhances the collaborative human-AI approach to personalized care [4, 10] (Table 5).

Table 5 Summary of case studies in personalized medicine

Study title	Methods	Task explained	Key insights	Context/ Outcomes
Allen (2024) [4]	Conceptual/ review; application of AI + XAI in precision medicine	Using AI to identify patient-specific patterns (genetics, environment, lifestyle) for personalized therapies	AI can match patients to optimal treatments; trust depends on transparency; XAI is essential to explain rationale	XAI bridges black-box AI and clinical reasoning, enabling trustworthy personalized care
Sadeghi (2024) [10]	Methodological review; categorization of XAI techniques (e.g., SHAP, LIME, Grad-CAM, concept-based models)	Explainability in high-risk domains, with focus on personalized/ clinical decision-making	Physicians require localized, patient-specific explanations; different XAI methods offer complementary strengths	XAI ensures transparency, interpretability, and adoption of AI in healthcare, especially in precision medicine

Drug Discovery and Genomics

Explainable AI (XAI) operationalizes transparency across the drug-discovery pipeline by revealing which molecular features, substructures, or clinical covariates drive model outputs, mitigating the "black-box" barrier to scientific and regulatory adoption. In early discovery, post-hoc attribution (SHAP, LIME, Integrated Gradients, DeepLIFT), attention over molecular graphs/SMILES, and gradient-based saliency tie predicted activity, binding affinity, and ADMET liabilities to concrete chemotypes—enabling rational hit prioritization and lead optimization rather than unguided screening. These explanations align computational predictions with medicinal-chemistry priors (e.g., toxicophore alerts, permeability determinants), improve candidate triage, and support audit-ready rationales for outcomes like poor absorption or metabolic instability. Beyond preclinical design, XAI strengthens decision support in drug–target/drug–drug interaction modeling and personalized medicine through interpretable DTI/DTA frameworks (e.g., DeFuseDTI, DTRE, ARGENT, DCGAN-DTA) that clarify drivers of predicted response and safety signals, surface bias for remediation, and aid cohort selection and clinical-trial optimization. Tools are increasingly expanding across multimodal and structure-aware platforms—ranging from AlphaFold 3 for structure and context, to graph-XAI systems like GraphIX, instruction-tuned molecular models like InstructMol, and end-to-end discovery platforms like PandaOmics. These advancements are helping to create mechanism-aware explanations and testable hypotheses. At the ecosystem scale, collaborative efforts like MELLODDY illustrate how privacy-preserving, federated learning can underpin explainable, industry-grade models without exposing proprietary data. Still, the field must navigate data scarcity, interpretability–accuracy trade-offs, and the requirement for scientifically meaningful, regulator-ready explanations. Collectively, XAI reframes AI from mere predictors into trustworthy, mechanism-anchored decision tools that can accelerate discovery while preserving rigor and reproducibility [35].

In high-dimensional genomics, XAI bridges the gap between model accuracy and biological/clinical understanding by either making architectures intrinsically more transparent or layering post-hoc explanations on existing models so that the contributions of genes, sequence motifs, and molecular pathways to predictions are explicit and trustworthy. A systematic mapping of the literature shows that feature-relevance methods and visual explanations are the dominant "common language" of interpretability in omics: CNNs on DNA sequences expose their internal logic through motif visualizations, DNNs on gene-expression data quantify gene/pathway contributions with SHAP-family and DeepLIFT analyses, and transformer models surface SNP–SNP interactions in complex diseases. This data–method alignment turns bare predictions into auditable mechanistic leads for biomarker selection, experimental design, and the interpretation of molecular signatures. At the same time, the research map highlights important gaps: XAI remains underused in areas such as 3D-genomics and several data modalities; local explanations and architecture simplification are applied less than their potential warrants; and the real-world impact of explanations on expert decision-making calls for study-designed evaluation. The most promising path is to combine multiple explanation methods within a single, coherent pipeline and to inject biological priors into model architectures, so explanations are both statistically persuasive and human-comprehensible; in this way, XAI advances from an add-on to a foundational mechanism that translates genomic predictions into mechanism-anchored evidence ready for clinical use [36].

Public Health and Epidemiology

In public health and epidemiology, explainable AI is becoming a valuable tool for gaining population-level health insights and supporting decision-making. AI systems have been suggested for tasks like disease surveillance, outbreak prediction, and health resource allocation, but their effectiveness depends on stakeholders' trust in the predictions they provide [17]. XAI can help by making epidemiological models more transparent—for example, by explaining how an AI model uses mobility data, infection rates, and other factors to predict the spread of an infectious disease [17]. During the COVID-19 pandemic, for example, researchers used XAI in predictive models to identify which factors (e.g., social distancing metrics, demographic factors) influenced predictions of case surges, helping shape public health interventions [17]. More broadly, explainable machine learning is used in disease risk modeling to explain why certain populations or individuals are identified as high-risk, helping to support preventative strategies [9]. A recent systematic review of XAI in disease prediction found that methods like SHAP and LIME were commonly used to interpret models predicting outcomes in areas like oncology and cardiology, highlighting XAI's role in making epidemiological predictions more understandable for health officials [9]. By offering human-understandable reasons for predictions—such as showing which environmental or social factors have the greatest impact on a community's health risk—XAI helps public health experts validate and communicate model results [9]. Ultimately, in public health and epidemiology, XAI enhances the credibility and accountability of AI-driven analyses. It helps shape data-driven health policies by providing clarity on why the AI suggests certain actions, ensuring that decision-makers can trust and effectively act on the algorithm's insights [10] (Table 6).

4 Conclusion

Explainable Artificial Intelligence (XAI) in health care is also critical since the stakes involve patient safety, clinic trust, and regulatory guidelines. This section shows, however, that even though AI models have proven wonderful predictive power in areas such as medical imaging, electronic health records (EHR), genomics, and personalized medicine, the fate of AI use will be dependent on how much interpretable and clinically meaningful explanations can be provided. Different classes of methods in XAI will cater to different application domains: gradient and activation-based methods (e.g. Grad-CAM, LRP) fit very well into medical imaging in which spatial localization of the relevant regions helps a radiologist to validate AI decisions;

Table 6 Summary of case studies in public health and epidemiology

Study (Year)	Task/Application	XAI method(s)	Main findings/results
Okada et al. (2023) [9, 17]	AI in emergency and public health—overview of XAI needs in emergency medicine and health system applications	Not specific (Conceptual review)	Noted that AI can aid epidemiological surveillance and resource allocation in public health, but explainability is vital for acceptance. Stressed that XAI allows clinicians and health officials to understand and trust AI recommendations in high-stakes settings [17]
Alkhanbouli et al. (2025) [9]	Disease prediction models—systematic review of XAI in various disease risk and outcome prediction studies	Commonly SHAP and LIME (most frequent in reviewed studies)	Across 30 studies, XAI improved transparency of predictive models in domains like oncology and epidemiology. The review found SHAP and LIME explanations frequently used to interpret model outputs, helping to identify key predictors in disease outcomes. It highlighted that while XAI builds trust in AI-driven predictions, challenges such as limited data diversity still need addressing for reliable public health AI [9]

perturbation-based and feature-importance methods (e.g. LIME, SHAP) fit very well into electronic health records and tabular clinical data where a risk factor ranking will provide actionable insights on prognosis; and rule-based or surrogate models are most useful in clinical decision support systems where transparent "if–then" causes trust, auditability, and alignment with guidelines. To incorporate applications across the board, XAI finds its strengths and limitations. In imaging, visual saliency maps enable quick and intuitive validation but may lack cause and be sensitive to model architecture. For risk prediction based on EHR, SHAP values can provide explanations on the level of detail provided by features, but the experience in terms of cost and degree of sensitivity when it comes to data imbalance may render its application less feasible for real-time scenarios. Model-agnostic approaches explain the process of revealing such interaction on features without identifying them as special; however, caution should be exercised in that the ability to create spurious explanations increases with the dimensionality. For personalized treatment, the same as for multimodal applications, the requirement for mixed heterogeneous data calls for a hybrid explanatory setup, in which multiple methods are orchestrated to achieve consistency across modalities. Across all these domains, the challenges of stability of explanations, actionability for clinicians, as well as fairness of explanation across subpopulations are still open.

References

1. Weiner EB, Dankwa-Mullan I, Nelson WA, Hassanpour S. Ethical challenges and evolving strategies in the integration of artificial intelligence into clinical practice. PLOS Digital Health [Internet]. Public Library of Science; 2025 [cited 2025 Sep 9];4:e0000810. https://doi.org/10.1371/JOURNAL.PDIG.0000810
2. Salahuddin Z, Woodruff HC, Chatterjee A, Lambin P. Transparency of deep neural networks for medical image analysis: A review of interpretability methods. Comput Biol Med. Elsevier Ltd; 2022. https://doi.org/10.1016/j.compbiomed.2021.105111
3. Hafeez Y, Memon K, AL-Quraishi MS, Yahya N, Elferik S, Ali SSA. Explainable AI in Diagnostic Radiology for Neurological Disorders: A Systematic Review, and What Doctors Think About It. Diagnostics 2025, Vol 15, Page 168 [Internet]. Multidisciplinary Digital Publishing Institute; 2025 [cited 2025 Sep 9];15:168. https://doi.org/10.3390/DIAGNOSTICS15020168
4. Hahn J, Giansanti D, Review S, Allen B. The Promise of Explainable AI in Digital Health for Precision Medicine: A Systematic Review. Journal of Personalized Medicine 2024, Vol 14, Page 277 [Internet]. Multidisciplinary Digital Publishing Institute; 2024 [cited 2025 Sep 9];14:277. https://doi.org/10.3390/JPM14030277
5. Baig MM, Hobson C, GholamHosseini H, Ullah E, Afifi S. Generative AI in Improving Personalized Patient Care Plans: Opportunities and Barriers Towards Its Wider Adoption. Applied Sciences 2024, Vol 14, Page 10899 [Internet]. Multidisciplinary Digital Publishing Institute; 2024 [cited 2025 Sep 9];14:10899. https://doi.org/10.3390/APP142310899
6. Startari A V. AI, Tell me your protocol: The Intersection of Technology and Humanity in the Era of Big Data. SSRN Electronic Journal [Internet]. Elsevier BV; 2025 [cited 2025 Sep 9]; https://doi.org/10.2139/SSRN.5260083
7. Nazir S, Dickson DM, Akram MU. Survey of explainable artificial intelligence techniques for biomedical imaging with deep neural networks. Comput Biol Med [Internet]. Pergamon; 2023 [cited 2025 Sep 9];156:106668. https://doi.org/10.1016/J.COMPBIOMED.2023.106668

8. Di Martino F, Delmastro F. Explainable AI for clinical and remote health applications: a survey on tabular and time series data. Artif Intell Rev [Internet]. Springer Nature; 2023 [cited 2025 Sep 9];56:5261–315. https://doi.org/10.1007/S10462-022-10304-3/TABLES/11
9. Alkhanbouli R, Matar Abdulla Almadhaani H, Alhosani F, Simsekler MCE. The role of explainable artificial intelligence in disease prediction: a systematic literature review and future research directions. BMC Med Inform Decis Mak [Internet]. BioMed Central Ltd; 2025 [cited 2025 Sep 9];25:1–17. https://doi.org/10.1186/S12911-025-02944-6/TABLES/2
10. Sadeghi Z, Alizadehsani R, CIFCI MA, Kausar S, Rehman R, Mahanta P, et al. A review of Explainable Artificial Intelligence in healthcare. Computers and Electrical Engineering [Internet]. Pergamon; 2024 [cited 2025 Sep 9];118:109370. https://doi.org/10.1016/J.COMPELECENG.2024.109370
11. Bienefeld N, Boss JM, Lüthy R, Brodbeck D, Azzati J, Blaser M, et al. Solving the explainable AI conundrum by bridging clinicians' needs and developers' goals. NPJ Digit Med [Internet]. Nature Research; 2023 [cited 2025 Sep 9];6:1–7. https://doi.org/10.1038/S41746-023-00837-4;SUBJMETA=689,692,700,706;KWRD=HEALTH+CARE,SOCIAL+SCIENCES
12. Freyer N, Groß D, Lipprandt M. The ethical requirement of explainability for AI-DSS in healthcare: a systematic review of reasons. BMC Med Ethics [Internet]. BioMed Central Ltd; 2024 [cited 2025 Sep 9];25:1–11. https://doi.org/10.1186/S12910-024-01103-2/FIGURES/4
13. Brankovic A, Cook D, Rahman J, Delaforce A, Li J, Magrabi F, et al. Clinician-informed XAI evaluation checklist with metrics (CLIX-M) for AI-powered clinical decision support systems. NPJ Digit Med [Internet]. Nature Research; 2025 [cited 2025 Sep 9];8:1–6. https://doi.org/10.1038/S41746-025-01764-2;SUBJMETA=308,479,575,648,692,706;KWRD=PUBLISHING,TRANSLATIONAL+RESEARCH
14. Joyce DW, Kormilitzin A, Smith KA, Cipriani A. Explainable artificial intelligence for mental health through transparency and interpretability for understandability. NPJ Digit Med [Internet]. Nature Research; 2023 [cited 2025 Sep 9];6:1–7. https://doi.org/10.1038/S41746-023-00751-9;SUBJMETA=114,631,692,700;KWRD=COMPUTATIONAL+BIOLOGY+AND+BIOINFORMATICS,HEALTH+CARE
15. Kyrimi E, McLachlan S, Wohlgemut JM, Perkins ZB, Lagnado DA, Marsh W, et al. Explainable AI: definition and attributes of a good explanation for health AI. AI and Ethics 2025 5:4 [Internet]. Springer; 2025 [cited 2025 Sep 9];5:3883–96. https://doi.org/10.1007/S43681-025-00668-X
16. Mesinovic M, Watkinson P, Zhu T. Explainability in the age of large language models for healthcare. Communications Engineering [Internet]. Springer Nature; 2025 [cited 2025 Sep 9];4:1–4. https://doi.org/10.1038/S44172-025-00453-Y;SUBJMETA=639,692,700,705;KWRD=HEALTH+CARE,MATHEMATICS+AND+COMPUTING
17. Okada Y, Ning Y, Ong MEH. Explainable artificial intelligence in emergency medicine: an overview. Clin Exp Emerg Med [Internet]. Korean Society of Emergency Medicine; 2023 [cited 2025 Sep 9];10:354. https://doi.org/10.15441/CEEM.23.145
18. Ihongbe IE, Fouad S, Mahmoud TF, Rajasekaran A, Bhatia B. Evaluating Explainable Artificial Intelligence (XAI) techniques in chest radiology imaging through a human-centered Lens. PLoS One [Internet]. Public Library of Science; 2024 [cited 2025 Sep 9];19:e0308758. https://doi.org/10.1371/JOURNAL.PONE.0308758
19. Abgrall G, Holder AL, Chelly Dagdia Z, Zeitouni K, Monnet X. Should AI models be explainable to clinicians? Crit Care [Internet]. BioMed Central Ltd; 2024 [cited 2025 Sep 9];28:1–8. https://doi.org/10.1186/S13054-024-05005-Y/TABLES/1
20. Jung J, Lee H, Jung H, Kim H. Essential properties and explanation effectiveness of explainable artificial intelligence in healthcare: A systematic review. Heliyon [Internet]. Elsevier Ltd; 2023 [cited 2025 Sep 9];9:e16110. https://doi.org/10.1016/J.HELIYON.2023.E16110/ASSET/1377163C-B7FF-4E3E-B6F4-170F9AD444D3/MAIN.ASSETS/GR1.JPG
21. Aboy M, Minssen T, Vayena E. Navigating the EU AI Act: implications for regulated digital medical products. NPJ Digit Med [Internet]. Nature Research; 2024 [cited 2025 Sep 13];7:237. https://doi.org/10.1038/S41746-024-01232-3

22. Marketing Submission Recommendations for a Predetermined Change Control Plan for Artificial Intelligence-Enabled Device Software Functions | FDA [Internet]. [cited 2025 Sep 13]. https://www.fda.gov/regulatory-information/search-fda-guidance-documents/marketing-submission-recommendations-predetermined-change-control-plan-artificial-intelligence. Accessed 13 Sep 2025
23. Art. 35 GDPR – Data protection impact assessment - General Data Protection Regulation (GDPR) [Internet]. [cited 2025 Sep 13]. https://gdpr-info.eu/art-35-gdpr/. Accessed 13 Sep 2025
24. WHO calls for safe and ethical AI for health [Internet]. [cited 2025 Sep 13]. https://www.who.int/news/item/16-05-2023-who-calls-for-safe-and-ethical-ai-for-health. Accessed 13 Sep 2025
25. for Medicair C, Services M, Learning Network M. HIPAA Basics for Providers: Privacy, Security, & Breach Notification Rules.
26. Muhammad D, Bendechache M. Unveiling the black box: A systematic review of Explainable Artificial Intelligence in medical image analysis. Comput Struct Biotechnol J [Internet]. Elsevier; 2024 [cited 2025 Sep 9];24:542–60. https://doi.org/10.1016/J.CSBJ.2024.08.005
27. Sun Q, Akman A, Schuller BW. Explainable Artificial Intelligence for Medical Applications: A Review. ACM Trans Comput Healthc [Internet]. ACMPUB27New York, NY; 2025 [cited 2025 Sep 9];6. https://doi.org/10.1145/3709367
28. Champendal M, Müller H, Prior JO, dos Reis CS. A scoping review of interpretability and explainability concerning artificial intelligence methods in medical imaging. Eur J Radiol [Internet]. Elsevier; 2023 [cited 2025 Sep 9];169:111159. https://doi.org/10.1016/J.EJRAD.2023.111159
29. Rosenbacke R, Melhus Å, McKee M, Stuckler D. How Explainable Artificial Intelligence Can Increase or Decrease Clinicians' Trust in AI Applications in Health Care: Systematic Review. JMIR AI [Internet]. JMIR Publications Inc.; 2024 [cited 2025 Sep 9];3:e53207. https://doi.org/10.2196/53207
30. Metsch JM, Hauschild AC. BenchXAI: Comprehensive benchmarking of post-hoc explainable AI methods on multi-modal biomedical data. Comput Biol Med [Internet]. Pergamon; 2025 [cited 2025 Sep 9];191:110124. https://doi.org/10.1016/J.COMPBIOMED.2025.110124
31. Pahud de Mortanges A, Luo H, Shu SZ, Kamath A, Suter Y, Shelan M, et al. Orchestrating explainable artificial intelligence for multimodal and longitudinal data in medical imaging. npj Digital Medicine 2024 7:1 [Internet]. Nature Publishing Group; 2024 [cited 2025 Sep 9];7:1–10. https://doi.org/10.1038/s41746-024-01190-w
32. Mohapatra RK, Jolly L, Dakua SP. Advancing explainable AI in healthcare: Necessity, progress, and future directions. Comput Biol Chem [Internet]. Elsevier; 2025 [cited 2025 Sep 9];119:108599. https://doi.org/10.1016/J.COMPBIOLCHEM.2025.108599
33. Nagendran M, Festor P, Komorowski M, Gordon AC, Faisal AA. Quantifying the impact of AI recommendations with explanations on prescription decision making. npj Digital Medicine 2023 6:1 [Internet]. Nature Publishing Group; 2023 [cited 2025 Sep 9];6:1–7. https://doi.org/10.1038/s41746-023-00955-z
34. Räz T, Pahud De Mortanges A, Reyes M. Explainable AI in medicine: challenges of integrating XAI into the future clinical routine. Frontiers in Radiology. Frontiers Media SA; 2025;5:1627169. https://doi.org/10.3389/FRADI.2025.1627169/BIBTEX
35. Qadri YA, Shaikh S, Ahmad K, Choi I, Kim SW, Vasilakos A V. Explainable Artificial Intelligence: A Perspective on Drug Discovery. Pharmaceutics [Internet]. 2025;17:1119. https://doi.org/10.3390/pharmaceutics17091119
36. Toussaint PA, Leiser F, Thiebes S, Schlesner M, Brors B, Sunyaev A. Explainable artificial intelligence for omics data: a systematic mapping study. Brief Bioinform. Oxford University Press; 2024. https://doi.org/10.1093/bib/bbad453

Explainability and Reliability of Large Language Models in Health Systems

Javad Jahangiri Darzehkonani, Azadeh Zamanifar, and Farhad Soleimanian Gharehchopogh

1 Introduction

The digital transformation in recent years has fundamentally transformed healthcare and medical systems. One of the most important developments is the emergence of Large Language Models (LLMs), models that are capable of understanding natural language, generating text, and interacting intelligently with users. It has quickly found a niche in medicine: from clinical documentation and extracting information in electronic health records to supporting clinical decision-making, helping diagnose diseases, assessing symptoms, and even educating patients [1, 2]. However, the arrival of LLMs in the healthcare field does not come with only opportunities. The "black box" nature of these models means that their internal processes, including the decision-making logic, the source of the data used, and the validity of the arguments, often remain opaque. It has become a significant concern in medicine, where decisions can directly affect patients' lives. Transparency, trust, and explainability are crucial in such an environment because medics and patients need to be able to understand and trust the reasoning behind a suggestion or result provided by the system.

In addition to the issue of transparency, challenges such as data bias, the production of incorrect or outdated information (AI illusions), as well as issues related to security and accountability in the use of LLMs, have made it more important to pay attention to explainability [3]. In the meantime, a new field called Explainable Artificial Intelligence (XAI) [4] is trying to build a bridge between algorithmic complexities and clinical needs, and gain the trust of human users [5].

J. J. Darzehkonani · F. S. Gharehchopogh (✉)
Department of Computer Engineering, Ur, C., Islamic Azad University, Urmia, Iran
e-mail: bonab.farhad@gmail.com

A. Zamanifar
Department of Computer Engineering, SR, C., Islamic Azad University, Tehran, Iran

A. Zamanifar et al. (eds.), *Explainable Large Language Models in Healthcare Applications*, Bio-IT and AI, https://doi.org/10.1007/978-3-032-15088-2_3

This book chapter examines the various dimensions of explainability, evaluation, and security of LLMs in health systems. First, the challenges arising from the black-box nature of these models and their clinical implications are reviewed, and then key technologies and tools for transparency are introduced. Next, existing approaches to evaluating the performance and security of LLMs will be reviewed, and finally, accountability, trust, and future directions in this field will be discussed.

2 The "Black Box" Challenge: The Challenge of Ambiguities in LLM Decisions

Large language models are often referred to as black box systems due to their complex architecture and large parameter sizes. This term means that although the model outputs are visible to the user, the exact path to those outputs is not transparent and traceable [6]. In areas such as machine translation or general text generation, this ambiguity may not be a serious threat. Still, in medicine, where every decision can have direct consequences for human health or life, the lack of transparency is a potential risk [7].

One of the most important consequences of this ambiguity is bias. Many LLMs are trained on internet data or public text sources that are themselves imbued with racial, gender, or cultural biases. If left unchecked in the clinical setting, these biases may lead to unfair or discriminatory decisions between patients and exacerbate existing inequalities [6]. Another challenge is hallucinations in LLMs that refer to the production of information that appears linguistically correct and even convincing, but has no scientific or clinical basis in reality. For example, an LLM may recommend a drug to treat a disease that does not exist or has not been supported by valid medical studies. Such errors in clinical settings can have serious consequences for patients [8]. Furthermore, the origin of the information and reasoning process of LLMs is often unclear. Clinicians who rely on these tools usually cannot tell whether a recommendation is based on valid scientific evidence or is simply the result of statistical patterns in the training data. This lack of traceability makes it difficult to trust the model's outputs [9].

A concrete clinical example: Imagine an LLM being used as a physician assistant in the emergency department to triage patients [10]. A patient presents with symptoms of severe headache and high blood pressure. Based on similar text data in its training database, the model suggests the possibility of "migraine" and recommends supportive treatment. However, in reality, the patient has a "cerebral hemorrhage" that requires immediate intervention. The lack of a clear explanation of why the model selected "migraine" can mislead the physician and have dangerous consequences for the patient. This example shows that without explainability, even plausible outputs can lead to severe clinical error. The hybridization of these factors makes it impossible to trust and accept LLMs in the health system without transparency of their internal processes. Therefore, scientific and industrial efforts have been directed towards the development of Explainable Artificial Intelligence (XAI). It aims to clarify the logic of decision-making and create human evaluation capabilities [11].

3 Explainability in Medicine: A Bridge Between Algorithm and Physician

For LLMs to be safely and effectively deployed in the healthcare system, it is not enough to have high accuracy; it is also necessary for physicians, patients, and regulatory agencies to understand why a model made a particular decision, which is where the concept of Explainable Artificial Intelligence (XAI) comes into play [12]. XAI refers to a set of methods and tools that aim to elucidate the internal logic of complex machine learning systems. Simply put, XAI attempts to answer the question: "Why did the model produce the result it did"?". In an environment like medicine, answering this question is crucial, as an inappropriate decision can directly harm the health or even the lives of patients. Figure 1 shows Conceptual distribution of "Transparency" and "Interpretability".

The benefits of explainability in multidimensional medicine are as follows:

- Building clinical trust: A physician will only trust an intelligent system if she can understand its logic and compare it with her medical knowledge [13].
- Reducing error and bias: By revealing the factors affecting the model's decision-making, hidden biases or incorrect data can be identified and corrected [14].
- Enhance teaching and learning: Explainable outcomes not only help physicians but can also be an effective educational tool for medical students [15].
- Support for shared human–machine decision-making: Clear explanations provide a foundation for safer collaboration between clinicians and AI systems [15].

For example, if an LLM answers the question "What is the cause of the patient's chest pain?" and gives the diagnosis "stable angina," a plausible explanation could show why symptoms such as pain on activity, a history of diabetes, or ECG changes contributed most to the outcome. Such transparency allows the clinician to validate or reject the model's decision in light of available clinical evidence [16]. In this way, XAI in medicine can be seen as a bridge between the algorithm and the clinician, a bridge without which the widespread and safe use of large language models in the healthcare system would be seriously challenged. Figure 2 shows a mind Map of various XAI methods.

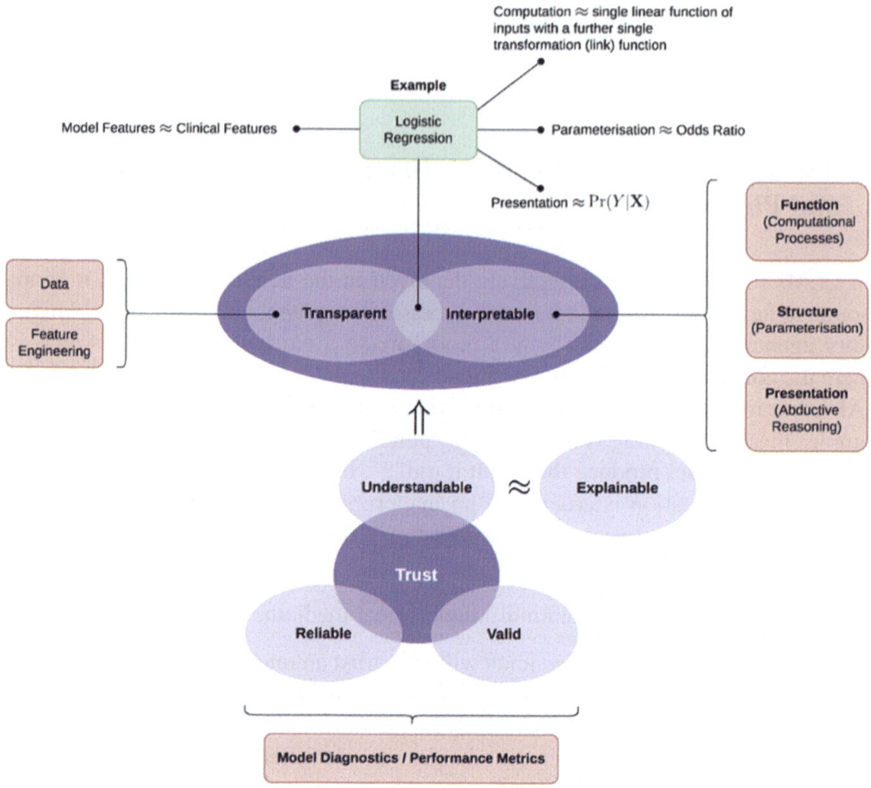

Fig. 1 Conceptual distribution of "Transparency" and "Interpretability"

4 Key Technologies and Making LLMs Transparent Approach

To disambiguate and increase trust in large linguistic models, a set of explainable artificial intelligence (XAI) methods and tools has been developed. Each of these technologies attempts to elucidate the internal logic of the model from a specific angle, allowing clinicians to better understand the reasons for the model's decisions [17].

4.1 SHAP, LIME and Local Explanation of Decisions

Shapley Additive exPlanations (SHAP) and Local Interpretable Model-agnostic Explanations (LIME) methods for explaining model decisions. SHAP is based on game theory and calculates the contribution of each input feature to the model's final

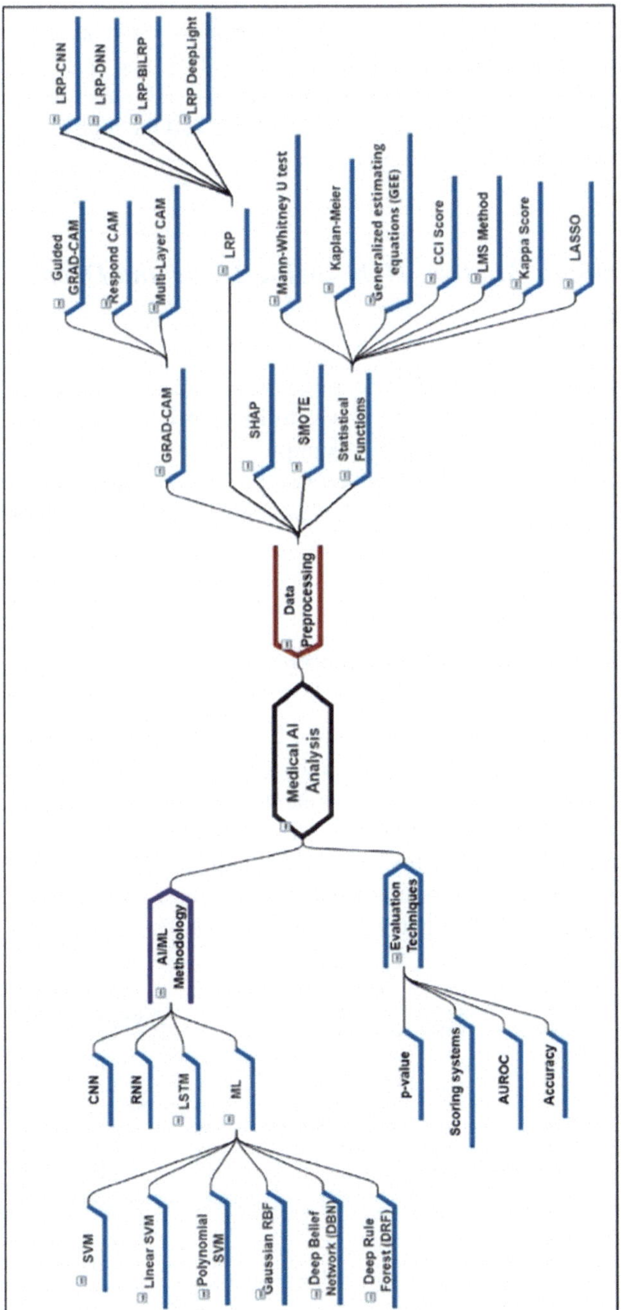

Fig. 2 A mind map of various XAI methods

decision. For example, in diagnosing diabetes, it can show that blood glucose levels contribute 40% and body mass index 20% to the model's outcome [18]. LIME, by building a simple, local model around a specific input, shows why the original model reached the conclusion it did. For example, it explains that "the presence of shortness of breath and chest pain" predicts heart disease [19]. Although these methods contribute to transparency, the stability of LIME results and the high computational cost of SHAP are their well-known limitations.

4.2 *Neurosymbolic Models: Hybridizing Logic and Deep Learning*

Neurosymbolic models are a hybrid of neural learning and rule-based logic. In this approach, neural networks provide the power to recognize complex patterns, and symbolic rules help to provide clarity. For example, if blood pressure is greater than 180 and there is a severe headache, the likelihood of a hypertensive crisis is high. These types of rules make the model's decisions more tractable for the physician [20].

4.3 *Chain-of-Thought (Cot) Argument*

One of the new approaches to engaging with LLMs is to have them explain their reasoning step by step. This technique is called CoT Prompting. Rather than just giving the final answer, the model also explains the path of its reasoning [21]. For example, instead of giving a straight answer, "The patient is having a heart attack" the model explains: So the probability of a heart attack is high → ECG shows ST changes → Has a history of diabetes → Patient has chest pain. Figure 3 shows difference between Chain of Thoughts (CoT) and Tree-of-Thought approaches.

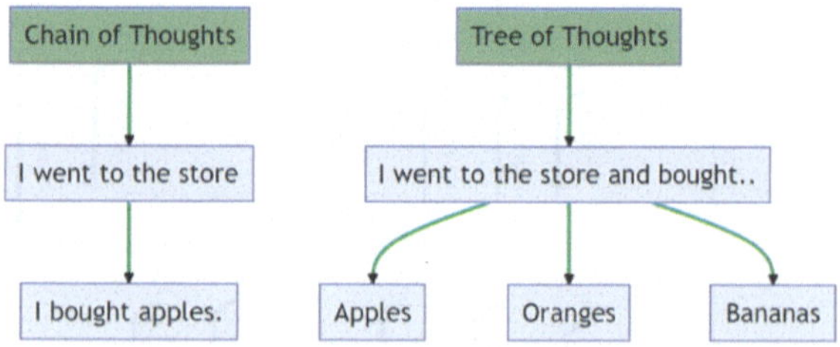

Fig. 3 Difference between chain of thoughts (CoT) and tree-of-thought approaches

5 Measuring the Performance of LLMs in a Clinical Environment

Evaluating the performance of large language models in the healthcare domain is of critical importance, as their decisions and recommendations can directly affect the fate of patients. Unlike many areas of technology, in medicine, accuracy alone is not enough; other dimensions, such as safety, empathy, comprehensiveness, and trustworthiness, must also be measured [22].

5.1 Limitations of Automatic Criteria

Traditional metrics in natural language processing, such as Accuracy, F1-score, BLEU, and ROUGE, are primarily designed for tasks such as machine translation or text summarization, but LLMs in medicine have a "generative" role; that is, they generate new text and suggestions, not just extract or classify. Therefore, these automated metrics cannot fully measure the actual quality of the model output in clinical scenarios. For example, an output can be linguistically correct, but scientifically or ethically completely wrong [23].

5.2 Human Assessment as the Gold Standard

For this reason, human evaluation remains the gold standard in healthcare. Clinicians and professionals should evaluate the outputs of LLMs in real-world scenarios to determine whether these systems can contribute to patient safety, reduce errors, and improve the quality of care [24]. These evaluations typically include:

- Correct understanding and interpretation of the clinical question (e.g., differentiating between a migraine headache and a cerebral hemorrhage).
- Accuracy and comprehensiveness of the answer (Have all important clinical possibilities been considered?)
- Clarity and empathy in the expression of the answer (Is the answer understandable to the patient or physician and presented in a human tone?)

5.3 QUEST Framework for Comprehensive Assessmen

One of the new frameworks in this field is QUEST, which stands for Quality, Understanding, Expression, Safety, Trust, and examines five main dimensions [24]:

- Q: Information quality—accuracy, completeness, timeliness.

- U: Comprehension and reasoning—the model's ability to interpret correctly and present logical arguments.
- E: Style and personality—clarity, empathy, and appropriateness of tone to the audience.
- S: Safety and harm—avoidance of bias, misinformation, or dangerous information.
- T: Trust and confidence—user (physician or patient) satisfaction and trust in the output.

This framework not only helps with qualitative measurement but can also serve as a basis for comparing different models in the health field.

5.4 Practical Methods of Evaluation

Practical assessment of LLMs in clinical settings typically involves the following steps [25]:

- Sample selection: Use diverse data, including different populations, different diseases, and real clinical scenarios.
- Clinical prompt design: Questions derived from medical records, emergency scenarios, or standard medical tests.
- Evaluation tools: Use questionnaires (such as Likert scales), comparisons with gold labels, and qualitative analyses by expert teams.
- Statistical analysis: Use statistical tests such as t-test, chi-square, and McNemar to compare the accuracy and reliability between model outputs and human decisions.

5.5 Continuous Monitoring in the Real World

Even if LLMs produce acceptable results in the laboratory, they face new challenges in the real world: new data, different languages, and unexpected clinical situations. Therefore, there is a need for continuous monitoring. This monitoring can include monitoring the inputs, outputs, biases, and even the level of uncertainty of the model. Only in this way can we ensure that LLMs remain safe, accurate, and reliable over time [26].

6 Security, Vulnerability, and Risk Mitigation Strategies in Medical LLMs

The introduction of large language models into the medical field brings with it new opportunities, as well as threats and risks. Due to their complex and data-driven nature, these models are vulnerable to fundamental vulnerabilities and unintended exploitation. Therefore, investigating security challenges and developing risk mitigation solutions is crucial to ensure patient safety and user trust [27].

6.1 Fundamental Problems: Data, Bias and Hallucinations

- Data quality: Many LLMs are trained on public internet data that may contain incomplete, outdated or unscientific information. This increases the risk of producing incorrect or misleading outputs [28].
- Bias: The presence of racial, gender or cultural biases in training data can lead to discriminatory decisions and exacerbate health inequalities [28].
- Hallucinations: LLMs may provide seemingly valid but fabricated information, such as referring to non-fiction articles or recommending unproven treatments. In a clinical setting, such errors can have serious consequences for patient health [28].

6.2 AI-CWE and Vulnerability Identification Frameworks

For structured vulnerability management, frameworks such as Artificial Intelligence—Common Weakness Enumeration (AI-CWE) have been proposed. These frameworks operate similarly to CWE in software security and list AI-specific weaknesses [29]. Examples of these weaknesses include:

- Over-reliance on limited training data.
- Individuals often lack self-awareness and warning mechanisms in situations of uncertainty.
- Vulnerability to data poisoning or prompt injection attacks.

Such frameworks allow for the systematic classification and monitoring of LLM vulnerabilities and help healthcare organizations take preventive measures.

6.3 New Solutions: RAG, AIBoM and Continuous Monitoring

- Retrieval-Augmented Generation (RAG): This approach augments the output of the LLM using reliable and up-to-date data sources. As a result, the risk of bias and outdated information is reduced. For example, an LLM can connect directly to the FDA's database of approved drugs when answering a drug query [30].
- AI Bill of Materials (AIBoM): Similar to SBOM in software, it helps to make the components of AI models transparent. This approach allows regulatory bodies to examine and, if necessary, trace the origin of data, algorithms, and dependencies [31].
- Continuous monitoring and supervision: The deployment of LLMs in the clinical setting should be accompanied by constant monitoring of outcomes. Indicators such as accuracy, fairness, correlation with clinical evidence, and level of uncertainty should be assessed periodically. This monitoring can be done by a combination of automated tools and human review to prevent clinical errors [32] (Table 1)

In conclusion, the security and reliability of LLMs are not possible without vulnerability management and the use of innovative solutions. Combining vulnerability identification frameworks with advanced techniques such as RAG and AIBoM, along with continuous human monitoring, can pave the way for the safe and responsible deployment of this technology in health systems..

Table 1 Challenges, consequences and risk reduction strategies in medical LLMs

Suggested solution	Clinical outcome	Fundamental challenge
Using reliable medical data, specialized databases, and RAG	Producing incomplete, outdated, or unscientific answers	Low data quality
Data diversification, human review, ethical frameworks	Discriminatory decisions exacerbate health inequalities	Bias in educational data
RAG to connect to trusted sources, continuous human monitoring	Production of fake articles and drugs, invalid treatments	Hallucinations
AI-CWE to identify weaknesses, increase sample diversity	Reduced the generalizability of the model across diverse populations	Over-reliance on limited data
Using AI-CWE, designing robust models	Possibility of malicious exploitation or bad decisions	Data/prompt manipulation vulnerability
Using AIBoM to clarify model components	Difficulty tracking errors and accountability	Lack of transparency of model components
Deploying automated monitoring systems + human review	Increased clinical errors without timely diagnosis	Lack of continuous monitoring

7 Summary and Evaluation

Large Language Models (LLMs) have provided significant opportunities for improving health systems in recent years, but they also have serious limitations and challenges. A review of recent studies shows that each of the proposed approaches has its own strengths and weaknesses in the areas of explainability, evaluation, and security.

Strengths of current approaches:

- Explainability methods (SHAP, LIME, Neurosymbolic, Chain-of-Thought) have taken an essential step towards transparency and allow clinicians to understand the decision-making logic of models better.
- Evaluation frameworks such as QUEST cover different dimensions of quality, safety and trust and are used as standard tools to review the clinical performance of LLMs.
- New security strategies (RAG, AIBoM, continuous monitoring) have been developed to reduce risk and prevent the generation of false information, opening new horizons for the secure deployment of LLMs.

Limitations and weaknesses:

- Many of the explainability methods are not yet stable and unified, and their results may vary for similar inputs.
- Human assessment, although considered the "gold standard", is expensive, time-consuming, and dependent on individual judgment.
- Risk reduction tools such as RAG require a strong and up-to-date data infrastructure, and in its absence, there remains the possibility of bias or outdated information.
- Legal regulations and ethical frameworks in many countries are still incomplete or under development, which makes widespread adoption and deployment uncertain.

Evaluation Summary:

Although significant progress has been made in improving the transparency, trustworthiness and security of LLMs, no single approach is sufficient. Hybridizing multiple tools and frameworks (e.g., using SHAP to explain decisions, QUEST for evaluation, and RAG to reduce errors) can be helpful in practice. In addition, the development of international guidelines and binding rules will be crucial to ensure trust and security in the future (Table 2).

8 Conclusion and Future Directions

Large Language Models (LLMs) have quickly established a niche in healthcare systems, transforming everything from clinical data documentation and extraction to decision support and patient and physician education. However, the black-box

Table 2 Comparison of key approaches of LLMs in the health system

Clinical application example	Weakness	Strength	Approach/Tools
Explaining the role of risk factors in predicting diabetes	High computation, difficulty in using big data	Determining the contribution of each feature to the final model decision, high transparency	SHAP
Explaining model selection in heart disease diagnosis	Instability of results, sensitive to input changes	Local simplification of the model decision, understandable to the user	LIME
Diagnosis of hypertensive crisis based on clinical symptoms and medical rules	Design complexity, need for strict rules	Combining explicit logic and deep learning, greater transparency	Neurosymbolic models
Explaining the diagnosis of heart attack by mentioning the clinical stages	Depending on the quality of the prompt, there is a possibility of errors in reasoning	Revealing the path of reasoning step by step	Chain-of-Thought (CoT)
Review of LLM response in emergency triage	Need for a human evaluator, time-consuming	Multidimensional assessment (quality, empathy, trust, safety)	QUEST Framework
Using credible scientific sources to answer pharmaceutical questions	Requires a strong and up-to-date database	Reducing illusions and using up-to-date data	RAG Recovery-Augmented Production
Tracing the origin of data and components used in the Medical LLM	In the early stages of development, implementation is limited in practice	Transparency in model components and dependencies	AIBoM

nature of these models, coupled with challenges such as bias, illusions, and security vulnerabilities, has made explainability and reliability key requirements for their secure deployment. This chapter showed that tools such as SHAP, LIME, neurosymbolic models, and Chain-of-Thought provide essential avenues for clarifying LLM decisions. The QUEST framework also highlights key dimensions such as accuracy, comprehensiveness, empathy, safety, and trust in evaluating the performance of models. On the other hand, approaches such as RAG, AIBoM, and continuous monitoring were introduced as novel steps to reduce risk and increase trust. Despite these advances, no single approach is sufficient. Hybrids of multiple technologies and frameworks, combined with human oversight, remain the best strategy for the safe use of LLMs in medicine. Furthermore, regulations such as the EU AI Act indicate that ethical rules and frameworks will increasingly play a role in ensuring trustworthiness and accountability.

Future directions of research and development:

- Dynamic and real-time XAI: Developing tools that can explain the reasoning behind a model's decision in real time.
- Hybrid models: Combining deep learning with explicit logic to increase transparency.
- Global assessment frameworks: Establishing international standards for measuring and monitoring medical LLMs.
- Interdisciplinary research: Collaboration between clinicians, data scientists, and lawyers to ensure safe and ethical deployment.

Strengthening human–AI teamwork: Designing systems that facilitate shared decision-making between clinicians and AI, rather than being a complete replacement. The future of LLMs in medicine is bright, but only if explainability, security, and accountability are considered as fundamental principles. The path forward requires a simultaneous scientific, technological, and regulatory effort to ensure that these models can safely and effectively serve patients and the medical community.

References

1. Maity, S. and M.J. Saikia, *Large Language Models in Healthcare and Medical Applications: A Review.* Bioengineering, 2025. **12**(6): p. 631.
2. Zhang, K., et al., *Revolutionizing health care: The transformative impact of large language models in medicine.* Journal of medical Internet research, 2025. **27**: p. e59069.
3. Qin, H. and Y. Tong, *Opportunities and challenges for large language models in primary health care.* Journal of Primary Care & Community Health, 2025. **16**: p. 21501319241312571.
4. Van der Velden, B.H., et al., *Explainable artificial intelligence (XAI) in deep learning-based medical image analysis.* Medical image analysis, 2022. **79**: p. 102470.
5. Mesinovic, M., P. Watkinson, and T. Zhu, *Explainability in the age of large language models for healthcare.* Communications Engineering, 2025. **4**(1): p. 128.
6. Li, H., J.-F. Fu, and A. Python, *Implementing Large Language Models in Health Care: Clinician-Focused Review With Interactive Guideline.* Journal of Medical Internet Research, 2025. **27**: p. e71916.
7. London, A.J., *Artificial intelligence and black-box medical decisions: accuracy versus explainability.* Hastings Center Report, 2019. **49**(1): p. 15-21.
8. Kim, Y., et al., *Medical hallucinations in foundation models and their impact on healthcare.* arXiv preprint arXiv:2503.05777, 2025.
9. Busch, F., et al., *Current applications and challenges in large language models for patient care: a systematic review.* Communications Medicine, 2025. **5**(1): p. 26.
10. Shool, S., et al., *A systematic review of large language model (LLM) evaluations in clinical medicine.* BMC Medical Informatics and Decision Making, 2025. **25**(1): p. 117.
11. Riedemann, L., M. Labonne, and S. Gilbert, *The path forward for large language models in medicine is open.* npj Digital Medicine, 2024. **7**(1): p. 339.
12. Saw, S.N., Y.Y. Yan, and K.H. Ng, *Current status and future directions of explainable artificial intelligence in medical imaging.* European journal of radiology, 2025. **183**: p. 111884.

13. Räz, T., A. Pahud De Mortanges, and M. Reyes, *Explainable AI in medicine: challenges of integrating XAI into the future clinical routine.* Frontiers in Radiology, 2025. **5**: p. 1627169.
14. Alkhanbouli, R., et al., *The role of explainable artificial intelligence in disease prediction: a systematic literature review and future research directions.* BMC medical informatics and decision making, 2025. **25**(1): p. 110.
15. Aziz, N.A., et al., *Unveiling Explainable AI in Healthcare: Current Trends, Challenges, and Future Directions.* medRxiv, 2024: p. 2024.08. 10.24311735.
16. Kalasampath, K., et al., *A Literature review on applications of explainable artificial intelligence (XAI).* IEEE Access, 2025.
17. Saarela, M. and V. Podgorelec, *Recent applications of Explainable AI (XAI): A systematic literature review.* Applied Sciences, 2024. **14**(19): p. 8884.
18. Salih, A.M., et al., *A perspective on explainable artificial intelligence methods: SHAP and LIME.* Advanced Intelligent Systems, 2025. **7**(1): p. 2400304.
19. Paul, J., *LIME and SHAP Interpretability for Medical AI Systems.* 2024.
20. Nawaz, U., M. Anees-ur-Rahaman, and Z. Saeed, *A review of neuro-symbolic AI integrating reasoning and learning for advanced cognitive systems.* Intelligent Systems with Applications, 2025: p. 200541.
21. Jeon, S. and H.-G. Kim, *A comparative evaluation of chain-of-thought-based prompt engineering techniques for medical question answering.* Computers in Biology and Medicine, 2025. **196**: p. 110614.
22. Budler, L.C., et al., *A Brief Review on Benchmarking for Large Language Models Evaluation in Healthcare.* Wiley Interdisciplinary Reviews: Data Mining and Knowledge Discovery, 2025. **15**(2): p. e70010.
23. Xiaolan, C., et al., *Evaluating large language models and agents in healthcare: key challenges in clinical applications.* Intelligent Medicine, 2025.
24. Tam, T.Y.C., et al., *A framework for human evaluation of large language models in healthcare derived from literature review.* NPJ digital medicine, 2024. **7**(1): p. 258.
25. Preiksaitis, C., et al., *The role of large language models in transforming emergency medicine: scoping review.* JMIR medical informatics, 2024. **12**: p. e53787.
26. Reddy, S., *Evaluating large language models for use in healthcare: A framework for translational value assessment.* Informatics in Medicine Unlocked, 2023. **41**: p. 101304.
27. Ferrag, M.A., et al., *Generative ai in cybersecurity: A comprehensive review of llm applications and vulnerabilities.* Internet of Things and Cyber-Physical Systems, 2025.
28. Mahajan, A., et al., *Cognitive bias in clinical large language models.* npj Digital Medicine, 2025. **8**(1): p. 428.
29. John, S., et al., *Owasp top 10 for llm apps & gen ai agentic security initiative.* 2025, OWASP.
30. Dong, G., et al. *Understand what LLM needs: Dual preference alignment for retrieval-augmented generation.* in *Proceedings of the ACM on Web Conference 2025.* 2025.
31. Stein, S., et al., *Leveraging MDS2 and SBOM data for LLM-assisted vulnerability analysis of medical devices.* Computational and Structural Biotechnology Journal, 2025.
32. Asgari, E., et al., *A framework to assess clinical safety and hallucination rates of LLMs for medical text summarisation.* npj Digital Medicine, 2025. **8**(1): p. 274.

Core Techniques for Explaining LLMs in Healthcare

Amirfarhad Farhadi and Reza Siasifar

1 Introduction

The impact of Artificial Intelligence (AI) on a wide range of aspects in our lives is steadily increasing. Among these aspects, healthcare is particularly significant, with many studies being conducted on various clinical tasks, many of which have achieved excellent results. Take a deep learning based autonomous dental treatment planning model as an example, which reached 100% precision in identifying damaged teeth and subsequently indicating a proper treatment plan [1]. Apart from clinical tasks, some administrative challenges in healthcare can also be addressed using AI, such as preventing insurance fraud by detecting semantic similarity between medical prescriptions [2].

Generative Artificial Intelligence (GenAI) is a type of AI that, when given a prompt, can generate different types of content, including text, images, and videos. These models have the potential to revolutionize healthcare by assisting in various fields [3]. One such example is in the pharmaceutical industry, where key challenges in drug discovery (i.e., cost reduction, precision, and accelerating the design process) can be managed using GenAI models [4]. As proposed in [5], GenAI can also be a solution for protecting patient privacy, a great concern in integrating AI into the healthcare domain due to the high sensitivity of medical information. Other creative solutions can also help address this concern. For example, smart contracts can be used to safeguard the clinical data exchange through Blockchain transactions, which can guarantee off-chain anonymity and on-chain transparency [6]. When the safety of the sensitive medical data communication is ensured, wearable and mobile

A. Farhadi (✉)
School of Computer Engineering, Iran University of Science and Technology, Tehran, Iran
e-mail: am_farhadi@mail.iust.ac.ir

R. Siasifar
Department of Computer Engineering, SRC, Islamic Azad University, Tehran, Iran
e-mail: reza.siasifar@iau.ac.ir

A. Zamanifar et al. (eds.), *Explainable Large Language Models in Healthcare Applications*, Bio-IT and AI, https://doi.org/10.1007/978-3-032-15088-2_4

medical sensors can also be integrated with AI to provide constant, real-time health status monitoring. However, this constant connectivity can cause challenges like high power consumption and mobility management, which need to be handled beforehand, possibly via customized communication protocols designed to address these challenges, such as EECASC [7].

It is safe to say that LLMs are the most conventional type of GenAI, which can offer support for a diverse range of tasks in healthcare, from low-risk tasks, such as administrative paperwork, to high-risk tasks, such as clinical diagnosis. Nevertheless, their integration into clinical environments, even for less-risky tasks, is still fraught with peril. The reason is that these models often function as "black boxes", which means that even though their outputs seem to have no grammatical or syntactical errors, they cannot be easily traced back to a clear rationale [8]. It is clear that in the healthcare field, where decisions profoundly affect human lives, the inability to ask "why" a particular recommendation was made poses an unacceptable risk, creating significant barriers to trust, patient safety, and regulatory compliance [9, 10].

Imagine a scenario where an LLM-powered diagnostic assistant predicts a patient who has previously had cancer as a high-risk candidate for recurrence, or suggests a preoperative guidance course; relying on these outputs in such applications, which were real-world examples of LLMs in healthcare that we will discuss further in the chapter, is unacceptable if the reasoning is not available or understandable. There are multiple expectations when provided with a health service as a patient or a layperson, some of which are obvious, such as the ability to trust, others more delicate and nuanced, such as the absence of bias of any kind. Both of these variations of expectations cannot be satisfied and will be prevented in case no interpretability is available. Clinicians also need to be able to pinpoint the errors and mistakes of these models if they are used as their assistants, verifying the source and cause, which can be from flaws in the input data (missing parameters from a laboratory test, for instance) or poor reasoning from the model. These desired explanations, however, are technically impossible to reach through direct internal examination of the modern LLMs, as they are multi-layered and complex systems with a stochastic and generative nature [9, 10].

As a response to this crucial necessity of explanations and interpretability, we seek to introduce, review, and analyze the core techniques that can be employed for explaining LLMs within the sensitive domain of healthcare. Figure 1 shows the categories we are going to review in this chapter. These methods comprise traditional explainable AI techniques that are widely applied to AI systems other than LLMs, in conjunction with more novel approaches, some of which are specifically for LLMs. The latter were mostly proposed in the past recent years, as the modern LLMs themselves are a rather new technology that has become mainstream quite recently. This chapter encompasses methods that are applied externally, together with intrinsic techniques that use concepts from the architecture of LLMs or prompt engineering to get transparent rationales in the intended output format. Understanding these techniques is not just an academic exercise; it is pivotal for making sure that the integration of LLMs into clinical workflows is done in a responsible way and ensuring

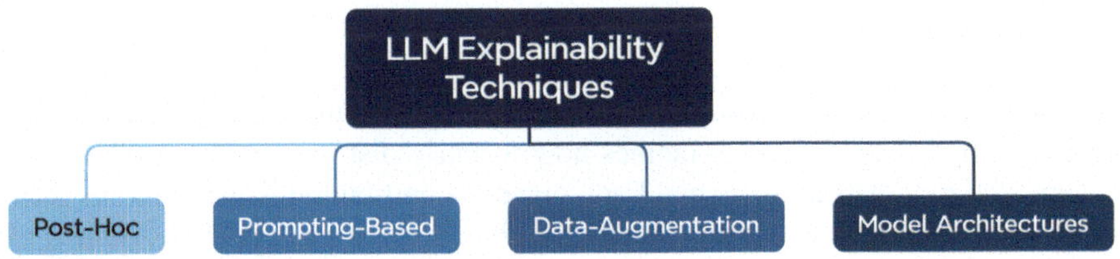

Fig. 1 Different categories of explainability techniques for large language models that are reviewed in this chapter

compliance with regulations to protect patients and other users from any kind of unwanted harm.

2 Post-Hoc Explanation Techniques

Post-hoc eXplainable AI (XAI) methods contain techniques that are applied after an AI model has been trained. The idea behind these methods is that the internal mechanisms of modern LLMs are too intricate to be examined directly; therefore, offering indirect solutions for analyzing a model's behavior externally can help circumvent this complexity [10]. Their strategy is to explain individual predictions, generally by measuring the contribution and relevance of different input features (such as specific words or pixels) to the final output. These clarifications can be crucial in the debugging process and guide the experts to pinpoint biases or limitations, as well as build trust with the provided insights, all achieved without the model itself being altered [11].

2.1 Feature Attribution Explanations

Feature attribution-based methods main goal is to discover the relevance of each input feature in the output that was received. We review two of the most widely used methods in XAI: SHAP and LIME [12], and also gradient-based methods.

SHAP (SHapley Additive exPlanations)

SHAP is a noticeable feature attribution method in the field of XAI and has been popularly explored in clinical XAI [9, 10]; it helps reach explainability by quantifying each input feature's contribution to the prediction of the model [8].

Theoretical Basis

This method uses a concept from cooperative game theory, the Shapley values developed by Lloyd Shapley, which defines how to fairly distribute a "payout" among players who took part in the game. Each player's contribution is measured as their marginal contribution across all possible combinations of players [13]. Applied to the LLMs, our input features, e.g., blood pressure or lab results, are assumed to be "players" and the output that we receive is considered the "payout".

It is important to know that SHAP is the only additive feature attribution method that complies with the following desirable criteria:

- **Local Accuracy**: The aggregate of feature attributions equals the model's output for a specific input. This ensures that the explanations are complete by forcing the whole prediction to be explained by the contributions of all individual features. This is critical in clinical decision support because it prevents uncertainty about what is actually influencing the output.
- **Missingness**: Missing or unobserved features do not have any sort of attribution or contribution in the explanation, and do not receive any fictional credit nor blame. Take an LLM-based diagnostic assistant, for example. If it predicts that a patient is likely to have pneumonia based on fever and cough features, and the patient's blood oxygen level is not mentioned (i.e., missing), missingness ensures that this feature is assigned with zero attribution in the explanation. Otherwise, clinicians might mistakenly believe that it has influenced the decision.
- **Consistency**: If a feature's importance increases after a model undergoes some changes (e.g., a newer version), the weight assigned to it in the explanation will not decrease. Suppose in the new version of the model from the previously mentioned example, the pneumonia probability in the presence of fever doubles; if the model is consistent, fever can never have a lower score in any explanation compared to the previous version. This is especially crucial when hospitals or regulators compare successive versions of a diagnostic model, as it ensures interpretability and logical coherence.

Being the only method of its kind with this trait distinguishes SHAP from the rest of the local surrogate models like LIME, which does not assure these properties. Additionally, the Shapley values are calculated using the complete dataset, unlike methods such as LIME, which take a local model approach for each prediction [11].

Local and Global Interpretability

The ability to explain a distinct prediction for a single input is called local interpretability. In SHAP, each feature gets assigned a unique SHAP value for the specific prediction to demonstrate how each factor has contributed to the result [13]. For instance, in a clinical model that predicts breast cancer recurrence, Liu et al. [14] utilized the Tree Explainer method in SHAP to explain why a patient was marked as high-risk by revealing the explicit contribution of some key factors, including tumor size, clinical stage III, number of lymph node metastases, and age.

Understanding the model's overall behavior is also a key necessity alongside local explanations for clinicians and researchers, which is referred to as global interpretability. SHAP achieves this by aggregating many local explanations through analyzing SHAP values across the entire dataset. Based on this analysis, summary plots can be generated to illustrate the most influential factors in the model's prediction on average [8]. This enables identifying the most influential features overall from various possible features, such as the key factors mentioned in the previous example, and uncovering the model's underlying decision-making patterns.

Applications in Healthcare

LLMs have not been an exception to SHAP's widespread integration in various clinical diagnostic applications. In the field of pathology, TraP-VQA, which is a vision–language transformer model for Pathology Visual Question Answering (PathVQA), uses SHAP for presenting textual explanations for its generated answers [10].

Furthermore, SHAP is being integrated directly into Clinical Decision Support Systems (CDSS). Most AI-powered CDSSs are now providing clinicians with SHAP plots to visualize how model predictions are achieved. However, SHAP outputs can be too complicated for non-technical users; that is why the studies show that the format of the explanation is significantly important, to the extent that combining them with a clinically-focused narrative explanation enhances clinicians' acceptance, trust, satisfaction, and usability of the AI recommendations. This has caused emerging approaches in which LLMs are used to translate quantitative SHAP values into more intuitive and comprehensible narratives.

Limitations

Computational complexity is a main limitation of SHAP, since the exact calculation of SHAP values is challenging, with a complexity that can be exponential relative to the number of features. This makes SHAP less practical for explaining LLMs with billions of parameters. Moreover, trying to manage the computational expense with approximation strategies can come with its own challenges and limitations. One of these challenges is choosing appropriate methods for feature removal and efficient value estimation. Another limitation is having to make assumptions that, in many real-world, complex cases, are not true, such as the model features being independent or the model being linear [11, 13]. Additionally, focusing on individual feature contributions can cause SHAP to overlook differences between samples, which can lead to errors [14].

It is also worth mentioning that studies about the development of multimodal LLMs in healthcare are limited, and further research on development, interpretability, and evaluation is needed to further advance this subject.

LIME (Local Interpretable Model-Agnostic Explanations)

Local Interpretable Model-agnostic Explanations, known as LIME, is a perturbation-based method that focuses on generating a local explanation by approximating the behavior of the original complex model around the specific input instance. It achieves this by fitting an interpretable, much simpler model, which is called a surrogate model. LIME subtly perturbs and alters the input features and then measures the effects of these small perturbations on the original model's prediction. After observing these output changes, the surrogate model determines the contribution of each input feature to the final prediction; therefore, the importance of each feature is revealed [11].

The explained mechanism shows us that LIME relies on observing consistent changes in output after repeatedly running the model with slightly modified inputs. However, LLMs do not produce fixed-dimensional outputs like most traditional machine learning models, and they have a generative and stochastic nature, which means the same input can generate a different output [9, 15]. This inherent probabilistic behavior and dynamic output structure violate the consistency required for this approach, as they impair the reliability of perturbation-based explanations and are difficult to handle for traditional local approximation methods such as LIME. This is an important limitation that can explain why LIME is not frequently adopted in prominent LLM studies or deployments in healthcare, in contrast to its broader use in other domains.

Gradient-Based Explanations

Gradient-based methods learn the importance of input features through calculating the gradient (partial derivatives) of the model output with respect to the input dimensions [11]. To exemplify, Grad-CAM (Gradient-weighted Class Activation Map) is a gradient-based visualization technique, which was implemented in the earlier-mentioned TraP-VQA model, to provide visual explanations as well as discussed textual explanations [10].

Saliency maps are another common example, which is a visualization that highlights the most impactful input features on the output, which can be pixels in an image or tokens in a text [11]. A vision transformer model for diagnosing COVID-19 from chest X-ray images used saliency maps to present visual explanations for the outcome diagnosis [10]. Despite the presence of this example, there were some reports of saliency maps failing when adopted for complex LLMs [9].

2.2 *Attention-Based Explanations*

Attention-based explanation methods are based on the attention concept, which is a core component of the transformer architecture; a revolutionary technology that became the foundation of the modern LLMs with billions of parameters [15]. An attention weight is assigned to each token that is being processed in the attention mechanism. These weights demonstrate the amount of contextual relationship between a series of tokens, which can then be visualized and interpreted to reveal the model's internal attention patterns, on the premise that higher weights align with higher significance of the input feature. However, several studies have demonstrated how the attention weights can miss the most important features or contain redundant information. This can reduce reliability and lead to misalignment with explanations from other methods [11]. Although this is an ongoing debate over attention, applications of this method in healthcare can still be found. The ClinicalBERT model, with the task of predicting hospital readmission, utilized attention weights to achieve interpretability and was evaluated through visualization of said weights [10].

2.3 Counterfactual Explanations

Counterfactual explanations describe decisions of a model by exposing it to multiple "what-if" scenarios and thereafter discovering the factors that would need to change in order to get a different result. This example-based approach is often more intuitive for humans to comprehend, compared to numerical attribution scores [11]. For example, consider a scenario in which you describe a patient with symptoms X, Y, and cough to a model to check if they have the disease Z, and receive a negative prediction. Subsequently, you apply a very minimal change and present the patient's new symptoms as X, Y, and dry cough, similar to asking "what if the coughs were dry?". If this small adjustment leads to a different output, which in this case would be a positive diagnosis, it reveals that the presence of dry cough is an influential symptom in the diagnosis of Z.

In a study performed to understand the effect of these explanations in human-AI collaborative decision making, Lee et al. [16] found that counterfactual explanations contributed to an optimization of both therapists' and laypersons' performance and agreement level when presented with correct AI outputs, and also reduced reliance on incorrect AI outputs at the same time. Moreover, it was shown that in this reduction, counterfactual explanations had 21% higher impact, compared to salient feature explanations.

Generating counterfactuals can be done through manual creation, paraphrasing, word replacement, or automated generators like Polyjuice, which create them using various sentence permutation methods, including negation, deletion, and shuffling. More sophisticated methods utilize two-stage approaches, initially finding the influential tokens to change using a feature attribution method such as gradient-based methods, and thereafter editing said tokens using another generative model (e.g., MICE, CREST) [11].

3 Prompting-Based Explanation Techniques

Prompting-based techniques focus on harnessing the LLM model itself to "show its work" by delicately engineering given prompts in a way that makes the model clarify the path it takes that leads to the final prediction or answer.

3.1 Chain-of-Thought (CoT) Prompting

Chain-of-Thought (CoT) prompting is a technique inspired by the human thought process, designed to instruct LLMs to break down the given task into a sequence of intermediate steps in order to evoke reasoning. This step-by-step rationale provides a human-readable explanation that is easy to understand and can also be used by clinicians to assess whether the model's output can be trusted or not. Furthermore, it also potentially enhances the outcome performance, especially when the tasks are sophisticated and require logical reasoning.

Implementation

To implement this method, we follow a two-step scheme. Initially, we cue the model to follow a step-by-step reasoning process in the first part of our prompt or even a separate one. Following that, we describe the main problem that we need its answer to be derived from that process. There are two main variations for triggering CoT in the first step of this scheme:

- **Zero-Shot CoT**: In this approach, we use a simple and brief instruction, such as "Let's think step-by-step".
- **Few-Shot CoT**: In this approach, we provide a few examples (shots) that demonstrate the specific pattern of rationales we are looking for, typically in a triplet format of question-explanation-answer [17].

Self-consistency prompting is another interesting strategy, especially for domains with complex reasoning paths like healthcare, in which multiple reasoning paths are generated and the most consistent one is chosen in the end, leading to noticeable improvements in some studies [18, 19].

Applications

Multiple medical question-answering (QA) tasks have been performed in studies to investigate the effectiveness of CoT. When CoT was applied in different LLM models like GPT-4 and Llama2, for benchmarking several standard medical datasets, such as MedQA-5 and Medbullets-5, improvements were shown on multiple different combinations, even with the simple zero-shot approach. However, in various other situations, including the benchmarks performed on more challenging cases, such as those from the JAMA Clinical Challenge, CoT did not lead to any improvements. This suggests that these tasks are relatively harder to reason about [20]. Self-consistency approach was also adopted for several datasets with multiple-choice questions, such as MedQA, MedMCQA, PubMedQA, and MMLU [19]. CoT was also used in several studies as a baseline or for general comparison, together with their newly proposed methods, which are mentioned later.

Limitations

Multiple downsides have also been spotted when adopting this method, in spite of the previously mentioned advantages. For instance, generated explanations through this method are not factually grounded and may even be unfaithful to the real decision-making process of the model [9]. Additionally, being susceptible to bias, such as when choice options in a few-shot prompt are reordered, is a significant negative aspect [9, 11]. Another disadvantage is the possibility of encountering new error types for multiple-choice questions, such as receiving "None of the above" from the model, or even a newly invented answer option after the reasoning process [20].

3.2 Specialized Diagnostic Reasoning Prompts

Specialized diagnostic reasoning prompts are an advanced form of CoT prompting, designed to induce LLMs to produce structured outputs by imitating the cognitive processes used by clinicians during the process of diagnosis. This allows the clinicians to evaluate the trustworthiness of the model's output and the reasoning behind it more effectively, as it becomes more interpretable and familiar to them. We review four different types of these methods proposed in [21] in Table 1.

All approaches above were followed with a pair of examples in the mentioned study to demonstrate the exact process and structure that is desired.

These approaches were performed on a modified MedQA USMLE (United States Medical Licensing Exam) dataset, using LLM models GPT-3.5 and GPT-4, alongside the traditional CoT as the control for the experiments. When applied in GPT-3.5, only the intuitive reasoning prompt enhanced the CoT correct responses percentage (from 46 to 48%), and the rest of the prompts led to lower percentages. On the other hand, all approaches except Bayesian inference helped raise GPT-4's correct responses percentage between 0.8 and 2.2% from the base 76% achieved by CoT. This shows us that as the LLM models continue to grow, the effect of specialized prompting strategies on them also seems to grow, alongside the overall enhancement in their performance. The differential diagnosis prompt was also applied solely in GPT-4 to the NEJM (New England Journal of Medicine) challenge question set, which decreased

Table 1 Four types of specialized diagnostic reasoning prompts

Prompt type	Implementation
Differential diagnosis prompt	First, we demand a step-by-step deduction for creating a list of possible diagnoses, i.e., differential diagnosis, and then following another step-by-step deduction for deciding the most likely one
Intuitive reasoning prompt	We ask the model to focus on making direct associations from a patient's symptoms, signs, and laboratory findings to possible diseases. This mimics a clinician's intuitive or "System 1" thinking
Analytic reasoning prompt	We try to mirror a clinician's analytical or "System 2" thinking by prompting the model to use analytic reasoning to reach a deduction of the physiologic or biochemical pathophysiology of the patient and subsequently identify the correct response step-by-step
Bayesian reasoning prompt	We ask the model to utilize step-by-step Bayesian inference, creating a prior probability that gets updated with each new information to finally reach a posterior probability that determines the diagnosis

CoT's 38% correct responses percentage by 4.2%, once again highlighting the shortcomings of CoT and CoT-based methods when faced with challenging medical cases. It was also observed that the incorrect diagnoses were substantially more likely to include rationales with logical errors of factual inaccuracies compared to the correct diagnoses.

3.3 Self-Questioning Prompting (SQP)

Self-Questioning Prompting (SQP) is a novel prompting strategy that encourages the LLM to be more aware of its own thinking process through a multi-step and human-like reasoning process. Throughout these steps, the model extracts key information from the problem in the form of questions, and then answers them itself. Stages are designed in a way that helps the user as well as the model itself to deeply comprehend the context and provides a clear path from which the output was reached. The process is broken down in [22] as follows:

1. We describe medical situations to the model and ask it to generate questions about them in order to guide the model into identifying the key elements.
2. The questions are generated by the model following the previous prompt, helping it to enhance its context comprehension.
3. We now ask the model to answer its own question using basic medical knowledge in our second prompt.
4. Finally, we prompt the model to perform the final task with the deep grasped understanding from previous stages.

The study performed both zero-shot and 5-shot prompting in standard prompting, CoT prompting, and SQP altogether across different LLM models, including Bard,

GPT-3.5, GPT-4, on eight datasets from the following six different tasks in the healthcare domain: named entity recognition, relation extraction, semantic textual similarity, natural language inference, document classification, and QA. SQP consistently outperformed standard prompting in all tasks except the document classification, leading to relative improvements ranging from 4.9% to 46.9%. It also had a close and competitive performance compared to CoT, leading to the best performance in most tasks except for the QA.

The mentioned exceptions, which caused inferior performance, could suggest that sometimes the complexity added by the self-questioning process is not guaranteed to contribute to a better understanding, or that the model's capacity for taking advantage of the extra information is insufficient. It is also worth mentioning that another valuable finding of the study was that 5-shot prompting generally, but not universally, improved performance across all tasks against zero-shot prompting.

3.4 *Ensemble Reasoning (EnsReas)*

Ensemble Reasoning (EnsReas) is another novel prompting strategy that attempts to improve the performance and consistency of LLMs with an iterative design that makes the LLM model self-critique and refine its own reasoning by presenting it with an ensemble of different reasoning paths that were previously generated for the problem at hand. The methodology is described in [18] as follows for each medical question:

1. We use Zero-Shot CoT with Self-Consistency (ZS-CoT-SC) prompt to generate a diverse set (e.g., a set of 10) of reasoning paths.
2. If all generated paths do not result in a consistent answer, we now use an EnsReas prompt. This prompt includes the original question, followed by the aggregated set of inconsistent reasonings, instructing the model to analyze the different perspectives and provide a final correct answer and rationale. It is also advisable to add an instruction for the model not to consider the majority vote as the definite correct answer, and to encourage critical analysis instead.
3. We iterate this process until a target minimum inconsistency percentage is met.

It is clear that this process will result in a very informative set of different thought processes and rationales that can widen the perspective of the clinicians in an easy-to-interpret way.

The study implemented EnsReas on GPT-3.5, GPT-4 turbo, and Med42-70B, which is an open-source clinical LLM, for answering a simple format of text-only, multiple-choice questions from all 3 levels or "Steps" of the public National Board of Medical Examiners (NBME) sample exam questions for the USMLE. EnsReas significantly improved the consistency of the final answers for all tested models, with the number of divergent responses shrinking after each iteration. In contrast, the results in diagnostic accuracy were mixed. In less powerful models like GPT-3.5 and Med42-70B, EnsReas invariably outperformed the strong ZS-CoT-SC and other less-complex tested prompting methods. For the more advanced GPT-4 turbo, however, EnsReas not only did not always cause improvement in the accuracy, but also performed slightly worse on some tasks. This implies that this method might not align with the increasing baseline capabilities of the newer LLMs. In addition, a possible trade-off between maximizing accuracy and maximizing consistency can be spotted, as it was shown that for some models, peak performance was achieved in the first iteration (EnsReas-1) while consistency continued to improve in the next iteration (EnsReas-2). The quality of correctness and helpfulness (i.e., offering new information or guidance) of a sample of the refined reasonings was also analyzed by an expert who scored them as "mostly correct" or "completely correct" across all steps, with no incorrect reasonings being provided for the correct answers. The reasonings were also evaluated to be insightful beyond basic medical knowledge and guiding towards the correct answer.

3.5 General Prompt Engineering

All of the introduced prompting-based methods are different examples of prompt engineering, a broad practice of designing and manipulating prompts to steer the LLM models toward intended outputs and behaviors. It is a discipline that seeks to optimize the performance of a general-purpose model for specific tasks, without the need to access or update the model weights [8]. Many different simple or complex forms of prompt engineering approaches can be used with the main goal of rationale elicitation or step-by-step progress, urging the model to generate justifications or intermediate steps for its output, thereby increasing transparency and interpretability. In Table 2, we review some applications of such approaches in the healthcare domain:

Despite these numerous use cases, some major drawbacks remain. First and foremost, crafting effective prompts is usually an iterative process that may involve plenty of trial and error, which can be time-consuming and require domain expertise. The LLMs are also highly sensitive to the exact structure and wording of prompts, where minor changes in phrasing can lead to variations in the quality and correctness of the output [28]. Additionally, even with detailed prompts, LLMs are still prone to hallucination. This means that they can generate plausible information that is not factually correct, making up imaginary references or clinical facts, for instance [26, 29].

4 Model Architecture and Data-Augmentation Techniques

In this section, we dive deep into several advanced model architectures and data augmentation techniques, which are designed to enhance the capabilities, explainability, and safety of LLMs in healthcare. Figure 2 demonstrates where each of the previously reviewed categories of techniques for explaining LLMs is applied,

Table 2 General prompt engineering applications in healthcare

Application	Details
Patient education	For developing RadGPT, a system for generating sets of simplified and patient-centered explanations of complex radiology reports, several engineered prompts were integrated to reach the desired outcome and ensure conciseness and maximum understandability [23]
Clinical decision support	Zero-shot performance of multiple reasoning and non-reasoning LLM models was put to test, with rationale elicitation being incorporated into prompts to also evaluate model transparency in three essential clinical tasks: primary diagnosis generation, ICD-9 (International Classification of Diseases, Ninth Revision) medical code prediction, and hospital readmission risk stratification [24]
Mental health	Yang et al. [25] trained MentaLLaMA, an open-source LLM for interpretable mental health analysis that generates human-level explanations, using few-shot prompting to generate explanations on their raw data in the training process
Medical education	LLMs can offer different possibilities in medical education, such as various personalized learning materials. Some examples of related prompts and their results can be found in [26], an interesting study that interviews ChatGPT about medical education. However, many challenges also come with these possibilities, such as algorithmic bias and overreliance [27]

along with data augmentation techniques, all compared to a conceptually illustrated architecture model.

4.1 Retrieval-Augmented Generation (RAG)

Retrieval-Augmented generation (RAG) is a technique that integrates data from external sources before generating a response to optimize the LLM's performance by producing more faithful and verifiable outputs. This technique is executed in two sequential steps: when given a query, an external data source is searched for the most relevant data, and subsequently, a response based on the retrieved data will be generated. Some new LLM models like Command-R from Cohere are now available, which use RAG to provide sources of information [9].

It is clear that RAG's mechanism makes it a powerful technique for addressing explainability challenges, especially in the healthcare domain, where a RAG-enabled model can back up its factual statements with citations to reliable sources like medical textbooks or clinical guidelines [9]. In a case study mentioned in [30], multiple standard LLMs, a RAG-enabled GPT-4 model, and four anesthesiologists with less than five years of medical experience were engaged in providing preoperative guidance within 14 diverse clinical scenarios. The GPT4-RAG model achieved the best accuracy with 91.4%, which was 5.1 percentage points higher than the answers generated by the physicians. Besides the performance, the GPT4-RAG model was also superior

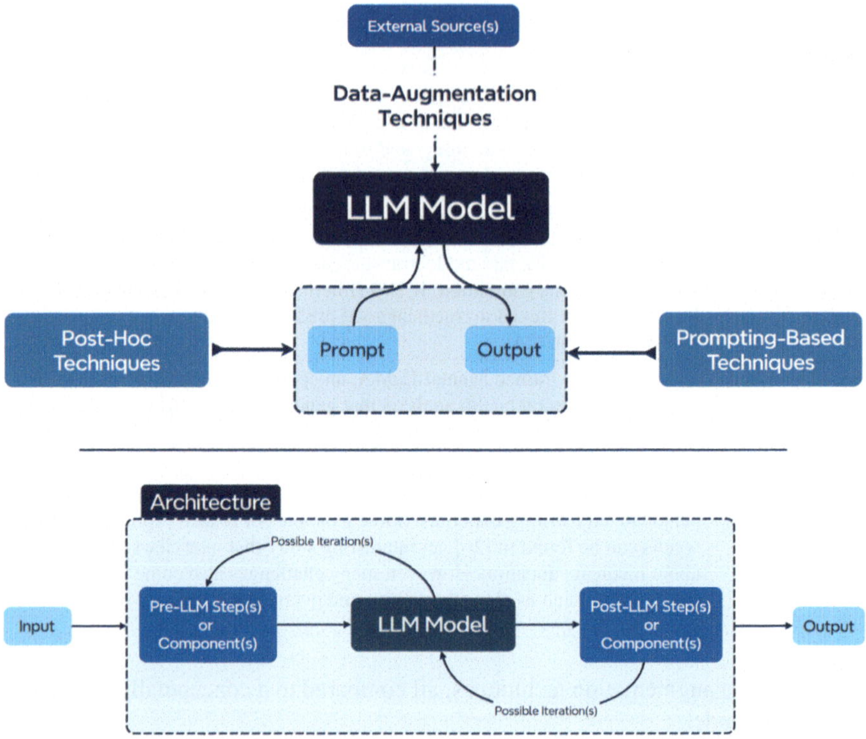

Fig. 2 Demonstrates where each category of techniques for explaining LLMs is applied, compared to a conceptually illustrated architecture model

in the aspect of time consumption, taking an average of 1 s for information retrieval and an average of 15–20 s for answer generation, while human physicians produced preoperative instructions in an average of 10 min.

Despite the mentioned upsides, it is important not to overlook some important weaknesses. The information retrieval step is not always guaranteed to work flawlessly, which can cause inaccuracies in the final output. Moreover, even though the system is grounded in external knowledge, it is still vulnerable to hallucinations [9].

4.2 Tree-of-Thoughts (ToT) and Graph-of-Thoughts (GoT)

Tree-of-Thoughts (ToT) and Graph-of-Thoughts (GoT) are novel and advanced reasoning frameworks with the purpose of unpacking and visualizing an LLM's internal reasoning process. ToT is a technique that represents LLM reasoning in multiple linear paths, each of them being similar to the outcomes of CoT prompting. GoT approaches to portray the LLM reasoning in an associative visual format by visualizing it as a graph, the nodes of which are representations of collections of data and information, and the edges demonstrate logical connections between them, and thus the impact of each part on the outcome. These methods imitate the human thought process of exploring numerous possible pathways and assessing them to arrive at the best solution and provide additional explainability through clarifying the information decisive to the final output [9].

There is a gap for future work to integrate these methods, since they were recently proposed and not many studies exist in the domain of healthcare that have utilized them, although some can be found. For instance, Yang et al. [31] proposed the Quantized Medical Tree of Thought (QM-ToT), claiming to be the first utilization of ToT for medical question answering tasks. This framework first employs ToT by expanding the question into a tree structure, where different chains of thought are generated. Following that, an incorporated evaluator module scores each generated reasoning path based on both reasoning quality and factual correctness. QM-ToT achieved significant accuracy enhancements over the MedQA-USMLE dataset with multiple LLM models, including LLaMA2-70b, Qwen2.5-72b, and LLaMA3.1-8b, over the CoT and CoT with self-consistency (CoT-SC) baselines, although performing 1.5% weaker with LLaMA3-70b compared to CoT-SC. In addition, substantial performance gains were exhibited by the framework for the questions categorized as medium and hard difficulty, in contrast to an occasional decline in accuracy for easy questions, which suggests its benefits in complex clinical reasoning.

4.3 Knowledge Graph (KG)

A medical Knowledge Graph (KG) serves as a medical knowledge repository that combines and shares biomedical information. In a KG, concepts are exhibited as nodes and semantic relationships. KGs are integrated into LLMs with the purpose of optimizing the model's diagnostic reasoning with the structured and contextually relevant medical information they provide. These augmented data help the LLM model to provide knowledge grounding and reduce the chances of hallucinations and factual mistakes by guiding it down the correct reasoning path, enhancing both performance and explainability.

Electronic Health Records (EHRs) were a major benefit for healthcare due to the improvement they caused in patient care and data management [5], but they can also overwhelm healthcare providers and consequently increase the risk of diagnostic inaccuracies. Despite the capacity of the transformer-based generative models like LLMs for deriving meaningful insights from EHRs [32], more advanced approaches like KG can also be implemented to address the challenge of the complexity and verbosity of EHRs, as was executed in [33]. They propose a system called DR. KNOWS (Diagnostic Reasoning Knowledge Graph System) to perform diagnosis prediction (or summarization) based on daily patient progress notes in SOAP format (Subjective, Objective, Assessment, and Plan). The specific task sought in this study was to predict the list of problems and diagnoses found in the "Plan" section using data from the other three sections. The system extracts the KGs through a complex process and then feeds them to the LLM model inside an engineered prompt template along with few-shot examples to obtain the desired output. It was shown that the augmented knowledge by KG not only led to a significantly better performance, but also 5 percentage points stronger agreement on the correctness of the model's rationale (compared to 50% in the absence of KGs), when presented to physicians as human evaluators. It is important to note that even if no data biases or flaws exist in the knowledge provided by KGs, which is a naïve assumption, the study exemplified that the LLM models still may hallucinate and generate misleading outputs. The study also found that the LLM models occasionally fail to leverage accurate knowledge paths that were provided to them.

5 Evaluating and Analyzing Explanations

Aiming towards explaining LLMs in healthcare goes beyond just technical solutions. Evaluation and analysis of the correctness and quality of the explanations is also a very important step, making sure of the trustworthiness and interpretability of the reasoning for all kinds of intended users [9]. In this section, we review different aspects of the four strategies shown in Fig. 3.

Fig. 3 Four main strategies for evaluating and analyzing explanations of the large language models

5.1 Human and Expert Evaluation

As expected, domain experts, such as clinicians, are the most reliable evaluators of LLMs in the healthcare domain. This is due to weak correlations of automated metrics and human assessments across multiple studies, and the various delicacies and critical dimensions of the healthcare domain, which automatic scores often come short of measuring [20, 34–36]. The recruitment of evaluators mostly depends on the task, with the tendency to select evaluators who represent the actual users, which can differ between physicians, residents, medical students, nurses, pharmacists, and non-expert users. Across the literature, the number of evaluators is most often reported as 20 or fewer [36]. Another key observation in the studies in regards to better adjudication is to ensure consistency by having several evaluators to assess each response, and in case of disagreement, using a third evaluator or a consensus meeting to determine the final judgment. On top of that, an effective way to reduce bias can be leaving the evaluators blinded to the source of the material they are evaluating, not knowing whether it was generated by a human or an AI. Nonetheless, a review of 142 studies found that this method was implemented in only 41 studies, and a majority of 56% do not explicitly report their blinding procedures [36].

Multiple types of human evaluation can be found in the reviewed studies, from validating diagnostic reasoning and summarization of medical evidence for factual correctness and thoroughness, to clinicians and occasionally patients themselves rating and evaluating patient-oriented materials, such as simplified reports and medical answers. These different varieties of AI outputs are evaluated for helpfulness and scientific truthfulness, along with different humanistic metrics like empathy or lack of potential for causing anxiety [9, 19, 34]. In addition, there are many studies throughout the literature that benchmark the performance of LLMs on medical licensing exams, in which they use medical experts to grade and assess their knowledge base and clinical reasoning against human benchmarks.

There is a gap, however, for the human evaluations to be completely reliable and generalizable, caused by a significant lack of standard guidelines and best practices [35, 36]. To bridge this gap, multiple specialized frameworks have been proposed or reviewed in the studies, some of which are more focused on explainability. For instance, QUEST (Quality, Understanding, Expression, Safety, Trust) is a framework specifically designed for human evaluation of LLMs in healthcare, which covers five core principles, broken down into 17 evaluation dimensions, involving accuracy, relevance, consistency, bias, etc. It is noteworthy that even with these frameworks, implementation can still vary and not capture every specific nuance of a particular use case [36]. Another solution can be fidelity metrics, i.e., using the guideline-provided answers in benchmark datasets of reference rationales for comparison with the explanations from LLM models. For example, testing to see if explanations for recommending a specific drug are consistent with a credible clinical guideline like NICE (National Institute for Health and Care Excellence). Similarly, functional tests or RAG-like approaches can use prompt-based explorations to test the alignment of generated explanations with reliable sources or guidelines [9].

5.2 *Red-Teaming*

Red-teaming, which is originally a cybersecurity term, is an adversarial evaluation method in which deliberate efforts are made to break the system or expose its vulnerabilities, tricking it to show unsafe or misleading behaviors before real-world deployments. The process is often performed by interdisciplinary experts and requires partnership with patients and clinicians to ensure clinical safety validation from their feedback. This is used to simulate high-stakes and difficult scenarios in healthcare to assess whether an LLM's explanations can be safe and faithful or not [9, 37].

5.3 *Uncertainty Quantification*

Uncertainty quantification refers to the ability of an LLM to assess and transmit its own uncertainty or confidence about its predictions [11]. This self-awareness in an LLM model of its own knowledge limitations is a critical aspect of metacognition in AI systems, a concept for improving interpretability and transparency of LLMs in healthcare [38].

Several approaches have been observed for this technique. Primarily, in consistency-based estimation, the LLM model is presented with the same question multiple times, or slightly altered versions of it, and then given a score based on observed consistency among the multiple responses. Secondly, in confidence elicitation, the LLM model is prompted to provide a qualitative verbal confidence or a numerical confidence score (e.g., on a 1–5 Likert scale or a percentage) along with the answer. Additionally, another approach is also mentioned, which relies on the fact that LLMs generate text by predicting each token based on a probability distribution. These token-level uncertainty methods calculate a confidence score for all tokens derived from their probabilities, and then aggregate them all to achieve the overall uncertainty [11]. The missing answer recall metric used with multiple choice questions was also an interesting approach, evaluating the model's capability to recognize when all provided options are incorrect. This was shown to be one of the most challenging metrics for the LLMs [38].

One application of these methods in healthcare can be selective prediction, where the model is designed to withhold the answers that their measured confidence falls below a certain threshold. Another application can be assessing an LLM's performance on medical QA simply by correlating model confidence for its answers with its accuracy. A major downside, however, is that LLMs commonly show poor metacognition and a tendency toward overconfidence, which can make some uncertainty quantification metrics useless. For example, most LLM models that were tested in a numerical confidence elicitation benchmark assigned a maximum level 5 confidence to their answers, leaving only GPT-4o that actually assigned confidence scores below 3 to its answers. Two other models, along with GPT-4o, were also observed to occasionally self-assign confidence scores 3 and 4 [38].

5.4 Watermarking and Labeling

If LLMs eventually make their way to broad deployment in healthcare, there will be situations where their diagnoses, answers, reports, or any other output of theirs are saved and kept along with patient records or attached to a clinician's diagnosis, for example. To make sure that LLM outputs and explanations are never missed in getting evaluated and being mistakenly assumed as expert-generated in such situations, labeling is a transparent way to explicitly mark any AI-generated content as such. Another approach is watermarking a digital identifier into AI-generated content, more often imperceptibly, so that it can be easily detected when checked via related tools. This method offers an advantage over labeling, as it remains robust even if any AI-generated content is accidentally or intentionally unlabeled. Even so, the effectiveness and reliability of the current watermarking technologies are uncertain [15].

6 Conclusion

Explainability for LLMs in healthcare is not a very simple goal to achieve, as it requires aligning technical methods, evaluation standards, and clinical workflows. In this chapter, we first reviewed variations of the traditional and widely used XAI methods, known as post-hoc techniques. Their theoretical strengths were mentioned along with their limitations, mainly not being very practical for complex modern LLMs with billions of parameters. We then moved on to the interesting and easily applicable prompting-based techniques, which harness the LLM model itself to "show its work" by simply or sophisticatedly engineered prompts, all designed to end up with a clear path that was taken to reach the final output. Crafting these prompts, however, can be an iterative, time-consuming process due to the LLMs being highly sensitive to exact prompt structure and wording, which also makes them prone to bias. These composed prompts may also sometimes become too complex, leading to unimproved or even diminished understanding and performance in some settings, compared to simple standard prompting. The final groups of techniques we discussed were model architectures, where multiple steps and/or components are implemented before and after LLMs to help them achieve better performance and interpretability, together with data augmentation, where citing external reliable sources makes the outputs more faithful and verifiable. Nevertheless, even information retrieval and model architectures are not guaranteed to work flawlessly and are liable to inaccuracies, misleading outputs, and hallucination, a major issue among many of the techniques that were discussed. In the end, we went through different aspects and approaches of evaluating and analyzing the LLMs' explanations, highlighting the significant lack of standard guidelines and best practices, along with the issue of poor metacognition of LLMs. This tendency toward overconfidence, coupled with widely observed hallucinations, shows us that while LLMs can be a very powerful tool in assisting both experts and patients in many different ways, there is a long road ahead for them to be ready for being implemented and used directly in the healthcare field, without any supervision from experts.

References

1. Rashidi Ranjbar F, Zamanifar A. Autonomous dental treatment planning on panoramic x-ray using deep learning based object detection algorithm. Multimed Tools Appl 2023;83:42999–3033. https://doi.org/10.1007/s11042-023-17048-4.
2. Parham Najafi Lataran, Amirfarhad Farhadi, Azadeh Zamanifar, Alireza Taheri. Developing a BERT-Enhanced Blockchain Model for HealthInsurance Fraud Detection. Journal of Information Systems Research and Practice 2024;2.
3. Rashidieranjbar F, Farhadi A, Zamanifar A. Revolutionizing Healthcare with Generative Artificial Intelligence Technologies, 2025, p. 189–221. https://doi.org/10.1007/978-3-031-76710-4_10.
4. Farhadi A, Zamanifar A, Faezipour M. Application of Generative AI in Drug Discovery. Application of Generative AI in Healthcare Systems, Cham: Springer Nature Switzerland; 2025, p. 155–74. https://doi.org/10.1007/978-3-031-82963-5_6.
5. Farhadi A, Taheri A. Application of GenAI in Synthetic Data Generation in the Healthcare System. Application of Generative AI in Healthcare Systems, Cham: Springer Nature Switzerland; 2025, p. 67–89. https://doi.org/10.1007/978-3-031-82963-5_3.
6. Shahparian J, Erfani SH, Zamanifar A. A secure and efficient authentication and key agreement protocol in blockchain-enabled VANETs. Computers and Electrical Engineering 2025;122:109947. https://doi.org/10.1016/j.compeleceng.2024.109947.
7. Zamanifar A, Nazemi E. EECASC: an energy efficient communication approach in smart cities. Wireless Networks 2020;26:925–40. https://doi.org/10.1007/s11276-018-1838-5.
8. Khullar D, Wang X, Wang F. Large Language Models in Health Care: Charting a Path Toward Accurate, Explainable, and Secure AI. J Gen Intern Med 2024;39:1239–41. https://doi.org/10.1007/s11606-024-08657-2.
9. Mesinovic M, Watkinson P, Zhu T. Explainability in the age of large language models for healthcare. Communications Engineering 2025;4. https://doi.org/10.1038/s44172-025-00453-y.
10. Nazi Z Al, Peng W. Large Language Models in Healthcare and Medical Domain: A Review. Informatics 2024;11. https://doi.org/10.3390/informatics11030057.
11. Zhao H, Chen H, Yang F, Liu N, Deng H, Cai H, Wang S, Yin D, Du M. Explainability for Large Language Models: A Survey. ACM Trans Intell Syst Technol 2024;15. https://doi.org/10.1145/3639372.
12. Hur S, Lee Y, Park J, Jeon YJ, Cho JH, Cho D, Lim D, Hwang W, Cha WC, Yoo J. Comparison of SHAP and clinician friendly explanations reveals effects on clinical decision behaviour. NPJ Digit Med 2025;8. https://doi.org/10.1038/s41746-025-01958-8.
13. Lundberg SM, Allen PG, Lee S-I. A Unified Approach to Interpreting Model Predictions. n.d.
14. Liu Y, Fu Y, Peng Y, Ming J. Clinical decision support tool for breast cancer recurrence prediction using SHAP value in cooperative game theory. Heliyon 2024;10. https://doi.org/10.1016/j.heliyon.2024.e24876.
15. Harrer S. Attention is not all you need: the complicated case of ethically using large language models in healthcare and medicine. EBioMedicine 2023;90. https://doi.org/10.1016/j.ebiom.2023.104512.

16. Lee MH, Chew CJ. Understanding the Effect of Counterfactual Explanations on Trust and Reliance on AI for Human-AI Collaborative Clinical Decision Making. Proc ACM Hum Comput Interact 2023;7. https://doi.org/10.1145/3610218.
17. Liévin V, Hother CE, Motzfeldt AG, Winther O. Can large language models reason about medical questions? Patterns 2024;5. https://doi.org/10.1016/j.patter.2024.100943.
18. Lucas MM, Yang J, Pomeroy JK, Yang CC. Reasoning with large language models for medical question answering. Journal of the American Medical Informatics Association 2024;31:1964–75. https://doi.org/10.1093/jamia/ocae131.
19. Singhal K, Azizi S, Tu T, Mahdavi SS, Wei J, Chung HW, Scales N, Tanwani A, Cole-Lewis H, Pfohl S, Payne P, Seneviratne M, Gamble P, Kelly C, Babiker A, Schärli N, Chowdhery A, Mansfield P, Demner-Fushman D, Agüera y Arcas B, Webster D, Corrado GS, Matias Y, Chou K, Gottweis J, Tomasev N, Liu Y, Rajkomar A, Barral J, Semturs C, Karthikesalingam A, Natarajan V. Large language models encode clinical knowledge. Nature 2023;620:172–80. https://doi.org/10.1038/s41586-023-06291-2.
20. Chen H, Fang Z, Singla Y, Dredze M. Benchmarking Large Language Models on Answering and Explaining Challenging Medical Questions. vol. 1. Long Papers; n.d.
21. Savage T, Nayak A, Gallo R, Rangan E, Chen JH. Diagnostic reasoning prompts reveal the potential for large language model interpretability in medicine. NPJ Digit Med 2024;7. https://doi.org/10.1038/s41746-024-01010-1.
22. Wang Y, Zhao Y, Petzold L. Are Large Language Models Ready for Healthcare? A Comparative Study on Clinical Language Understanding. vol. 219. 2023.
23. Herwald SE, Shah P, Johnston A, Olsen C, Delbrouck JB, Langlotz CP. RadGPT: A System Based on a Large Language Model That Generates Sets of Patient-Centered Materials to Explain Radiology Report Information. Journal of the American College of Radiology 2025;22:1050–9. https://doi.org/10.1016/j.jacr.2025.06.013.
24. Naliyatthaliyazchayil P, Muthyala R, Gichoya JW, Purkayastha S. Evaluating the Reasoning Capabilities of Large Language Models for Medical Coding and Hospital Readmission Risk Stratification: Zero-Shot Prompting Approach. J Med Internet Res 2025;27. https://doi.org/10.2196/74142.
25. Yang K, Zhang T, Kuang Z, Xie Q, Huang J, Ananiadou S. MentaLLaMA: Interpretable Mental Health Analysis on Social Media with Large Language Models. WWW 2024 - Proceedings of the ACM Web Conference, Association for Computing Machinery, Inc; 2024, p. 4489–500. https://doi.org/10.1145/3589334.3648137.
26. Eysenbach G. The Role of ChatGPT, Generative Language Models, and Artificial Intelligence in Medical Education: A Conversation With ChatGPT and a Call for Papers. JMIR Med Educ 2023;9. https://doi.org/10.2196/46885.
27. Abd-Alrazaq A, AlSaad R, Alhuwail D, Ahmed A, Healy PM, Latifi S, Aziz S, Damseh R, Alrazak SA, Sheikh J. Large Language Models in Medical Education: Opportunities, Challenges, and Future Directions. JMIR Med Educ 2023;9. https://doi.org/10.2196/48291.
28. Hager P, Jungmann F, Holland R, Bhagat K, Hubrecht I, Knauer M, Vielhauer J, Makowski M, Braren R, Kaissis G, Rueckert D. Evaluation and mitigation of the limitations of large language models in clinical decision-making. Nat Med 2024;30:2613–22. https://doi.org/10.1038/s41591-024-03097-1.
29. Shen Y, Heacock L, Elias J, Hentel KD, Reig B, Shih G, Moy L. ChatGPT and Other Large Language Models Are Double-edged Swords. Radiology 2023;307. https://doi.org/10.1148/RADIOL.230163.
30. Wang D, Zhang S. Large language models in medical and healthcare fields: applications, advances, and challenges. Artif Intell Rev 2024;57. https://doi.org/10.1007/s10462-024-10921-0.
31. Yang Z, Qian J, Huang Z-A, Tan KC. QM-ToT: A Medical Tree of Thoughts Reasoning Framework for Quantized Model 2025.
32. Mahalegi HSM, Farhadi A, Balogh G, Nagy E. Application of Generative AI in Resource Optimization in Healthcare Systems. Application of Generative AI in Healthcare Systems, Cham: Springer Nature Switzerland; 2025, p. 175–95. https://doi.org/10.1007/978-3-031-82963-5_7.

33. Gao Y, Li R, Croxford E, Caskey J, Patterson BW, Churpek M, Miller T, Dligach D, Afshar M. Leveraging Medical Knowledge Graphs Into Large Language Models for Diagnosis Prediction: Design and Application Study. JMIR AI 2025;4. https://doi.org/10.2196/58670.
34. Tang L, Sun Z, Idnay B, Nestor JG, Soroush A, Elias PA, Xu Z, Ding Y, Durrett G, Rousseau JF, Weng C, Peng Y. Evaluating large language models on medical evidence summarization. NPJ Digit Med 2023;6. https://doi.org/10.1038/s41746-023-00896-7.
35. Maslinski J, Grasfield R, Awasthi R, Mishra S, Mahapatra D, Mathur P. Understanding Large Language Models in Healthcare: A Guide to Clinical Implementation and Interpreting Publications. Cureus 2025. https://doi.org/10.7759/cureus.82397.
36. Tam TYC, Sivarajkumar S, Kapoor S, Stolyar A V., Polanska K, McCarthy KR, Osterhoudt H, Wu X, Visweswaran S, Fu S, Mathur P, Cacciamani GE, Sun C, Peng Y, Wang Y. A framework for human evaluation of large language models in healthcare derived from literature review. NPJ Digit Med 2024;7. https://doi.org/10.1038/s41746-024-01258-7.
37. Chang CT, Farah H, Gui H, Rezaei SJ, Bou-Khalil C, Park YJ, Swaminathan A, Omiye JA, Kolluri A, Chaurasia A, Lozano A, Heiman A, Jia AS, Kaushal A, Jia A, Iacovelli A, Yang A, Salles A, Singhal A, Narasimhan B, Belai B, Jacobson BH, Li B, Poe CH, Sanghera C, Zheng C, Messer C, Kettud DV, Pandya D, Kaur D, Hla D, Dindoust D, Moehrle D, Ross D, Chou E, Lin E, Haredasht FN, Cheng G, Gao I, Chang J, Silberg J, Fries JA, Xu J, Jamison J, Tamaresis JS, Chen JH, Lazaro J, Banda JM, Lee JJ, Matthys KE, Steffner KR, Tian L, Pegolotti L, Srinivasan M, Manimaran M, Schwede M, Zhang M, Nguyen M, Fathzadeh M, Zhao Q, Bajra R, Khurana R, Azam R, Bartlett R, Truong ST, Fleming SL, Raj S, Behr S, Onyeka S, Muppidi S, Bandali T, Eulalio TY, Chen W, Zhou X, Ding Y, Cui Y, Tan Y, Liu Y, Shah N, Daneshjou R. Red teaming ChatGPT in medicine to yield real-world insights on model behavior. NPJ Digit Med 2025;8. https://doi.org/10.1038/s41746-025-01542-0.
38. Griot M, Hemptinne C, Vanderdonckt J, Yuksel D. Large Language Models lack essential metacognition for reliable medical reasoning. Nature Communications 2025;16. https://doi.org/10.1038/s41467-024-55628-6.

Designing Trustworthy and Explainable Clinical Decision Support Systems

Deniz Noor Mohammadzadeh Maleki, Azadeh Zamanifar, Amirfarhad Farhadi, and Majid Sorouri

Abbreviations

AI	Artificial Intelligence
ML	Machine Learning
CDSS	Clinical Decision Support System
EHR	Electronic Health Record
XAI	Explainable Artificial Intelligence
SA	Situation Awareness
AUC	Area Under the Curve
FL	Federated Learning
DP	Differential Privacy
PETs	Privacy-Enhancing Technologies
MLOps	Machine Learning Operations
CBN	Causal Bayesian Network
PDP	Partial Dependence Plot
ICE	Individual Conditional Expectation
TCAV	Testing with Concept Activation Vectors

D. N. M. Maleki
Department of Computer Engineering, Ta.C., Islamic Azad University, Tabriz, Iran
e-mail: Deniz.Noormohammadzadehmaleki@iau.ir

A. Zamanifar
Department of Computer Engineering, SR.C, Islamic Azad University, Tehran, Iran
e-mail: azamanifar@iau.ac.ir

A. Farhadi
School of Computer Engineering, Iran University of Science and Technology, Tehran, Iran
e-mail: am_farhadi@mail.iust.ac.ir

M. Sorouri (✉)
Department of Electronic Engineering, Maynooth University, Maynooth, Co. Kildare, Ireland
e-mail: majid.sorouri@mu.ie

A. Zamanifar et al. (eds.), *Explainable Large Language Models in Healthcare Applications*, Bio-IT and AI, https://doi.org/10.1007/978-3-032-15088-2_5

SHAP	Shapley Additive explanations
LIME	Local Interpretable Model-Agnostic Explanations
HCD	Human-Centered Design
FDA	Food and Drug Administration
HTA	Health Technology Assessment

1 Introduction

Clinical decision support systems (CDSS) augment clinician judgment by combining patient data with curated clinical knowledge to generate patient-specific recommendations at the point of care. These systems can improve safety and quality when they are tightly integrated with workflows and evaluated beyond algorithmic accuracy to include clinical utility and impact [1]. Yet history shows that predictive performance alone rarely guarantees better outcomes; rigorous external validation, calibration, updating, and impact evaluation are indispensable to avoid well-calibrated but clinically unhelpful tools. Methodological reviews emphasize that even strongly discriminative models can be clinically useless if predictions fall outside actionable decision thresholds or are not maintained against dataset shift [2].

Trustworthiness in clinical AI depends on transparent intent, robust data quality, performance under distribution shift, accountable governance, and human-factors-aware implementation. A recent systematic review proposes a structured, medical-AI-specific data quality framework (METRIC) that links dataset properties to fairness, robustness, and interpretability foundational preconditions for trustworthy CDSS [3]. Explainability is central because most modern CDSS employ complex machine-learning models whose decision logic is otherwise opaque to clinicians and regulators [4]. Comprehensive surveys show that model-embedded and post-hoc XAI methods (e.g., saliency, feature attribution, prototypes) can improve clinician comprehension, error detection, and calibrated reliance when evaluated with human-centered studies rather than proxy technical metrics alone [5]. Despite potential benefits, CDSS can degrade safety through human–automation interaction failures such as automation bias over-reliance on suggestions that suppresses independent verification. Systematic evidence across health informatics shows that automation bias is amplified by verification complexity, time pressure, and cognitive load, underscoring the need for design features (e.g., uncertainty displays, rationale exposure, and reversible recommendations) that promote appropriate skepticism [6].

Alerting mechanisms, a dominant interaction pattern in CDSS raise their own challenges: bibliometric analyses indicate a shift from patient-safety emphasis toward system usability, with "alert fatigue" now a key threat to effectiveness and clinician trust [7]. Prior meta-research therefore recommends tuning relevance thresholds, suppressing low-value alerts, and embedding learning loops that continuously recalibrate triggers to local practice. Fairness is equally foundational: biases introduced during data collection, labeling, modeling, and deployment can propagate inequities across patient groups, making formal bias assessment and mitigation requirements for clinical AI [8]. Broad reviews of fairness in healthcare AI highlight dataset representativeness, appropriate performance stratification, and post-deployment monitoring as essential safeguards for trustworthy CDSS [9]. Regulatory and standards ecosystems now expect transparency and human oversight for AI-enabled decision support, including clear statements of intended use, data provenance, performance claims, and conditions of safe deployment [10]. Complementary governance expectations from peer-reviewed and standards-aligned literature stress lifecycle risk management and post-market surveillance to maintain trust over time [1].

At the frontier, large language models (LLMs) introduce natural-language interfaces, code-and-knowledge grounding, and explanation generation that can wrap predictive analytics, but their reliability hinges on robust data quality, guardrails, and human-in-the-loop validation [11]. Current surveys caution that LLM-augmented CDSS must be audited for hallucinations, provenance of evidence, calibrated uncertainty, and fair performance across subgroups before clinical deployment [5]. Consequently, designing trustworthy and explainable CDSS requires an end-to-end socio-technical approach that links dataset quality (METRIC), rigorous validation and impact evaluation, human-factors-aware interaction design, fairness safeguards, and explicit governance and monitoring [3]. This chapter provides a comprehensive and systematic examination of the design principles, governance frameworks, and emerging technologies underpinning trustworthy and explainable Clinical Decision Support Systems (CDSS), elucidating how transparency, fairness, privacy-preserving learning, and human-centered design collectively transform algorithmic models into accountable, interpretable, and clinically reliable partners for evidence-based decision-making across modern healthcare. Figure 1, is show framework illustrating the core pillars of trustworthy CDSS, where explainability, fairness, privacy, robustness, and human-centered design collectively ensure clinical safety and reliable decision support.

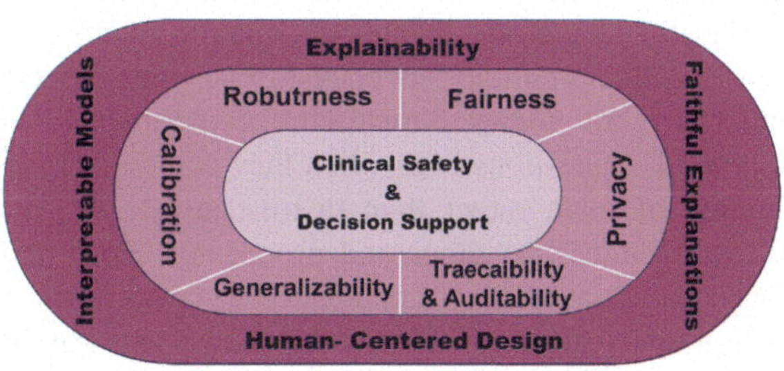

Fig. 1 Framework for achieving clinical safety and reliable decision support through explainability, fairness, privacy, and human-centered AI design

2 The Trust Problem in Clinical AI

The rapid expansion of artificial intelligence (AI) in healthcare has generated unprecedented expectations regarding diagnostic precision, efficiency, and personalization. Yet, despite decades of progress in predictive analytics, deep learning, and clinical decision support systems (CDSS), widespread clinical adoption remains restrained by a persistent and multifaceted "trust problem." Clinicians, patients, and healthcare institutions hesitate to rely on models whose reasoning, safety, and accountability are not transparent or continuously verifiable [12]. Confidence in clinical AI is not a simple yes or no but a scaled relationship between humans and machines, based on empirical evidence, clarity of understanding, and the compatibility of technology with professional standards. When systems lack credible validation or disrupt established decision-making workflows, clinicians appropriately exercise skepticism. This aligns with human–AI collaboration research, which conceptualizes trust as "calibrated reliance" neither blind acceptance nor complete rejection but reliance proportionates to demonstrable [13].

A fundamental barrier is the lack of transparency in machine-learning algorithms. Many top-performing deep-learning models function as statistical black boxes, complicating the process of understanding how particular inputs affect outputs. In medicine, where accountability and explainability are ethical imperatives, this opacity undermines cognitive trust and raises medicolegal uncertainty [12]. Clinicians may reasonably fear that an inscrutable model could misclassify a case or conceal bias, and they lack the means to interrogate its rationale. Another factor eroding confidence is dataset shift the degradation of model performance when real-world data diverge from training data. Differences in patient demographics, sensor calibration, or clinical documentation practices can silently alter predictive [14]. Without explicit monitoring, drift detection, and retraining protocols, even initially validated models can become unsafe, making dynamic oversight a precondition for sustained trust.

Equally crucial are human–automation interaction errors, particularly automation bias the cognitive tendency to over-rely on computer recommendations and under-weight contrary clinical cues. Experiments demonstrate that clinicians may accept AI suggestions even when they contradict observable data [6]. This miscalibration of trust, rather than distrust, creates a paradox where well-intentioned reliance reduces vigilance and propagates algorithmic errors directly to patients. Conversely, under-trust a reluctance to engage with CDSS despite proven benefit leads to under-utilization of valid guidance and wasted analytical potential. Striking a balance between automation bias and unjustified skepticism therefore requires iterative trust calibration: adaptive feedback, transparent rationale, and interface designs that encourage appropriate verification [15]. Clinical workflow integration is another determinant of justified trust. When CDSS outputs are context-irrelevant, interruptive, or misaligned with clinician reasoning, they generate cognitive friction and alert fatigue, eventually prompting users to ignore all notifications including life-critical [7].

Trust, therefore, emerges not only from model accuracy but also from interaction quality, timing, and perceived usefulness in clinical context. Beyond interface design, data provenance and quality form the epistemic foundation of trustworthy AI. Biased or incomplete data yield models that generalize poorly across subpopulations, perpetuating healthcare disparities [8]. Fairness audits, demographic performance stratification, and transparent reporting of dataset composition are thus moral and scientific imperatives. When equity is clearly integrated, trust reaches beyond single users to encompass communities and regulators. Explainability has been praised as a solution for distrust, but empirical evidence shows a more complex truth. Explainable-AI (XAI) techniques such as saliency maps, feature attribution, or counterfactual reasoning can enhance interpretability but may also oversimplify or mislead if unvalidated for clinical comprehension [5].

Effective explanation is not merely technical transparency but causability the user's ability to form accurate causal mental models of system behavior [16]. Furthermore, explainability alone cannot compensate for lack of validation or governance. Trustworthy CDSS require end-to-end lifecycle management; from data acquisition through post-deployment monitoring, encompassing calibration, uncertainty quantification, bias mitigation, and documentation [3]. Transparent reporting frameworks such as TRIPOD-AI, CONSORT-AI, and DECIDE-AI operationalize this lifecycle by specifying documentation requirements for model development, validation, and clinical evaluation [17]. At the ethical and institutional level, accountability and liability concerns strongly influence clinician trust.

When there is a lack of clarity about the decision-making authority between human and algorithm, practitioners may be subject to moral or legal responsibility for inscrutable outputs of the system [18]. Clear roles by the identification of AI as assistive, not autonomous helps maintain health professional agency and public confidence. Likewise, trust in the institution requires transparent governance and compliance with regulations along with communication. Health systems that release validation data, update cycles and handling protocols engender confidence across clinical teams and patients [19].

Conversely, secrecy or proprietary opacity can provoke suspicion, even when models perform well technically. Recent advances such as large-language-model (LLM)-augmented CDSS introduce new trust dynamics. LLMs can generate natural-language explanations, summarize guidelines, and synthesize evidence, but they also hallucinate facts, misattribute sources, and obscure uncertainty. These risks make human-in-the-loop validation and provenance tracing indispensable [11]. From an engineering standpoint, Kondylakis, H et al. framework translate abstract ethics into actionable design principles fairness, universality, traceability, usability, robustness, and explainability providing developers with operational criteria for measurable trustworthiness [20]. Incorporating these principles into certification and procurement policies can establish trust as an institutional standard rather than viewing it as a personal feeling. Post-market monitoring is the last and frequently overlooked component. Continuous auditing for data drift, error clusters, and adverse-event reports ensures that trust, once earned, is not lost through complacency. Public

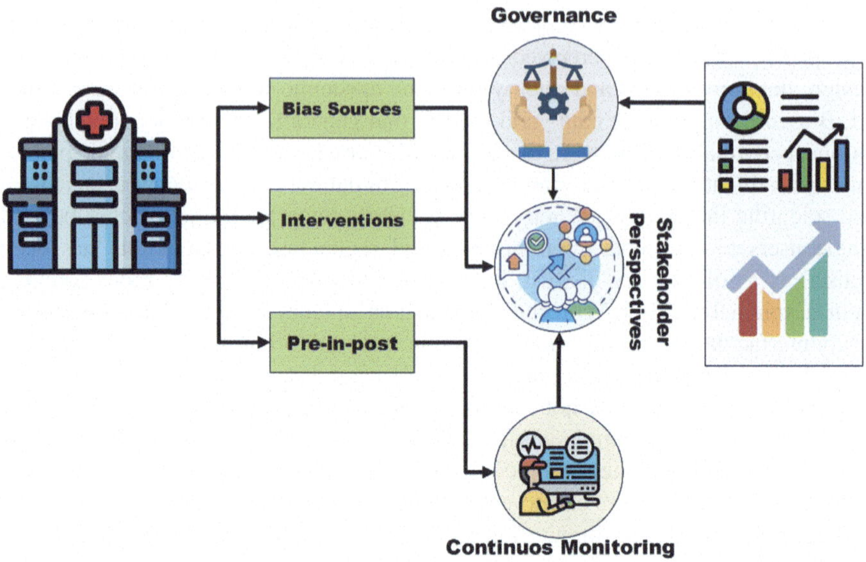

Fig. 2 Framework for bias identification, intervention, and continuous monitoring in clinical decision systems, emphasizing governance, stakeholder perspectives, and performance metrics

registries of deployed algorithms and open feedback mechanisms have been proposed as accountability infrastructures [21].

Ultimately, the trust challenge in ClinicalAI is as much socio-technical as computational. It is the product of interactions between human cognition, institutional governance, ethical norms and design of the system. It therefore cannot be solved by engineers alone, as also referred to in [22], but has to be tackled interdisciplinarily between engineers, clinicians, ethicists, regulators and patients. Each party adds its own layer of trust: technical, relational and moral within the network. The developers of trustworthy CDSS must hence adopt trust-by-design as part of the systems' life cycle: strict data wrangling, intelligible modeling, equitable assessment, human-centred interaction, transparent accountability and adaptive governance. Only by integrating these dimensions can AI systems progress from theoretical promise to dependable clinical partners [23] (Fig. 2).

3 Trustworthy and Explainable (XAI)

Clinical decision support systems (CDSS) promise measurable improvements in diagnostic accuracy, timeliness, and consistency, but their routine adoption is constrained by a pervasive trust deficit that arises from opaque model behavior and insufficiently transparent validation; clinicians and institutions therefore insist on intelligible justification, auditable provenance, and demonstrable coherence with medical logic and practice, not mere discrimination metrics [12, 16]. Explanation in medicine is an epistemic requirement: it must enable scrutiny, contestability, and shared accountability so that clinicians can form correct causal mental models or "castability" from algorithmic outputs, which is why interpretable-by-design approaches are often preferred and post-hoc explanations must be validated for fidelity and stability when used [5, 12, 16]. The XAI toolkit in healthcare broadly comprises global model-level interpretability, local instance explanations, concept- and prototype-based methods, and counterfactual reasoning, each presenting trade-offs among fidelity, stability, and cognitive usability; when embedded into a lifecycle of assurance dataset curation, calibration, external validation, uncertainty quantification, and continuous monitoring these methods reveal whether predictions rely on clinically meaningful signals or on spurious, non-transportable associations [5, 12, 14]. Dataset shift is a primary risk to justified trust because model performance and explanatory concordance can degrade as deployment data diverge from training distributions; consequently, explanation stability must be systematically audited across sites and over time, with predetermined governance triggers for recalibration, rollback, or retraining when XAI surfaces non-transportable relationships [14]. Human automation interaction evidence shows that well-designed explanations can appropriately calibrate reliance reducing over-trust under high uncertainty and supporting uptake when evidence is robust whereas poorly constructed explanations produce miscalibration and can exacerbate automation bias, particularly when verification is cognitively costly; therefore, XAI must reduce rather than increase clinicians' cognitive burden by delivering concise, context-aware, and actionable rationales [6, 25]. To operationalize explainability as an obligation rather than an aspiration, reporting standards and evaluation frameworks mandate lifecycle transparency and empirical assessment of interaction, usability, and safety (for example, DECIDE-AI), which together promote reproducibility, open audit trails, and independent verification of both predictive performance and explanation stability [12, 17, 26]. Fairness is an essential target of XAI because explanation methods expose proxy variables and skewed feature usage that can produce disparate outcomes across subgroups; routine subgroup-specific explanation profiling, paired with stratified performance reporting, enables targeted mitigation of inequitable behavior before it becomes entrenched in clinical workflows [8]. Workflow fit moderates the utility of explanations: excessive, poorly contextualized alerts produce fatigue and signal disregard, whereas brief, patient-specific explanations that connect model outputs to guideline concordance or plausible physiological mechanisms improve verification and clinical actionability [7]. At the modality level, counterfactual explanations often align naturally with clinical reasoning by identifying minimal, plausible changes that would alter predicted risk and thereby support counselling and shared decision-making, while concept- and prototype-based methods map statistical patterns to medical semantics (e.g., lesion types, waveform motifs) to facilitate multidisciplinary review and teach-back; crucially, these approaches must be constrained by clinical feasibility and safety to avoid recommending unattainable or harmful interventions [5, 12]. The arrival of large-language-model–augmented CDSS intensifies these imperatives because LLMs introduce hallucination and provenance opacity risks; trustworthy integration therefore requires explicit citation of evidence, calibrated uncertainty estimates, stepwise reasoning that clinicians can verify, and human-in-the-loop adjudication checkpoints [11, 27]. Finally, because trust must be continually earned and maintained, post-deployment XAI monitoring should track explanation drift, subgroup disparities, and adverse-event signals with prespecified governance actions for retraining or withdrawal; when XAI is co-designed with clinicians and patients and embedded across data stewardship, modeling, validation, interface design, and surveillance, explainability becomes an operational cornerstone that converts predictions into reasons, supports fairness audits, enables accountability, and anchors calibrated human reliance in high-stakes care [16, 26, 28] (Fig. 3).

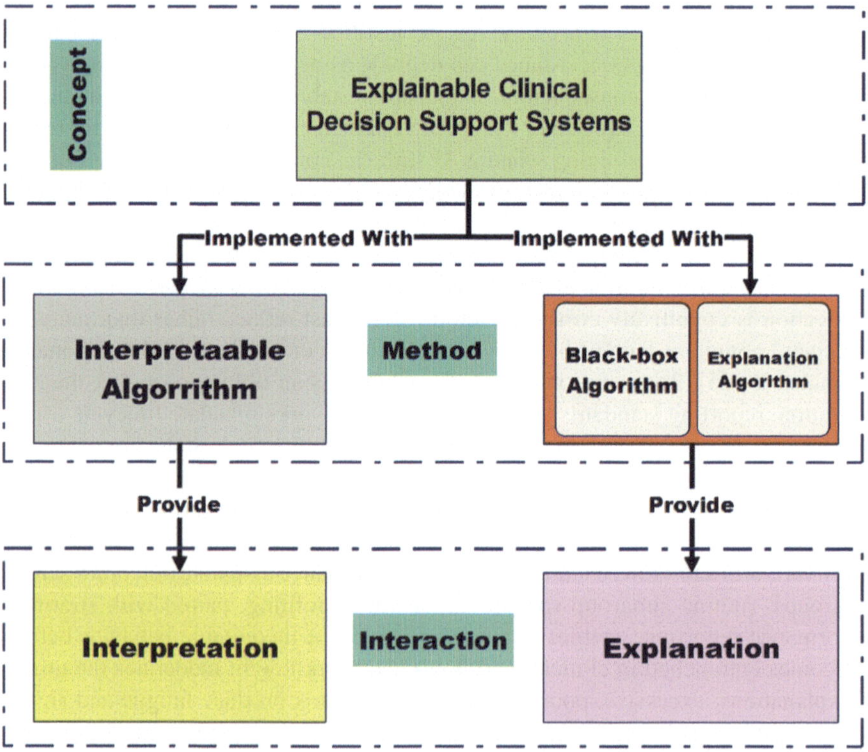

Fig. 3 Conceptual framework of trustworthy and explainable clinical decision support systems (X-CDSS)

4 The Evolution of Trustworthy and Explainable Clinical Decision

Clinical decision support systems (CDSS) have evolved over the past fifty years from transparent, rule-based expert systems into hybrid, data-driven platforms that reflect both increased computational capacity and a deeper appreciation of healthcare as a socio-technical domain. Despite this evolution, the core objective remains constant: to supply reliable, transparent, and clinically appropriate recommendations that support but do not supplant human judgement [29, 30]. In high-risk clinical settings, explainability and trustworthiness are therefore not optional virtues but preconditions for deployment.

Early CDSS (e.g., MYCIN, INTERNIST-I) relied on symbolic IF–THEN logic and excelled at provenance: each recommendation could be traced to explicit rules and evidence, enabling audit, challenge, and integration with workflow triggers such as reminders or order sets [29, 30]. Standardization efforts (Arden, and later FHIR-based computable forms and EBMonFHIR) preserved that explainability chain-linking bedside logic back to source evidence and made transparency a technical design constraint rather than mere documentation [29, 30]. However, rule-based systems proved brittle in practice. Knowledge elicitation and maintenance were onerous; rules often failed outside their original context due to drifting terminology, mapping errors, or outdated interfaces. Operational studies find ubiquitous failures from broken triggers to stale data connections, recasting reliability as a lifecycle governance problem rather than a pure logic problem [31, 32]. Concurrently, poor human–computer interaction most notably alert fatigue driven by ill-timed or excessive notifications revealed that many failures were engineering and ergonomic, not conceptual, lapses [32].

The proliferation of EHRs and high-dimensional clinical data (time series, imaging, free text) accelerated adoption of machine learning (ML), deep learning, and large language models (LLMs) capable of detecting complex, nonlinear patterns beyond hand-coded rules [33]. But these models introduced new trust challenges: statistical "black boxes" whose internal reasoning is opaque to clinicians, prompting the rise of explainable AI (XAI) as a discipline committed to making model behaviour interpretable, accountable, and aligned with medical reasoning [24, 34, 35]. Contemporary CDSS therefore favor hybrid architectures that combine rule-based guardrails (safety constraints, eligibility, workflow triggers) with ML components for stratification and pattern detection preserving explainability while capturing empirical complexity [34, 35]. FHIR-linked computable guidelines further enable traceability from model outputs back to evidence artifacts, sustaining the rule-based heritage of provenance into learning-based epochs [35]. With learning models come operational responsibilities; MLOps practices (data versioning, reproducible pipelines) and continuous surveillance are required to detect drift, calibration decay, and emerging biases; validation must become vigilance [36, 37]. Successful deployment is fundamentally socio-technical requiring workflow redesign, clinician training,

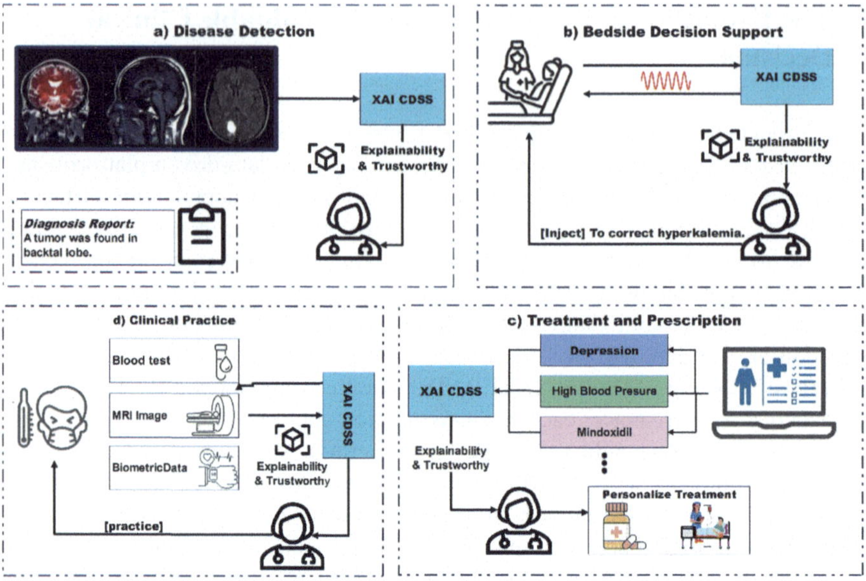

Fig. 4 Rule-based expert systems to hybrid, learning-driven CDSS, ensuring explainability, provenance, continuous MLOps governance, and human-in-the-loop oversight is essential for trustworthy bedside decision support

and interface-level explainability because accuracy alone does not guarantee clinical value or adoption [37, 38]. LLMs expand capabilities for narrative reasoning and evidence synthesis, but they also bring hallucination and provenance risks that mandate human-in-the-loop verification and tightly coupled evidence retrieval [35]. Figure 4 Explain transparent rule-based expert systems to hybrid, learning-driven platforms, the evolution of CDSS compels explainability, provenance, continuous MLOps-backed governance, and human-in-the-loop socio-technical design to ensure reliable, auditable bedside decision support (Table 1).

5 Designing Trustworthy CDSS

Reliability is that characteristic that separates experimental AI proof-of-concepts from clinical decision-support systems (CDSS) that must be trusted to perform in operation [37, 47]. For a CDSS to gain trust, it must show reliability, safety, robustness, and fairness not as independent goals but as interconnected aspects of a comprehensive assurance program that manages design, validation, deployment, and lifecycle oversight. Relying solely on technical accuracy, such as high AUC or F1 does not guarantee clinical acceptance or regulatory approval without proven,

Table 1 Comparative summary of leading XAI techniques for CDSS scope, outputs, exemplar uses, strengths, and deployment caveats in clinical practice

Technique/tool	Scope (global/local)	Typical output	Example clinical use	Strengths for CDSS	Key caveats
SHAP (tabular)	Local and global	Feature attributions per case; global importance summaries	Risk drivers for sepsis readmission from EHR features	Additive, theoretically grounded; aggregates well for dashboards	Sensitive to background distribution; correlated features complicate attribution [5]
LIME (local linear surrogate)	Local	Sparse linear explanation around a prediction	Explaining an individual chest-X-ray triage score to a radiologist	Model-agnostic; human-parsable coefficients	Instability across runs/seed; locality kernel choices affect faithfulness [39]
Integrated Gradients (IG)	Local (DL)	Pixel/voxel attributions along gradients	Highlighting infarct-relevant regions in brain MRI classifiers	Implementation-invariant; less noisy than plain gradients	Still saliency-based; can highlight non-causal textures [40]
Counterfactual explanations (model-agnostic)	Local (actionable)	Minimal, plausible changes to flip a prediction	What if Cr fell by 0.3 mg/dL—would AKI alert clear?"	Action-oriented; supports treatment planning and consent	Must enforce plausibility/feasibility; can be adversarial [41]
Causal/CBN-based counterfactuals	Local and global	Do-calculus-grounded counterfactuals and path analyses	Governance review of surgical decisions under alternatives	Encodes causal structure; policy-relevant "what-if"	Requires valid causal graph; data/assumption heavy [42]
TCAV (concept-based)	Local and global (DL)	Concept sensitivity scores (e.g., "consolidation", "edema")	EEG abnormality detection explained by clinical concepts	Bridges human concepts ↔ features; bias checks	Needs curated concept sets; concept leakage risk [43]
SurvSHAP(t) (survival XAI)	Local and global	Time-dependent attributions for time-to-event models	Explaining evolving drivers of BCR after prostatectomy	Adds the time dimension clinicians need	More complex to communicate; method still maturing [44]

(continued)

Table 1 (continued)

Technique/tool	Scope (global/local)	Typical output	Example clinical use	Strengths for CDSS	Key caveats
PDP/ICE (global effect plots)	Global (and individualized via ICE)	Marginal effect curves (1–2 features)	Showing dosage–response shape for a dosing CDSS	Simple; detects non-linearities and interactions	Assumes feature independence; can mislead with correlated covariates [45]
SHAP versus LIME (comparative perspective)	Methods overview	Guidance on when to use which	Selecting an explainer for tabular CDSS (labs/vitals)	Practical comparison; deployment-oriented	Neither guarantees causal faithfulness [46]

consistent reliability across various patients, institutions, and time periods [37, 47–49]. Practically, reliability requires that the model predictions have to be accurately well calibrated and not drift in real-life ways: temporal drift when there is case-mix, clinical practice, or site variation. Along with discrimination measures, drift in calibration, temporal validation, and prospective monitoring of performance are necessary because empirical experiences in imaging and EHR applications show that otherwise correct models can silently degrade after deployment. As such, reliability is enabled as a practice of continuous monitoring, timely re-tuning, and version-controlled versioning of data, models, and pipelines the MLOps and post-deployment monitoring practice which are now the core quality-assurance layers [6, 31, 32, 50]. Safety broadens this perspective from algorithmic activity to the sociotechnical setting in which CDSS operate. The vast majority of actual-world safety accidents originate not from algorithmic failure but from incorrect inputs, workflow incompatibility, poor human–machine interfaces, and human cognitive risks such as automation bias and complacency. Sound safety engineering therefore combines rigorous pre-market testing with system-level hazard analysis (e.g., bow-tie modelling), explicit escalation procedures, and human-in-the-loop controls for marginal or unclear cases. Cerebral alarm semantics firing windows, recurrence rules, and contextual triggers are needed to avoid alert fatigue while preserving clinical sensitivity. Organizational processes (root-cause analysis, structured remediation, incident reporting) are as critical to safety as model logic itself [51]. Robustness addresses predictability under stress: distributional changes, missing or corrupted modalities, sensor noise, and intentional adversarial perturbations. Medical imaging and signal research has already shown vast amplification of small, rational perturbations into large, clinically important output variations. Robust CDSS should therefore be thoroughly stress-tested under off-nominal and worst-case conditions, and mitigation techniques such as adversarial training, tight data hygiene, runtime anomaly detection, and formally defined safety arguments that favor graceful deterioration over catastrophic loss. In addition to calibration stability and dependable runtime behavior, robustness makes up the technical triad of reliable design [52].

Fairness equitable sharing of predictive performance and clinical benefit across subgroups is key to trust. CDSS can either enhance or worsen current health disparities based on its design and monitoring.

Fairness is context-specific, goals vary for screening, triage, resource distribution, or guidance, and no single numerical value captures all ethical considerations [53]. Blind reliance on global precision can mask significant subgroup differences; also, equity standards (e.g., calibration parity vs. balanced error rates) may not be compatible. Misguided optimization can produce "leveling down" harming all groups; fairness interventions should instead "level up" by improving performance for under-privileged groups without lowering overall safety [1, 52, 54]. Bias reduction must accordingly be multi-faceted. Pre-processing strategies (reweighting, enrichment, deduplication) address dataset imbalance; in-processing methods include fairness constraints or adversarial losses in optimization; post-processing moves thresholds to meet operational equity requirements. In domains where decisions have significant utility impact (e.g., specialist triage, organ allocation), fairness evaluation must also consider real-world benefit–harm trade-offs across populations and over time [52, 55]. Causal fairness improves statistical parity by identifying and altering causal pathways leading to unfairness so that mitigation takes place at the mechanism of inequity rather than superficial correlations. These strategies attempt to preserve balanced reasoning and auditable decision logic that remain clinically justifiable to targeted subgroups [56]. Commitment to organization and governance is crucial.

The rivalry between race-aware and race-neutral modeling highlights an ethical nuance: Excluding protected traits may obscure risks for minority communities; including them could unintentionally reinforce systemic bias. Practical, contingent solutions memoing repeatedly when and why sensitive attributes are used, using their application in combination with subgroup validity monitoring, and publishing mitigation findings allow therapists to conduct themselves responsibly [53]. Achieving fairness in actual CDSS requires more than technological remedies. Establishing routine disclosure of subgroup performance, remediation efforts, and monitoring outcomes as an institution instills transparency and accountability into organizational culture and not just into regulation as an afterthought [37, 47, 56].

6 Human Centered Design

Human-centered design (HCD) views clinical decision support systems (CDSS) as socio-technical solutions that should align with clinicians' cognitive workflows, collective situation awareness, and task requirements rather than just enhance predictive accuracy. This involves gathering user objectives and limitations across various roles, iteratively creating prototypes of interaction patterns, and verifying if the content, timing, and method of assistance genuinely lower cognitive load and coordination expenses during care delivery. Evidence from human-AI teaming research shows that transparency and calibrated explainability must be designed to sustain clinicians' situation awareness (what the model "knows," "is doing," and "will do next") rather than added post hoc; otherwise, explanations can distract, induce overreliance, or fail to surface model limits when it matters [57]. At the micro-interaction level, alert design illustrates HCD in action: aligning notifications with clinically meaningful temporal landmarks (e.g., order entry vs. administration vs. monitoring windows) can curb alert burden and improve actionability, because "when" an alert fire is as consequential as "what" it says [31].

Operationalizing HCD for CDSS typically combines contextual inquiry, workflow mapping, participatory co-design, and formative usability testing (think scenario-based simulations with representative users) before live deployment. Studies that followed this path report measurable gains in adoption and clinician performance. For example, a user-centered CDSS to promote early peanut introduction embedded prompts and education into well-child visits; after iterative usability refinement, implementation was associated with more consistent counseling and better alignment of actions with guideline intent an effect attributable to fitting the tool to real-world pediatric workflows [58].Beyond single sites, HCD also encourages continuous learning loops capturing use patterns (e.g., alert dwell times, acknowledgment latencies) and qualitative feedback to update triggers, presentation, and escalation pathways as practices evolve [31].

HCD connects directly to trust and explainability: explanations should be role-sensitive (clinician vs. pharmacist vs. nurse), purpose-specific (verification, justification, education), and evaluated with end users using agreed criteria rather than solely technical proxies. Recent syntheses emphasize that explainability which is not anchored in user needs rarely improves safety or adoption; they advocate standardized, multi-stakeholder evaluation and reporting during early clinical studies so that utility, risks, and workflow effects are visible to decision-makers [3, 58].

7 Trustworthy Federated and Continuous Learning in CDSS

As clinical data remain geographically scattered and extremely confidential, federated learning (FL) and privacy-preserving machine learning have been the front-runners for the future generation of CDSS. FL does not centralize patient records but synchronizes model training across institutions, keeping raw data intact and reducing regulation- and ethics-based risks of confidentiality breaches [59]. In reality, sites or hospitals train local model updates with their own data and share gradients or parameters through a trusted aggregator. This paradigm is able to enable data sovereignty requirements, alleviate cross-border data transfer reluctance, and potentially accelerate multi-site collaboration without mandating a lowest-common-denominator dataset. It also better captures real-world clinical heterogeneity, since models can learn across various EHR systems, imaging protocols, and care pathways without convening the data into a single warehouse [59]. Dual-goal complementary privacy technologies also protect this environment. Differential privacy restricts what can be inferred from published parameters, adding calibrated noise to updates such that membership or attribute inference about any individual patient is provably difficult. Homomorphic encryption and secure aggregation safeguard model updates at rest and in transit, allowing for global model computation without revealing site-level contributions, while secure enclaves (TEE) may provide runtime protection for the aggregator. Together, these methods ensure that even if communications are intercepted by an attacker or access to the global model is gained, reverse engineering sensitive patient-level data is still hard in practice [60]. Balancing privacy budgets and computing overhead so that training is up to date for clinical deployment involves making the statistical and cryptographic privacy layers clinically feasible. However, systematic reviews today point out that FL in healthcare is not the silver bullet. Generalizability benefits over heterogeneous populations are countered by non-IID data, labeling skewness, and site-specific documentation idiosyncrasies all of which obstruct or destabilize convergence [31]. Communication efficiency is another bottleneck: frequent synchronization of big models can be expensive on hospital networks; compression, scarification, or partial participation strategies are generally necessary. Security is still an open problem: poisoning attacks on a subset of clients can cause the global model to move toward harmful behaviors, and gradient leakage or membership inference are still feasible without strong defenses. Robust aggregation (e.g., median-based or Byzantine-resilient policies), client update anomaly detection, and contribution provenance tracking have to be top-class requirements, not an afterthought, for FL-enabled CDSS [31]. The vast majority of legacy CDSS are static and degrade with evolving clinical practice, patient cohorts, and data distributions. Continuous learning systems aim to achieve this by continuously updating models with new data and monitoring for drift, fairness, and calibration virtually in real time [48]. In oncology and critical care, initial experiments using continuous pipelines sometimes a combination of batch retraining with online updating suggest that adaptive retraining can preserve accuracy across shifting cohorts, updated treatment protocols, and shifting imaging protocols [49]. Critically, ongoing learning can support local personalization (e.g., ward- or device-specific tuning) without losing a shared backbone, achieving a balance between global generalizability and local fit [48, 49]. But ongoing learning also poses new governance challenges. Unchecked updates can propagate local biases throughout the system, lower performance for under-represented subgroups, or erode reproducibility if versioning is lax. To counteract these dangers, human-in-the-loop oversight must be constructed across multiple levels; review gates for model updates, approval by clinicians in shadow mode, and A/B testing aimed at chosen roll-outs. Automated monitors would observe predictive stability (e.g., AUROC/AUPRC trends), calibration (e.g., ECE), fairness statistics amongst protected features and care facilities, as well as multiple forms of drift (covariate, label, concept). Clear rollback procedures, signed artifacts, immutable experiment history, and reproducible pipelines are table stakes to safe clinical iteration [37]. Responsible adoption from the MLOps perspective involves an auditable lifecycle: data lineage and consent provenance, declarative pipelines, environment capture, and strict separation of development, validation, and production [37]. Release candidates must be constructed to fulfill pre-established acceptance tests, including external validation on hold-out sites, stress testing under simulated shift, and adversarial red-team testing. Monitoring must be followed up post-deployment with alerting thresholds and "circuit breakers" that stop or quarantine a model automatically when specified safety, equity, or calibration thresholds are breached. Documentation artifacts model cards, data sheets, and change logs support accountability and clinician trust through the recording of intended use, performance envelopes, monitoring plans, and known limitations [37]. Transition from pilots to standard deployment depends not just on technological advances but on responsibility, accountability, and governance in adoption processes. Effective programs create multi-stakeholder governing boards clinicians, developers, ethicists, patient representatives, security officers, and regulators to deliberate on trade-offs among performance and transparency, global generality and local equity, speed and safety

[40]. The boards ought to assume responsibility for policy for consent reuse, data minimization, and site participation; create escalation procedures for incidents; and have continuous contact with front-line clinicians who can reveal failure modes embedded in aggregate metrics [40]. Incentive alignment is also crucial: contribution credit to FL members, shared benchmarks, and transparent reporting reduce competitive barriers and encourage long-term collaboration. Finally, compliance mandates such as the EU Artificial Intelligence Act (2024) and evolving FDA guidelines make it evident that explainability, post-market surveillance, and liability provisions must be core aspects of CDSS especially if powered by adaptive or federated algorithms [61]. This encompasses strict risk stratification, detailed reporting of model refreshments, traceable decision-making that is accessible to clinicians, and required monitoring with specific reporting mandates. Combining methods privacy-preserving federated learning for multi-site training, continuous monitoring for drift and equity, strong evaluation frameworks, and human-focused governance offers a realistic solution. Accountable adoption is not just an agreement to anticipated performance but to consistent, safe, and ethical decision-making assistance that will survive the operating, legal, and social pressures of contemporary health care [61].

The work collectively highlights a pragmatic path toward merging augmented machine learning, trustworthy data foundations, and generative AI into clinical decision support and prioritizing the governance, security, and validation practices that drive trustworthy, explainable systems [62]. Experiments in automated image-based treatment planning (e.g., deep-learning object detection for panoramic dental X-rays) demonstrate both the clinical promise of task-specific vision models as well as the concurrent need for robust external validation, calibrated uncertainty prediction, and interpretable outputs before deployment [62]. Protocol-level technical analyses of blockchain-protected authentication and BERT-based ledger schemes demonstrate mutually supporting technological trends for ensuring data integrity, supporting auditable provenance, and enabling cross-institutional workflows in a blanket-free centralization capability that is critical in scenarios where CDSS must balance distributed sovereignty over data with traceability and responsibility [63]. Infrastructure and communications studies (e.g., energy-aware smart-city networking) focus on the system dependability on effective and robust network and edge structures that enable real-time CDSS operation under constrained resources and dynamic connectivity [64]. Furthermore, the recent Springer book chapters on generative AI on synthetic data generation and resource optimization recognize high-value, practical levers (privacy-preserving synthetic data, GenAI-augmented augmentation, and operational optimization) that can propel CDSS development at speed with a need for explicit protection for fidelity, bias prevention, and auditability [65, 66]. Together, these papers argue that reliable, explainable CDSS design is not just about high-performance models but also integrated technical controls (communications resilience, privacy, provenance), formal verification and recalibration schemes, and human-explanation mechanisms to make model outputs actionable, auditable, and institutionally acceptable [67, 68].

8 Conclusion

Clinical Clinical decision support systems only come to establish a long-term place in care when they are designed, regulated, and tested as socio-technical systems instead of as separate algorithms. All along in this chapter, we have argued that trustworthiness is an emergent property bridging reliability and resilience, fairness and privacy, and end-to-end monitoring while explainability is a design constraint which must be balanced against clinical tasks, cognitive load, and workflow. When these pieces are in alignment, CDSS move from "better forecasts" to quantifiably safer, more dependable, and accountable decisions. When they are not, the same magnitude of scale and velocity that make CDSS so powerful have the potential to amplify bias, automation error, and operational risk. The working principle is simple: performance measures are needed, but without calibrated explanations, uncertainty communication, and lifecycle monitoring, they are never sufficient. A trustworthy CDSS life cycle begins with evidence-based stewardship of information and bias analysis; moves forward through model development with explicit reliability, justice, and privacy objectives; and concludes in deployment practices that impose post-market surveillance as routine clinical quality assurance. Continuous monitoring for drift, calibration degradation, and subgroup performance failure along with circuit breakers, rollback plans, and auditable change logs turns safety into a commitment-on-going from barrier-at-once. Human-centered design connects the lifecycle together: descriptions and alerts must increase situation awareness, reduce cognitive load, and help clinicians make distinctions between trusting and distrusting. In this framing, "building trust" is a question of tuning trust with interfaces, instruction, and feedback systems that tune human judgment to match model ability and skepticism.

Going forward, the most promising advances in the domain federated and privacy-preserving learning, continuous adaptation, and selective use of large language models will only be feasible under explicit governance. Federated learning can potentially enhance representativeness without data transfer, but at the expense of PETs, strong aggregation, and contribution provenance. Continuous learning can preserve accuracy under shift, but only if complemented with human-in-the-loop review gates and reproducible MLOps. LLMs can generalize CDSS to free text and dialogue, but require evidence-linked constraints, open data governance, and uncertainty-sensitive output to forestall assertive error. In all of these, new regulatory frameworks endorse what good practice already indicates: simple accountability, traceability, and stakeholder-appropriate transparency are prerequisites not an afterthought of clinical use. sound and open CDSS are built where technical safeguards (strength, calibration, fairness, privacy), human-factors excellence (workflow fit, cognitively efficient explanations, uncertainty communication), and institutional governance (risk management, monitoring, auditability, and transparent liability) intersect. The roadmap applied is to pair multi-site, privacy-preserving learning with systematic testing and post-deployment surveillance; to frame explanation as a problem of interaction design associated with clinical decisions; and to codify surveillance that preserves equity and safety over the long term. Well done, CDSS become transparent, equitable, and clinically actionable partners predictive systems that justify and learn responsibly, improving care while preserving the trust on which medicine depends.

References

1. Sutton, R. T., Pincock, D., Baumgart, D. C., Sadowski, D. C., Fedorak, R. N., & Kroeker, K. I. (2020). An overview of clinical decision support systems: benefits, risks, and strategies for success. NPJ digital medicine, 3(1), 17.
2. Binuya, M. A. E., Engelhardt, E. G., Schats, W., Schmidt, M. K., & Steyerberg, E. W. (2022). Methodological guidance for the evaluation and updating of clinical prediction models: a systematic review. BMC Medical Research Methodology, 22(1), 316.
3. Schwabe, D., Becker, K., Seyferth, M. et al. The METRIC-framework for assessing data quality for trustworthy AI in medicine: a systematic review. npj Digit. Med. **7**, 203 (2024). https://doi.org/10.1038/s41746-024-01196-4
4. Chaddad, A., Peng, J., Xu, J., & Bouridane, A. (2023). Survey of Explainable AI Techniques in Healthcare. Sensors, 23(2), 634. https://doi.org/10.3390/s23020634
5. Sadeghi, Z., Alizadehsani, R., Cifci, M. A., Kausar, S., Rehman, R., Mahanta, P.,... & Pardalos, P. M. (2024). A review of Explainable Artificial Intelligence in healthcare. Computers and Electrical Engineering, 118, 109370.
6. Lyell, D., & Coiera, E. (2017). Automation bias and verification complexity: a systematic review. Journal of the American Medical Informatics Association, 24(2), 423-431.
7. Chien, S. C., Chen, Y. L., Chien, C. H., Chin, Y. P., Yoon, C. H., Chen, C. Y.,... & Li, Y. C. (2022, March). Alerts in clinical decision support systems (CDSS): a bibliometric review and content analysis. In Healthcare (Vol. 10, No. 4, p. 601). MDPI.
8. Cross, J. L., Choma, M. A., & Onofrey, J. A. (2024). Bias in medical AI: Implications for clinical decision-making. PLOS Digital Health, 3(11), e0000651.
9. Ueda, D., Kakinuma, T., Fujita, S., Kamagata, K., Fushimi, Y., Ito, R.,... & Naganawa, S. (2024). Fairness of artificial intelligence in healthcare: review and recommendations. Japanese journal of radiology, 42(1), 3-15.
10. US Food and Drug Administration. (2023). Clinical decision support software: guidance for industry and food and drug administration staff. September, 2022.
11. Noor, A. A., Manzoor, A., Mazhar Qureshi, M. D., Qureshi, M. A., & Rashwan, W. (2025). Unveiling Explainable AI in Healthcare: Current Trends, Challenges, and Future Directions. Wiley Interdisciplinary Reviews: Data Mining and Knowledge Discovery, 15(2), e70018.
12. Amann, J., Blasimme, A., Vayena, E., Frey, D., Madai, V. I., & Precise4Q Consortium. (2020). Explainability for artificial intelligence in healthcare: a multidisciplinary perspective. BMC medical informatics and decision making, 20(1), 310.
13. Quinn, T. P., Senadeera, M., Jacobs, S., Coghlan, S., & Le, V. (2021). Trust and medical AI: the challenges we face and the expertise needed to overcome them. Journal of the American Medical Informatics Association, 28(4), 890-894.
14. Finlayson, S. G., Subbaswamy, A., Singh, K., Bowers, J., Kupke, A., Zittrain, J.,... & Saria, S. (2021). The clinician and dataset shift in artificial intelligence. New England Journal of Medicine, 385(3), 283-286.
15. Rosenbacke, R., Melhus, Å., McKee, M., & Stuckler, D. (2024). How explainable artificial intelligence can increase or decrease clinicians' trust in AI applications in health care: systematic review. Jmir Ai, 3, e53207.
16. Holzinger, A., Langs, G., Denk, H., Zatloukal, K., & Müller, H. (2019). Causability and explainability of artificial intelligence in medicine. Wiley interdisciplinary reviews: data mining and knowledge discovery, 9(4), e1312.

17. Vasey, B., Nagendran, M., Campbell, B., Clifton, D. A., Collins, G. S., Denaxas, S.,... & McCulloch, P. (2022). Reporting guideline for the early stage clinical evaluation of decision support systems driven by artificial intelligence: DECIDE-AI. bmj, 377.
18. Alonso, M., Astobiza, A.M. & Ortega Lozano, R. AI-mediated healthcare and trust. A trust-construct and trust-factor framework for empirical research. Artif Intell Rev **58**, 337 (2025). https://doi.org/10.1007/s10462-025-11306-7
19. Steerling, E., Siira, E., Nilsen, P., Svedberg, P., & Nygren, J. (2023). Implementing AI in healthcare—the relevance of trust: a scoping review. Frontiers in health services, 3, 1211150.
20. Kondylakis, H., Osuala, R., Puig-Bosch, X., Lazrak, N., Diaz, O., Kushibar, K.,... & Lekadir, K. (2025). A review of methods for trustworthy AI in medical imaging: The FUTURE-AI Guidelines. IEEE Journal of Biomedical and Health Informatics.
21. Subasri, V., Krishnan, A., Kore, A., Dhalla, A., Pandya, D., Wang, B.,... & Dolatabadi, E. (2025). Detecting and Remediating Harmful Data Shifts for the Responsible Deployment of Clinical AI Models. JAMA Network Open, 8(6), e2513685-e2513685.
22. Sagona, M., Dai, T., Macis, M., & Darden, M. (2025). Trust in AI-assisted health systems and AI's trust in humans. npj Health Systems, 2(1), 10.
23. Bürger, V. K., Amann, J., Bui, C. K., Fehr, J., & Madai, V. I. (2024). The unmet promise of trustworthy AI in healthcare: why we fail at clinical translation. Frontiers in Digital Health, 6, 1279629.
24. Adadi, A., & Berrada, M. (2018). Peeking inside the black-box: a survey on explainable artificial intelligence (XAI). IEEE access, 6, 52138-52160.
25. Rivera, S. C., Liu, X., Chan, A. W., Denniston, A. K., Calvert, M. J., Ashrafian, H.,... & Yau, C. (2020). Guidelines for clinical trial protocols for interventions involving artificial intelligence: the SPIRIT-AI extension. The Lancet Digital Health, 2(10), e549-e560.
26. Collins, G. S., Moons, K. G., Dhiman, P., Riley, R. D., Beam, A. L., Van Calster, B.,... & Logullo, P. (2024). TRIPOD+ AI statement: updated guidance for reporting clinical prediction models that use regression or machine learning methods. bmj, 385.
27. McDermott, M. B., Wang, S., Marinsek, N., Ranganath, R., Foschini, L., & Ghassemi, M. (2021). Reproducibility in machine learning for health research: Still a ways to go. Science translational medicine, 13(586), eabb1655.
28. Rudin, C. (2019). Stop explaining black box machine learning models for high stakes decisions and use interpretable models instead. Nature machine intelligence, 1(5), 206-215.
29. Papadopoulos, P., Soflano, M., Chaudy, Y., Adejo, W., & Connolly, T. M. (2022). A systematic review of technologies and standards used in the development of rule-based clinical decision support systems. Health and Technology, 12(4), 713-727.
30. Ohno-Machado, L., Jiang, X., Kuo, T. T., Tao, S., Chen, L., Ram, P. M.,... & Xu, H. (2023). A hierarchical strategy to minimize privacy risk when linking "De-identified" data in biomedical research consortia. Journal of biomedical informatics, 139, 104322.
31. Canovas-Segura, B., Morales, A., Juarez, J. M., & Campos, M. (2023). Meaningful time-related aspects of alerts in Clinical Decision Support Systems. A unified framework. Journal of Biomedical Informatics, 143, 104397.
32. Thayer, J. G., Franklin, A., Miller, J. M., Grundmeier, R. W., Rogith, D., & Wright, A. (2024). A scoping review of rule-based clinical decision support malfunctions. Journal of the American Medical Informatics Association, 31(10), 2405-2413.
33. Biesheuvel, L. A., Workum, J. D., Reuland, M., van Genderen, M. E., Thoral, P., Dongelmans, D., & Elbers, P. (2025). Large language models in critical care. Journal of Intensive Medicine, 5(02), 113-118.
34. Li, J., Zhou, Z., Lyu, H., & Wang, Z. (2025). Large language models-powered clinical decision support: enhancing or replacing human expertise?. Intelligent Medicine, 5(1), 1-4.
35. Kierner, S., Kucharski, J., & Kierner, Z. (2023). Taxonomy of hybrid architectures involving rule-based reasoning and machine learning in clinical decision systems: A scoping review. Journal of biomedical informatics, 104428–104428.
36. Anjali Rajagopal, M. B. B. S. (2024). Machine Learning Operations in Health Care: A Scoping Review.

37. Sahiner, B., Chen, W., Samala, R. K., & Petrick, N. (2023). Data drift in medical machine learning: implications and potential remedies. The British Journal of Radiology, 96(1150), 20220878.
38. Lacson, R., Eskian, M., Licaros, A., Kapoor, N., & Khorasani, R. (2022). Machine learning model drift: predicting diagnostic imaging follow-up as a case example. Journal of the American College of Radiology, 19(10), 1162-1169.
39. Del Ser, J., Barredo-Arrieta, A., Díaz-Rodríguez, N., Herrera, F., Saranti, A., & Holzinger, A. (2024). On generating trustworthy counterfactual explanations. Information Sciences, 655, 119898.
40. Kyrimi, E., Mossadegh, S., Wohlgemut, J. M., Stoner, R. S., Tai, N. R., & Marsh, W. (2025). Counterfactual reasoning using causal bayesian networks as a healthcare governance tool. International journal of medical informatics, 193, 105681.
41. Baniecki, H., Sobieski, B., Szatkowski, P., Bombinski, P., & Biecek, P. (2025). Interpretable machine learning for time-to-event prediction in medicine and healthcare. Artificial Intelligence in Medicine, 159, 103026.
42. Salih, A. M., Raisi-Estabragh, Z., Galazzo, I. B., Radeva, P., Petersen, S. E., Lekadir, K., & Menegaz, G. (2025). A perspective on explainable artificial intelligence methods: SHAP and LIME. Advanced Intelligent Systems, 7(1), 2400304.
43. Berrett, J., de Kruiff, A., Pedell, S., & Reilly, A. (2022). Augmented assistive technology: the importance of tailoring technology solutions for people living with dementia at home. International Journal of Human-Computer Studies, 165, 102852.
44. Brenner, A., Knispel, F., Fischer, F. P., Rossmanith, P., Weber, Y., Koch, H.,... & Kutafina, E. (2024). Concept-based AI interpretability in physiological time-series data: Example of abnormality detection in electroencephalography. Computer Methods and Programs in Biomedicine, 257, 108448.
45. Farhadi, A., Zamanifar, A., Faezipour, M. (2025). Application of Generative AI in Drug Discovery. In: Zamanifar, A., Faezipour, M. (eds) Application of Generative AI in Healthcare Systems. Springer, Cham. https://doi.org/10.1007/978-3-031-82963-5_6
46. Taheri, A., Farhadi, A., Zamanifar, A. (2025). Application of GenAI in Clinical Administration Support. In: Zamanifar, A., Faezipour, M. (eds) Application of Generative AI in Healthcare Systems. Springer, Cham. https://doi.org/10.1007/978-3-031-82963-5_4
47. Lee, S., Yin, C., & Zhang, P. (2023). Stable clinical risk prediction against distribution shift in electronic health records. Patterns, 4(9).
48. Rajagopal, A., Ayanian, S., Ryu, A. J., Qian, R., Legler, S. R., Peeler, E. A.,... & Kawamoto, K. (2024). Machine learning operations in health care: A scoping review. Mayo Clinic Proceedings: Digital Health, 2(3), 421-437.
49. Rohren, E., Ahmadzade, M., Colella, S., Kottler, N., Krishnan, S., Poff, J.,... & Ghasemi-Rad, M. (2025). Post-deployment Monitoring of AI Performance in Intracranial Hemorrhage Detection by ChatGPT. Academic Radiology.
50. Khera, R., Simon, M. A., & Ross, J. S. (2023). Automation bias and assistive AI: risk of harm from AI-driven clinical decision support. Jama, 330(23), 2255-2257.
51. Sorin, V., Soffer, S., Glicksberg, B. S., Barash, Y., Konen, E., & Klang, E. (2023). Adversarial attacks in radiology–A systematic review. European journal of radiology, 167, 111085.
52. Huang, Y., Guo, J., Chen, W. H., Lin, H. Y., Tang, H., Wang, F.,... & Bian, J. (2024). A scoping review of fair machine learning techniques when using real-world data. Journal of biomedical informatics, 151, 104622.
53. Wang, Y., Song, Y., Ma, Z., & Han, X. (2023). Multidisciplinary considerations of fairness in medical AI: A scoping review. International Journal of Medical Informatics, 178, 105175.
54. Meerwijk, E. L., McElfresh, D. C., Martins, S., & Tamang, S. R. (2024). Evaluating accuracy and fairness of clinical decision support algorithms when health care resources are limited. Journal of Biomedical Informatics, 156, 104664.
55. Piccininni, M., Wechsung, M., Van Calster, B., Rohmann, J. L., Konigorski, S., & van Smeden, M. (2024). Understanding random resampling techniques for class imbalance correction and their consequences on calibration and discrimination of clinical risk prediction models. Journal of biomedical informatics, 155, 104666.

56. Ladin, K., Cuddeback, J., Duru, O. K., Goel, S., Harvey, W., Park, J. G.,... & Kent, D. M. (2024). Guidance for unbiased predictive information for healthcare decision-making and equity (GUIDE): considerations when race may be a prognostic factor. npj Digital Medicine, 7(1), 290.
57. Endsley, M. R. (2023). Supporting Human-AI Teams: Transparency, explainability, and situation awareness. Computers in Human Behavior, 140, 107574.
58. Omar, M., Sorin, V., Collins, J.D. et al. Multi-model assurance analysis showing large language models are highly vulnerable to adversarial hallucination attacks during clinical decision support. Commun Med **5**, 330 (2025). https://doi.org/10.1038/s43856-025-01021-3
59. Teo, Z. L., Jin, L., Liu, N., Li, S., Miao, D., Zhang, X.,... & Ting, D. S. W. (2024). Federated machine learning in healthcare: A systematic review on clinical applications and technical architecture. Cell Reports Medicine, 5(2).
60. Khalid, N., Qayyum, A., Bilal, M., Al-Fuqaha, A., & Qadir, J. (2023). Privacy-preserving artificial intelligence in healthcare: Techniques and applications. Computers in Biology and Medicine, 158, 106848.
61. Van Kolfschooten, H., & Van Oirschot, J. (2024). The EU artificial intelligence act (2024): implications for healthcare. Health Policy, 149, 105152.
62. Rashidi Ranjbar F, Zamanifar A. Autonomous dental treatment planning on panoramic x-ray using deep learning based object detection algorithm. Multimedia Tools and Applications. 2024 Apr;83(14):42999-3033.
63. Shahparian J, Erfani SH, Zamanifar A. A secure and efficient authentication and key agreement protocol in blockchain-enabled VANETs. Computers and Electrical Engineering. 2025 Mar 1;122:109947.
64. Zamanifar A, Nazemi E. EECASC: an energy efficient communication approach in smart cities. Wireless Networks. 2020 Feb;26(2):925-40.
65. Lataran, P. N., Farhadi, A., Zamanifar, A., & Taheri, A. (2024). Developing a BERT-enhanced blockchain model for HealthInsurance fraud detection. Journal of Information Systems Research and Practice, 2(3), 56-61.
66. Farhadi, A., Taheri, A. (2025). Application of GenAI in Synthetic Data Generation in the Healthcare System. In: Zamanifar, A., Faezipour, M. (eds) Application of Generative AI in Healthcare Systems. Springer, Cham. https://doi.org/10.1007/978-3-031-82963-5_3
67. Rashidieranjbar, F., Farhadi, A., Zamanifar, A. (2025). Revolutionizing Healthcare with Generative Artificial Intelligence Technologies. In: Vajjhala, N.R., Roy, S.S., Taşcı, B., Hoque Chowdhury, M.E. (eds) Generative Artificial Intelligence (AI) Approaches for Industrial Applications. Information Systems Engineering and Management, vol 24. Springer, Cham. https://doi.org/10.1007/978-3-031-76710-4_10
68. Mahalegi, H.S.M., Farhadi, A., Balogh, G., Nagy, E. (2025). Application of Generative AI in Resource Optimization in Healthcare Systems. In: Zamanifar, A., Faezipour, M. (eds) Application of Generative AI in Healthcare Systems. Springer, Cham. https://doi.org/10.1007/978-3-031-82963-5_7

Transfer Learning for Explainable AI in Clinical LLMs

Amirfarhad Farhadi and Nasser Mozayani

Abbreviations

LLM	Large Language Model
XAI	Explainable AI
LSTM	Long Short-Term Memory
CNN	Convolutional neural network
FCN	Fully Connected Network
EHR	Electronic health record
CDSS	Clinical decision support system
HIPAA	The Health Insurance Portability and Accountability Act of 1996
GDPR	General Data Protection Regulation
ABCM	Attention-Based Cross-Modal
ABTL	Adversarial-based transfer learning
QA	Question answering
NER	Named entity recognition
CVD	Cardiovascular disease
IID/non-IID	(Independent and identically distributed)/(not IID)
CoT	Chain-of-Thought
VLP	Vision-Language Pretrained (models)
MLLM	Multimodal Large Language Model
PEFT	Parameter-efficient fine-tuning
DP	Differential privacy
CL	Continual learning
WAM	Wavelet Attribution Method

A. Farhadi · N. Mozayani (✉)
School of Computer Engineering, Iran University of Science and Technology, Tehran, Iran
e-mail: mozayani@iust.ac.ir

A. Farhadi
e-mail: am_farhadi@mail.iust.ac.ir

A. Zamanifar et al. (eds.), *Explainable Large Language Models in Healthcare Applications*, Bio-IT and AI, https://doi.org/10.1007/978-3-032-15088-2_6

CEF	Correlational Explanatory Faithfulness
FT	Full Fine-Tuning
KD	Knowledge distillation

1 Large Language Models

Large Language Models (LLMs), which are advanced AI systems designed to process and generate human-like texts, are now the most widely used and the perhaps the most prominent AI tool, with many applications in almost any field and topic, ranging from general topics to more sophisticated, technical ones, such as law, finance, education, medicine and countless many other areas. In finance, they are employed for fraud detection, risk assessment, and customer service automation, reducing operational costs and improving accuracy [1]. Education benefits from LLMs through personalized learning experiences and automated content generation, while cybersecurity applications include vulnerability assessments and threat detection, as seen in the medical device industry [1]. In healthcare, LLMs assist in diagnosing diseases, personalizing treatment plans, and automating patient data management, enhancing efficiency and decision-making [1, 2]. Emerging applications also integrate LLMs with other technologies. For instance, multimodal models analyze video streams in agriculture, enabling real-time livestock monitoring and anomaly detection [3]. Similarly, LLMs enhance scientific discovery by automating data analysis and hypothesis generation [4]. LLMs' capabilities can be categorized as important, interesting, and challenging tasks, such as text generation, translation, code generation, summarization, and query answering. Unlike many other AI tools and techniques, their transformative power makes them versatile systems; they can be easily retrained (tuned) to adapt to new domains. Earlier LLMs were based on statistical techniques, such as n-grams; these techniques suffered from an inability to capture long-term dependencies [5], a limitation often observed in natural language texts. Some neural network architectures, such as Long Short-Term Memory (LSTM), mitigated the problem of sequence modeling but suffered from scalability and parallelization issues [6]. Contemporary Large Language Models, at their heart, are based on transformer architecture; the transformer architecture, which revolutionized LLMs, leverages two innovations: self-attention and positional encoding. Self-attention enables the model to estimate how words are related while predicting; positional encoding is responsible for capturing word order and sequential patterns [7, 8]. These innovations laid the foundation for models like BERT and GPT. BERT employs a bidirectional encoder to understand context from both directions, excelling in tasks requiring deep comprehension [9]. In contrast, GPT employs a unidirectional decoder, making it highly effective for generative tasks such as text creation [10, 11]. More recently, models like Llama have optimized transformer designs for efficiency, focusing on reducing computational costs while maintaining performance [12]. These Architectural variations highlight the trade-offs between bidirectional understanding and generative capabilities, shaping the diverse applications of LLMs across domains [13, 14]. Advancements in the realm of LLMs are not limited to what was just mentioned, and there is an ongoing interest in this topic. As mentioned before, LLMs are broadly applied in medicine, aiding practitioners (either physicians or researchers) and patients. We will introduce some of them in the following sections, but our primary goal is to explore the methods leveraged to tackle one of the main problems they suffer from: a lack of transparency. Transparency is closely related to interpretability and explainability. Generally speaking, despite their transformative potential, LLMs encounter challenges such as ethical concerns, biases, and computational demands that persist, necessitating careful implementation and ongoing research [1, 15]. Their challenges are not limited to what is mentioned. We should add hallucinations (where LLMs generate plausible but inaccurate information) [16], environmental costs (due to their high computational demands), privacy risks (such as data leakage) [17], security vulnerabilities (such as prompt stealing and context injection, expose models to adversarial manipulation [18, 19] and our topic of interest in this chapter: the lack of transparency. Like many other AI models, LLMs suffer from a lack of transparency, which, among the aforementioned challenges, is one of the most serious ones. This problem may result in a lack of trustworthiness. Considering the widespread applications of LLMs, the issue of lack of transparency is more severe compared to other AI models, necessitating more care and attention. Some efforts are dedicated to making these models transparent, and still more researchers are trying to push the boundaries toward more transparent, trustworthy LLMs. In the following section, after a discussion on the topic of explainable AI, we briefly mention some of the methods to make LLMs more transparent; obviously, a comprehensive treatment of the subject is not possible here, and the reader should consult other, more relevant resources.

2 Explainable AI

With the advent of AI, increasingly complex, black-box algorithms and methods were invented and applied, at the cost of decreased transparency. What mattered most was model performance in predicting unseen patterns, and vagueness was often not considered a significant issue to address. The issue of lack of transparency (i.e., vagueness) is not a trivial one; it can degrade the model's interpretability and explainability, resulting in a lack of trustworthiness, especially in more sensitive tasks. Explainable AI (XAI) is a relatively recent field of study and research that emerged as a response to the increasing complexity of machine learning models, particularly deep learning systems, trying to mitigate the problem of the opaqueness of AI models. It aims to make some existing opaque methods more explainable without altering their architecture [20], and also provides frameworks for designing interpretable models from scratch. Meanwhile, the existing models that cannot be explained in some ways are being gradually eliminated, because explainability is of great importance! The following Fig. 1 depicts the trend of 'Explainable AI' over

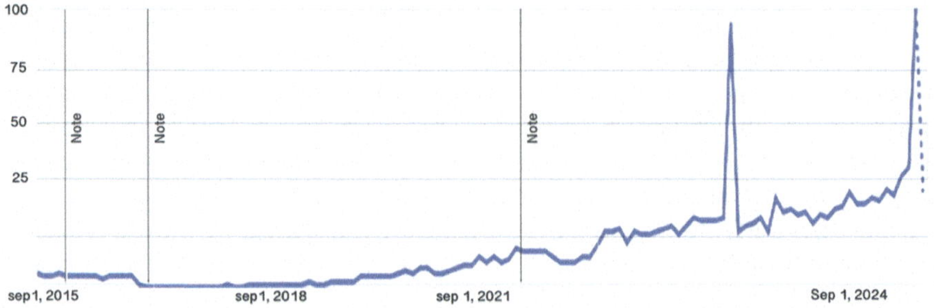

Fig. 1 The trend of explainable AI over the last 10 years

the last 10 years in Google Trends. As the trend indicates, there is a growing interest in this topic, which deserves further elaboration.

Explainable AI (XAI) has its roots in the early development of symbolic reasoning and expert systems, which prioritized human-understandable decision-making processes. These systems, prominent in the 1970s and 1980s, were designed to emulate human expertise in specific domains through rule-based logic and knowledge representation. Notable examples include MYCIN and DENDRAL, which used symbolic reasoning to provide interpretable outputs in medical diagnosis and chemical analysis, respectively [20, 21]. The emphasis on transparency in these systems stemmed from their reliance on explicit rules and logical structures. This approach allowed users to trace the reasoning behind decisions, improving trust and usability. However, these systems faced limitations, such as a lack of crispness and poor scalability, which hindered their application to more complex and dynamic problems [21, 22]. The transition from symbolic AI to hybrid paradigms, such as neural-symbolic systems, marked a significant evolution. These approaches sought to combine the interpretability of symbolic reasoning with the adaptability of neural networks, addressing the shortcomings of early expert systems while retaining their focus on explainability [23, 24]. This foundational work laid the groundwork for modern XAI, bridging historical principles with contemporary machine learning advancements. Early efforts in explainable AI (XAI) focused on rule extraction, feature importance measures, surrogate models, and prototype-based explanations, laying the foundation for modern interpretability methods. These techniques aimed to address the opacity of machine learning models by providing insights into their decision-making processes [25, 26]. Rule extraction methods translated complex models into simpler, rule-based representations, while feature importance measures quantified the contribution of individual features to model predictions. Surrogate models, such as decision trees, approximated black-box models to make their behavior more interpretable. Prototype-based explanations identified representative examples to illustrate model behavior [25–28]. The transition from local to global explanation techniques marked a significant evolution in XAI. Local methods, such as LIME and SHAP, provided instance-specific explanations, while global methods aimed to capture the overall behavior of models. LIME approximated predictions with interpretable models in the vicinity of a specific instance, whereas SHAP leveraged Shapley values from cooperative game theory to fairly attribute feature importance [29, 30]. These advancements have significantly enhanced model transparency, enabling their application in more critical domains, such as healthcare and finance, where trust and accountability are of great importance.

Recent advancements in explainable AI (XAI) focus on enhancing interpretability through innovative methods and addressing challenges in fairness, robustness, and usability. Gradient-based attribution techniques, for instance, have seen significant improvements, such as filtering high-frequency noise to enhance explanation clarity and faithfulness [31]. Additionally, novel approaches like IG2 integrate counterfactual gradients to reduce attribution noise and improve baseline selection, offering more reliable feature attributions [32].

Causal explanation frameworks are gaining traction for their ability to identify uncertainty sources and mitigate their impact on predictions, particularly in Bayesian deep learning [33]. Multimodal interpretability methods, such as the Wavelet Attribution Method (WAM), extend gradient-based techniques to diverse data types like images, audio, and 3D shapes, providing unified and context-aware explanations [34]. Human-centered evaluation protocols are increasingly emphasized to ensure explanations are meaningful and actionable for diverse users. These protocols assess dimensions such as trust, usability, and fairness, while addressing the variability in user expertise and preferences [35, 36]. However, challenges persist, including standardizing evaluation methodologies and aligning explanations with ethical principles like fairness and accountability [37, 38]. These trends highlight the ongoing efforts to make XAI systems more transparent, equitable, and user-aligned. Explainability in transformer-based large language models (LLMs) faces unique challenges due to their scale, architecture, and training paradigms. These challenges vary across encoder-only, decoder-only, and encoder-decoder architectures, each of which exhibits distinct strengths and limitations. Encoder-only models, such as BERT, excel in understanding tasks like classification but struggle with generative tasks. Decoder-only models, like GPT-3, are highly effective for text generation but often lack interpretability in their decision-making processes. Encoder-decoder models, such as T5, balance understanding and generation capabilities, making them versatile but computationally intensive [9, 11]. Explainability techniques for LLMs are typically categorized into fine-tuning and prompting paradigms. Fine-tuning involves training models on labeled datasets to generate explanations, while prompting leverages pre-trained models with carefully designed input prompts to elicit explanations. Prompting is particularly advantageous for few-shot learning scenarios but may produce fewer faithful explanations compared to fine-tuned models [39, 40]. Evaluation metrics for LLM explanations include faithfulness, plausibility, and human alignment. Recent advancements, such as Correlational Explanatory Faithfulness (CEF), aim to quantify explanation fidelity by measuring shifts in model predictions under counterfactual conditions, offering a more nuanced assessment of explanation quality [41, 42]. These metrics are crucial for ensuring trustworthy and actionable explanations in high-stakes applications. Previously, we mentioned that the terms transparency, explainability, and interpretability are closely related; however, we now argue that

Fig. 2 Distinctions between interpretability and explainability

these terms carry different meanings and serve different purposes, aiming at various aspects of transparency. While interpretability focuses on understanding how a model works internally, explainability emphasizes making its predictions comprehensible to users. Clarifying these concepts is vital for fostering transparency, usability, and ethical AI adoption. In other words, interpretability and explainability are distinct yet interconnected concepts in Explainable AI (XAI), with interpretability focusing on understanding a model's internal mechanics and explainability addressing the reasons behind specific predictions. Interpretability refers to the degree to which a human can comprehend how an AI model processes inputs to produce outputs, often linked to simpler, inherently transparent models like decision trees or linear regression [43]. Explainability, on the other hand, emphasizes post-hoc methods that clarify why a model made a particular decision, such as feature attribution techniques like SHAP or LIME [29].

In summary, the distinction between explainability and interpretability in Explainable AI (XAI) lies in their focus and application: interpretability emphasizes understanding a model's internal mechanics, while explainability centers on making predictions comprehensible to users. This report examined their definitions, methodologies, metrics, and ethical considerations, highlighting their complementary roles in promoting transparency and trust in AI systems (Fig. 2).

3 Transfer Learning

Transfer learning, as its name may imply, is a machine learning paradigm in which pre-trained, usually general models for some source tasks are transferred to different but related tasks. This approach enables models to leverage knowledge obtained from one domain to improve the performance in another domain; this reduces the reliance on large, labeled datasets, especially when obtaining reliable data is difficult and expensive, if not impossible. It also reduces the computational complexity accompanying training a new, massive model (for example, a deep neural network) from scratch [44]. Another essential feature of transfer learning is that, unlike many traditional machine learning approaches, it does not assume identical feature spaces or data distributions. These two limiting assumptions prevent the reuse of knowledge obtained from one task to generalize to other similar tasks [44]. The core principle of transfer learning lies in identifying and transferring shared knowledge between the source and target domains. Their feature spaces and data distributions define domains, while tasks are characterized by their label spaces. Transfer learning can be categorized into two types: homogeneous transfer learning, where source and target domains share the same feature space, and heterogeneous transfer learning, where feature spaces differ [45]. A primary challenge in transfer learning is how to distinguish beneficial knowledge in a source domain from the inherent cross-domain noise, due to the varied distributions, and transfer it to a target domain [45]. Other challenges include domain misalignment (due to feature distribution shifts, label space mismatches, and so on), and negative transfer—where knowledge transfer undesirably degrades performance on the target domain—remain active areas of research [46].

Finally, in addition to the challenges mentioned, data privacy is another concern; transfer learning requires access to data from the source domain. In some cases, for example, in medical applications, access to healthcare data is severely limited. Researchers proposed some methods using federated learning, which mitigates the problem of raw data privacy while training the model. Still, these methods suffer from some issues, mainly due to non-IID (data points that violate the assumption of being independent and identically distributed, which results in model bias and reduced generalization) data distributions [47, 48].

After discussing what transfer learning is, we now turn our attention to some methods and approaches to transfer knowledge.

3.1 Instance-Based Transfer Learning

Instance-based transfer learning refers to reusing instances from the source domain to improve performance on the target domain, especially when labeled data is scarce in the target domain. This approach is practical mainly when the two source and target domains share similar distributions, but in this case, covariate shift (according to [49], "...where the marginal distribution over the input covariates differs in the source and the target domains, while the conditional distribution of the output given the input covariates is similar across the two domains...") is a problem to be addressed [44, 49] One method to tackle the issue of covariate shift is to pre-train on the source data and fine-tune based on the target data, as proposed in [49]. Some other methods, such as importance weighting and density ratio estimation, are basically based on re-weighting the source samples to align them with the target distribution [50, 51]; the methods just mentioned are some examples, and a deeper discussion of these methods is out of the scope of this book.

3.2 Feature-Based Transfer Learning

As its name implies, this technique is based on adapting or transforming the learned features from the pre-trained model of the source domain to the general features of the specific target task [52]. Especially when the model is gigantic and complex, as in CNNs (convolutional neural networks), one can only fine-tune the last layers, usually FCNs (Fully Connected Networks), which represent more high-level details, and leave the first convolutional layers, which represent low-level details, unchanged. This makes it possible to fine-tune the model without any specialized, high-performance hardware (for example, GPU/TPU), and one can even retrain the model using a regular CPU. This task is achieved by identifying a shared latent feature space -usually using dimensionality reduction techniques or neural networks, for example, convolutional autoencoders and spectral feature alignment—or by re-weighting the features to make the source and target domains similar, which results in a reduction in distribution discrepancy and enhances transferability. Briefly speaking, in this approach, the features extracted from the source domain are made relevant and useful for the target domain, even if the two domains do not share similar distributions or feature spaces [53–56]. It is worth noting that this approach does have some disadvantages and limitations, some of which include high computational complexity of transferring feature spaces or learning latent spaces [55], dependency on feature quality in terms of the quality of the features extracted and the similarity between domains [53], and limited scalability, especially when the domains are very different. The labeled data in the target domain is the case [56].

3.3 Parameter-Based (Model-Based) Transfer Learning

This approach is based on transferring parameters or weights from a pre-trained model (for example, a neural network) from a source domain to a target domain. This approach assumes the learned weights or parameters can be modified to be transferred to the target domain successfully [57, 58]. This approach provides excellent scalability and flexibility by making it possible to reuse large-scale, pre-trained models such as transformers in different tasks and diverse domains such as image classification, natural language processing, and so on [59, 60]. Some drawbacks of this approach may include the risk of overfitting due to fine-tuning on a small target dataset, especially if the source and target domain are not much relevant [58] and limited interpretability due to transferred parameters; the model may work fine in the target domain but the model may suffer from transparency; in other words, we will not know why the model is working well [58].

3.4 Adversarial-Based Transfer Learning

Adversarial-based transfer learning (ABTL) leverages adversarial training, where a feature extractor and a domain discriminator are trained in opposition. The feature extractor tries to confuse the domain discriminator, making it unable to distinguish between source and target domain features [61, 62]. This approach tries to address the problem of domain shift. Domain shift occurs when the data distributions in the source and target domains differ, which can degrade model performance when transferring the knowledge [44, 61]. Adversarial-based transfer learning faces several significant challenges, including training instability [63, 64], high computational costs [65, 66], and sensitivity to hyperparameters [67, 68]. Additionally, the adversarial methods used in this approach may degrade its interpretability; it may be challenging to understand their decision-making processes. This lack of transparency limits their application in more sensitive tasks such as healthcare and autonomous systems [69, 70].

3.5 Domain Adaptation

This approach addresses the specific challenge of adapting pre-trained models to new domains with different data distributions but shared tasks (e.g., classification or regression). This is crucial in scenarios where labeled data in the target domain is scarce or expensive to obtain. By mitigating performance degradation caused by distribution shifts, domain adaptation enhances model robustness and applicability across diverse fields.

As previously stated, unlike general transfer learning, where tasks may differ, domain adaptation assumes the tasks remain the same but the data distributions vary between domains. The source domain refers to the dataset where labeled data is available. In contrast, the target domain represents the dataset where the model is applied, often with limited or no labeled data. The primary goal is to mitigate the performance degradation caused by distribution shifts, which can manifest as covariate shift (differences in input distributions), prior shift (differences in label distributions), or concept shift (differences in conditional distributions) [44, 71, 72]

Domain adaptation is commonly categorized into supervised, semi-supervised, and unsupervised approaches, depending on the availability of labeled target data. For instance, unsupervised domain adaptation assumes no labeled target data, relying on techniques such as domain-invariant feature learning or adversarial training to align distributions [61, 73] This framework is crucial for applications where collecting labeled data in the target domain is expensive or impractical, such as adapting

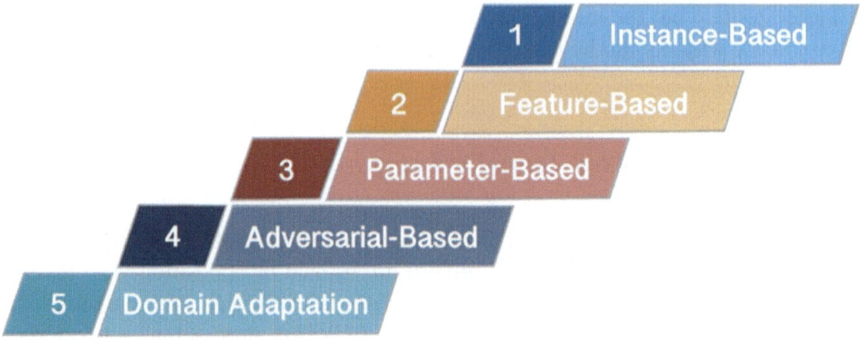

Fig. 3 Transfer learning approaches

models trained on synthetic data to real-world scenarios [74, 75]. Domain adaptation methods can be broadly categorized into four methodological families: instance re-weighting (this approach adjusts the importance of source domain samples to match betterto match better the target domain distribution- examples: density ratio estimation and iterative re-weighting), feature transformation (these methods seek to learn domain-invariant feature representations—example: Transfer Component Analysis), parameter sharing (this strategy involves sharing model parameters across domains while fine-tuning domain-specific components), and adversarial approaches (these methods employ adversarial training to align domain distributions- example: Domain-Adversarial Neural Networks). Each family addresses the challenge of aligning source and target domains differently, leveraging unique strategies to mitigate distributional shifts [44, 73, 76–78]. Domain adaptation faces several critical limitations that hinder its effectiveness and reliability, including negative transfer (when knowledge from the source domain degrades performance in the target domain, often due to mismatched label spaces or irrelevant source data), large domain gaps, lack of theoretical guarantees, and evaluation inconsistencies [61, 62, 79–81]. The following Fig. 3 depicts the transfer learning approaches.

After a brief discussion about the preliminaries, we now turn our attention to the main topic of this chapter: transfer learning for explainable AI in medical LLMs. In the following section, we begin with a discussion on LLMs' applications in healthcare (i.e., medical LLMs); then, we provide some specific, successful implementations. After that, we explore some challenges regarding implementing LLMs in healthcare. Finally, we end the section by making medical LLMs explainable using transfer learning. It goes without saying that our treatment cannot be comprehensive; what we aim to do is to conduct a brief survey, covering the most recent and interesting items. Again, we invite the eager reader to consult many of the invaluable resources that are available (Table 1).

Table 1 Summary of transfer learning methods in biomedical NLP

Method	Core idea	Strengths	Weaknesses/risks	Examples
Instance-based	Reweights source samples to match target distribution (covariate shift)	Simple, effective with few target labels	Sensitive to domain mismatch	Importance weighting, fine-tuning
Feature-based	Learns shared latent representations between domains	Reduces domain gap; efficient fine-tuning	Computationally heavy; depends on feature quality	Autoencoders, Spectral Feature Alignment
Parameter-based	Transfers model weights/ parameters from pre-trained models	Scalable; reuses large models (e.g., transformers)	Overfitting on small data; low interpretability	Fine-tuning BERT-like models
Adversarial-based	Aligns domains via adversarial training (feature extractor vs. discriminator)	Improves cross-domain generalization	Unstable training; high compute cost	DANN, GAN-based alignment
Domain adaptation	Adjusts source model to target domain with different data distribution	Robust to distribution shifts	Negative transfer, large domain gap risk	TCA, feature transformation, adversarial DA

4 Models of Medical LLMs

As discussed before, Large Language Models (LLMs) are transforming medicine and healthcare greatly by utilizing advanced natural language processing capabilities to address complex biomedical challenges. Built on transformer architectures and fine-tuned for domain-specific tasks (i.e., medicine and healthcare), these models are enabling breakthroughs in clinical decision support, medical research, patient aid, and administrative efficiency. Generally, LLMs are pre-trained on vast corpora to learn general language patterns and then fine-tuned on domain-specific datasets to address specialized tasks; as discussed before, the process of fine-tuning a pre-trained model on a domain-specific task is what we referred to as transfer learning. In this section, after exploring some examples of LLMs applied in medicine, we discuss some of the challenges associated with utilizing LLMs in healthcare. If we want to provide some more famous, successful samples of LLMs applied in medical settings, we can list the following cases.

4.1 BioBERT

BioBERT, like BERT, is a domain-specific language representation model developed based on a similar architecture and further pre-training on large-scale biomedical corpora such as the PubMed abstracts and PubMed Central full-text articles. Beginning from BERT's general-domain weights (being initially trained from English Wikipedia and BooksCorpus), it is further pre-trained on the biomedical corpus for better capturing the domain-specific terminologies and semantics. Minor architectural modification allows BioBERT to be fine-tuned for major biomedical text mining applications such as named entity recognition (NER), relation extraction (RE), and question answering (QA). Experimentation demonstrates that BioBERT significantly outperforms both BERT and previously best models of known performance. The reported results show that BioBERT improved NER by an additional 0.62% F1, RE by 2.80% F1, and QA by 12.24% MRR. They also emphasize their assertion that pre-training from a biomedical corpus is essential to comprehend complex biomedical texts, citing this as the reason for the BioBERTBASE that was also pre-trained using the eight NVIDIA V100 GPUs for not less than 23 days employing the BERT WordPiece tokenizer for compatibility purposes. Everything, thus, ends up confirming that domain-specific pre-training of BERT is very crucial to effective biomedical text mining [82].

4.2 Med-PaLM

The use of sensitivity training has been achieved via designing Med-PaLM as the domain-adapted version of the instruction prompt tuning method using a data- and parameter-efficient instruction prompt tuning method for the large language model Flan-PaLM with 540 billion parameters. MultiMedQA benchmark using seven medical datasets, such as MedQA, MedMCQA, and PubMedQA, indicates that the system was able to achieve top scores: at 67.6% for MedQA and 79.0% at PubMedQA, outperforming previous biomedical models such as BioGPT and PubMedGPT. References to a study in human evaluations indicated that measures of 92.6% of written answers from Med-PaLM had a favorable match with scientific consensus and reduced the proportion of possibly harmful responses from 29.7 to 5.9%, almost at par with those generated by clinicians. Major improvements have been made in medical understanding, reasoning, and knowledge retrieval, but lapses and occasional wrong errors are still witnessed. The authors conclude that instruction prompt tuning is a successful technique in aligning general-purpose LLMs towards the medical domain, with a significant narrowing of the gap between AI systems and clinicians. However, it is emphasized that Med-PaLM is not yet deployable for real-world clinical purposes but needs further safety, fairness, and ethical reliability work [83].

4.3 ClinicalBERT

ClinicalBERT is more specialized in the domain than BERT and is fine-tuned on clinical notes of the MIMIC-III database, capturing the vocabulary and context nuances in medical language. It has shown the potential in achieving insight extraction from unstructured electronic health records (EHR) and has leveraged transfer learning to perform effective clinical tasks with minimal additional training. The paper, however, noted some challenges regarding the use of large language models in healthcare, which include high computational cost, noise sensitivity, data bias, limited interpretability, privacy concerns, and regulatory ambiguity-these are the factors that continue to bar large-scale clinical implementation [84]. Various organizations around the world have continuously sought effective methods to prevent or control this problem. HI-BERT (Health Insurance Bidirectional Encoder Representations from Transformers), a model built upon BERT with modifications made to its [85].

4.4 GPT-3 and GPT-4

Generative Artificial Intelligence (GenAI) is a technology capable of creating diverse forms of content—such as text, images, videos, and code—based on a given prompt, and GPT-4 is one of the most popular models of this [86]. These well-known, general-purpose LLMs have been significantly adapted for healthcare tasks like generating patient-friendly medical reports, summarizing clinical notes, and answering medical questions. For example, GPT-4 has been used to create interpretive pathology reports that improve patient understanding [87]. It should be noted that these kinds of LLMs are general domain LLMs, and one may not expect their performance to be comparable with some other specific-task LLMs, but they still can be fine-tuned to transfer their knowledge to more dedicated tasks.

4.5 HELIOT

HELIOT, employing an RAG architecture, is a large language model-based clinical decision support system for managing adverse drug reactions (ADRs). This architecture retrieves information from extensive drug databases and integrates it with the analysis of clinical notes to produce more accurate and personalized alerts to users. HELIOT uses GPT-4o to process drug-related information from PubChem and normalize clinical terms into English to reduce linguistic bias. In evaluations on 1,000 expert-validated synthetic clinical scenarios, HELIOT attained perfect inter-rater agreement (Fleiss' kappa = 1.00) and an average F1-score of 0.9869 in clinical classification. The system achieved more than a 50% reduction in interruptive alerts compared with traditional CDSSs and had a mean response time of approximately 2.8 s per patient on average. The authors concluded that HELIOT performs with high accuracy and efficiency in controlled conditions, but it now requires validation in a real-world clinical environment before clinical implementation [88].

4.6 DeepSeek-V3

The Model DeepSeek-V3 assesses clinical decision making in diagnosis and treatment recommendation tasks-together with the other models DeepSeek-R1, GPT-4o, and Gemini-2.0 Flash Thinking Experimental (Gem2FTE). The study comprised 125 curated cases with real patients across five specialties: internal medicine, neurology, surgery, gynecology, and pediatrics, thus covering both common and rare diseases. Results indicate that DeepSeek-V3 performed at par with GPT-4o and DeepSeek-R1, having an average value of 4.43 out of 5 for the diagnosis and 4.31 for treatment; there was no significant statistical difference compared to DeepSeek-R1 ($P = 1$, rrb $= 0.03$). All the new models, including DeepSeek-V3, outperformed the previous benchmarks GPT-4, GPT-3.5, and Google Search. The model consistently exhibited the same performance irrespective of frequency and medical specialty, thus indicating robustness. The authors underline that DeepSeek-V3 and other open-source LLMs offer a sensible alternative to their proprietary counterpart, GPT-4o: it is privacy-compliant and cost-effective, being able to be deployed in-house within hospital IT infrastructures. They stress that the clinical accuracy is high while underlining that further validation and governance frameworks are needed before deploying these systems to real-world clinical practice [89].

4.7 GatorTron

GatorTron is a clinical language model of large scale, pre-trained from scratch on more than 90 billion words, including over 82 billion de-identified texts from the UF Health system along with PubMed (6 B), Wikipedia (2.5 B), and MIMIC-III (0.5 B). This model follows the architecture of BERT using masked language modeling (MLM) and sentence order prediction (SOP) as main objectives, and employs byte-pair encoding (BPE) vocabulary. A total of three sizes of models were trained- base (345 M), medium (3.9 B), and large (8.9 B) parameters. It took around 6 days for GatorTron-large to train on 992 NVIDIA A100 GPUs. This model was evaluated on five clinical NLP tasks: clinical concept extraction (NER), medical relation extraction (MRE), semantic textual similarity (STS), natural language inference (NLI), and medical question answering (MQA)—benchmark datasets for this study included i2b2 2010, i2b2 2012, n2c2 2018, and MedNLI. GatorTron thus proved to be the state-of-the-art model, with huge improvements made over BioBERT, ClinicalBERT, and BioMegatron. For instance, GatorTron achieved F1 scores of 0.8996/0.8091/0.9000 on NER datasets, F1 = 0.9627 on MRE, Accuracy = 0.9020 on NLI, and for MQA achieved EM = 0.3155, F1 = 0.9719 (Medication), and EM = 0.9310, F1 = 0.9719 (Relation). Scaling up model size consistently improved performance across four of the five tasks (except STS, where the 3.9B model performed best) [90].

4.8 BioMistral

This is an open-source Large Language Model tailored explicitly for the biomedical domain. It is built on the Mistral foundation model and further pre-trained on biomedical texts, such as those from PubMed Central, to enhance its performance in medical applications. BioMistral is designed to address challenges in adapting general-purpose LLMs to the medical domain. It excels in tasks like medical question answering (QA), named entity recognition (NER), and structured information extraction. BioMistral has been tested for multilingual generalization by translating and evaluating its performance across seven languages, marking it as one of the first LLMs to undergo large-scale multilingual evaluation in the medical domain [91]. BioMistral can be deployed locally in healthcare settings, offering advantages in data privacy and compliance with regulations like HIPAA (The Health Insurance Portability and Accountability Act of 1996) and GDPR [92] (General Data Protection Regulation). Finally, while BioMistral shows promise, it has limitations in handling complex reasoning tasks and achieving the same level of accuracy as larger proprietary models like GPT-4. For example, its responses in patient communication tasks were found to be less comprehensive and empathetic compared to some other models, such as GPT-4 [93]. The provided list is not comprehensive, but it suffices for our purpose to provide some famous examples. Now, we turn our attention to some challenges regarding applying LLMs in healthcare (Table 2).

5 Challenges in Using LLMs in Healthcare

Contrary to some success stories about LLMs' applications in medical and healthcare settings, this section aims to explore the challenges associated with using LLMs in these domains. Considering the limited pages of this chapter, this would not be a comprehensive treatment, and we focus mainly on the challenges regarding explainability, but this does not imply that the other aspects do not matter!

Table 2 Overview of LLMs models applied in medical settings

Model	Base/type	Main data	Main task(s)	Key result/metric	Main point
BioBERT	BERT-based	PubMed abstracts, PMC full texts	NER, RE, QA	+ 0.62% F1 (NER), + 2.8% (RE), + 12.24% MRR (QA)	Domain pretraining crucial for biomedical text mining
Med-PaLM	Flan-PaLM (540B) tuned with instruction prompts	7 medical QA datasets (MultiMedQA)	QA, reasoning	67.6% (MedQA), 79% (PubMedQA), 92.6% consensus match	Aligns general LLMs to medicine; not yet deployable
ClinicalBERT	BERT fine-tuned on MIMIC-III	Clinical notes (EHR)	EHR text analysis	Qualitative improvement	Sensitive to noise, privacy, and interpretability issues
GPT-3/GPT-4	General LLM (OpenAI)	General-domain data	Summarization, QA, report generation	Improved patient comprehension	General models adaptable but not domain-optimized
HELIOT	RAG-based CDSS (uses GPT-4o)	Drug DB + 1000 synthetic cases	ADR management	$F1 = 0.9869$; $\kappa = 1.00$; > 50% fewer alerts	Accurate, fast (≈2.8 s), needs real-world validation
DeepSeek-V3	Open-source LLM	125 clinical cases (5 specialties)	Diagnosis, treatment	4.43 (Dx), 4.31 (Tx); $P = 1$	Matches GPT-4o; privacy-compliant; needs regulation
GatorTron	BERT-style (trained from scratch)	90 B words (UF Health + PubMed + Wiki + MIMIC-III)	NER, MRE, STS, NLI, MQA	Best overall (F1 ≈ 0.9–0.97)	Large-scale model improves complex reasoning
BioMistral	Mistral-based biomedical LLM	PubMed Central + multilingual sets	QA, NER, info extraction	Strong multilingual performance	Privacy-friendly but weaker reasoning vs GPT-4

5.1 *Data Privacy, Security, Regulatory and Legal Challenges, and Ethical Concerns*

LLMs often require access to sensitive, private patient data, raising concerns about privacy and compliance with regulations like HIPAA and GDPR. Ensuring secure data handling and storage is critical to prevent breaches [94]. Data breaches may encompass severe consequences and may not be considered just a hypothetical problem; they can potentially influence many patients! Some techniques, like differential privacy, are being explored to mitigate these risks [95]. As mentioned before, federated learning is another technique to address the challenge of privacy, but this technique suffers from some complexities, mainly regarding its implementation. Also, some challenges regarding ethical concerns, such as the potential for misuse, lack of accountability, and the impact on clinician-patient relationships, need to be addressed carefully [96, 97]

5.2 *Bias and Fairness*

LLMs learn from data, so they can inherit biases from the datasets they are trained on, potentially amplifying healthcare disparities such as those related to race, gender, and age. For example, underrepresented populations (due to the scarcity of training samples) may receive less accurate or less practical recommendations [96]. To mitigate this issue, we should utilize techniques available to learn from imbalanced data or provide sufficient data to represent all categories almost equally, which may not be feasible. This is a real challenge to overcome, waiting for more investigations.

5.3 *Hallucinations and Factual Accuracy*

LLMs are prone to generating plausible (i.e., seeming likely to be true but necessarily not accurate!) but incorrect or misleading information, known as hallucinations. This raises significant risks in clinical decision-making where accuracy is of great importance [96]. One way to identify and mitigate such answers is to check them against real-world data by human experts.

5.4 *Computational Demands*

Training and deploying LLMs require substantial computational resources, making them practically expensive and environmentally harmful. This limits their accessibility severely, especially in resource-constrained settings [84]. As discussed before, one effective way to overcome this challenge is to leverage transfer learning; some pre-trained models can be fine-tuned to be used in new domains.

5.5 Integration with Clinical Workflows

Despite their transformative nature, integrating LLMs into existing healthcare systems, such as electronic health records (EHRs), may be challenging. Poor integration can lead to inefficiencies and user resistance [94]. In other words, these tools must be implemented in a way that is not in conflict with existing workflows and complements them [98].

5.6 Interpretability and Transparency

As mentioned before several times, the 'black-box' nature of LLMs makes it difficult for clinicians to understand how decisions are made, reducing trust and limiting adoption in high-stakes environments [94, 97]. Explainable AI (XAI) frameworks are being developed to address this issue, but are not yet widely implemented [96]. We primarily focus on this issue in this chapter and try to explore some existing solutions intended to tackle it. After discussing challenges related to using LLMs in healthcare, we now turn our attention to the main topic of this chapter: Transfer Learning for Explainable AI in Medical LLMs.

Now, we are ready to explore the main topic of this book chapter. What follows is a brief discussion on the application of transfer learning to make medical LLMs more explainable (Table 3).

6 Transfer Learning for Explainable AI in Medical LLMs

In this final section, we discuss the applications of transfer learning in making medical LLMs more explainable, thereby increasing their trustworthiness. Making LLMs explainable cannot be achieved straightforwardly and, like many real-world problems, encompasses serious challenges. Any discussion on methods without mentioning the challenges and limitations would not be complete and realistic. Before we explore the techniques, we mention some of the difficulties ahead.

Table 3 Challenges and mitigation strategies for medical LLMs

Challenge	Description	Example impact	Potential mitigation/approach
Data privacy and security	Sensitive clinical data risks leakage; compliance with HIPAA/GDPR	Data breaches affecting patient trust	Federated learning, differential privacy
Bias and fairness	Models inherit social or sampling biases	Underdiagnosis in minority groups	Balanced data, bias auditing
Hallucination/inaccuracy	Generates plausible but wrong outputs	Misleading clinical advice	Human-in-the-loop validation, factual grounding
Computational demands	High training and inference costs	Limited access in low-resource settings	Parameter-efficient fine-tuning (PEFT), cloud optimization
Integration with workflows	Hard to embed into EHR/CDSS pipelines	Clinician resistance, inefficiency	Interoperable APIs, usability testing
Interpretability/transparency	Black-box decisions reduce clinician trust	Reduced adoption in hospitals	Explainable AI (XAI), SHAP, CoT reasoning
Regulatory and ethical concerns	Ambiguous legal framework	Accountability and liability risks	Governance frameworks, ethical AI standards

6.1 Main Challenges in Integrating Transfer Learning with Explainable AI

The integration of transfer learning with explainable AI (XAI) in medical applications faces several significant challenges:

Data Heterogeneity and Domain Shifts

Medical datasets often vary significantly in terms of quality, format, and population representation. This heterogeneity complicates the generalization of transfer learning models across different clinical settings. Domain shifts, such as variations in imaging protocols or disease prevalence, further complicate this issue, reducing the external validity of models [99, 100].

Trade-Offs Between Interpretability and Accuracy

Enhancing interpretability through XAI methods, such as SHAP or Grad-CAM, can sometimes compromise the accuracy of transfer learning models. This trade-off creates a tension between achieving high performance and ensuring transparency, which is critical in clinical applications [101, 102].

Preferability of Model Explanations

There is evidence that XAI methods vary in their alignment with clinicians' needs and preferences; in other words, the issue of preferability of explanations by clinicians is subjective, and there is not a universally agreed-upon set of criteria for that [103, 104]. The current XAI methods primarily meet the requirements in principle, but they are too inflexible and do not fully fulfill clinicians' expectations and needs [105]. As [104] argues, SHAP and Araucana-XAI are two promising candidates for improving the application of XAI in clinical decision support systems (DSSs); the study also shows the importance of clinicians' expertise and specialty [104].

Need for Contextual and User-Centric Explanations

The literature emphasizes that explanations need to be context- and user-dependent to be effective. Current XAI methods often lack the flexibility to provide answers that align with the dynamic and nuanced requirements of clinical decision-making [105, 106]. As [106] argues, "... prior work considers explanations as stationary entities with no account for the temporal dynamics of patient care" [106].

The mentioned items collectively support the idea that some XAI methods, while technically robust, provide explanations that are not always easy for clinicians to understand or apply in practice. This highlights the need for user-centric, context-aware XAI approaches tailored to the specific needs of medical professionals. In other words, Clinicians' perceptions play a pivotal role in the adoption of XAI methods in healthcare. Factors such as understandability, actionability, trust, training, and workflow integration significantly influence whether XAI tools are embraced or rejected. To ensure successful adoption, developers must prioritize user-centered design, provide targeted training, and align XAI outputs with clinicians' needs and expectations. After a brief discussion on some challenges regarding integrating transfer learning with explainable AI, in the following sub-section, we explore some relevant methods and techniques. Now that we have discussed the challenges, we will explore some of the methods and techniques themselves.

6.2 Attention-Based Transfer Learning Frameworks and Knowledge Transfer in Explainable LLMs

Transfer learning and explainable AI (XAI) are crucial in advancing medical large language models (LLMs), addressing challenges related to data scarcity and interpretability. Transfer learning reduces the need for extensive labeled data by adapting pre-trained models to specialized medical tasks, while XAI enhances trust by making AI decisions interpretable. Together, these approaches address critical challenges in diagnostics, treatment planning, and patient care.

Attention-Based Cross-Modal (ABCM) Transfer Learning Framework

Attention-Based Cross-Modal (ABCM) Transfer Learning aims to bridge distinct data modalities, such as text, images, and audio, by leveraging attention mechanisms to enable effective knowledge transfer. This approach addresses challenges such as sparse data and domain discrepancies by aligning heterogeneous modalities in a shared latent space. Unlike traditional unimodal models, ABCM frameworks dynamically integrate complementary features across modalities, enhancing task performance and interpretability.

Cross-modal transfer learning facilitates knowledge sharing across different data modalities, including text, images, and audio, to enhance performance in tasks with limited labeled data. Unlike traditional transfer learning, which assumes the source and target domains share the same modality, cross-modal transfer learning bridges distinct modalities by leveraging shared semantic spaces or latent representations. This approach has been applied to tasks like image-text retrieval, multimodal human–computer interaction, and medical diagnostics, demonstrating its potential to address challenges like sparse data and domain discrepancies [56, 107–109].

The Attention-Based Cross-Modal (ABCM) framework integrates modality-specific encoders, attention-guided fusion layers, cross-attention blocks, and task-specific decoders to enable effective multimodal learning and transfer. Each component plays a distinct role in processing, aligning, and integrating information across modalities.

Modality-Specific Encoders extract features unique to each modality, such as text, images, or audio. They employ architectures like convolutional neural networks (CNNs) or transformers to capture modality-specific patterns. Attention-Guided Fusion Layers integrate features from multiple modalities by dynamically weighting their contributions. Attention mechanisms, such as channel and spatial attention, ensure that salient features are emphasized while redundant information is suppressed. Cross-Attention Blocks facilitate interaction between modalities by aligning their features in a shared latent space. Cross-attention mechanisms enhance the transfer of complementary information, enabling robust multimodal representation learning. Task-Specific Decoders translate the fused and aligned features into outputs tailored to specific tasks, such as classification or prediction. They are designed to leverage the enriched multimodal representations for improved performance [110–116].

As a case study, in cardiovascular disease (CVD) prediction, the ABCM framework outperformed traditional models by effectively combining clinical and imaging data. Extensive testing revealed a marked improvement in prediction accuracy and interpretability, offering more profound insights into disease mechanisms [109]. Although [109] states that the integration of cross-modal data by means of attention-based mechanisms provides a deeper understanding of cardiovascular diseases, enabling more informed clinical decision-making and personalized patient care, it does not argue how this integration results in deeper understanding (i.e., explainability). Still, utilizing several modalities and forcing the LLM to train on them may mitigate some plausible-seeming outputs (i.e., hallucinations), which are usually observable in unimodal LLMs. Additionally, when data scarcity occurs in some modalities, which may result in model bias, this shortage can be compensated for by data from different modalities.

Obviously, this method still suffers from some challenges; it shares some challenges with other methods, such as sensitivity to data scarcity, domain shift, and computational overhead. The challenge of computational overhead may be more severe in this case due to the application of several modalities. Additionally, while attention mechanisms provide some level of interpretability, they often lack transparency in cross-modal interactions, making it difficult to fully understand decision-making processes [110, 117]. Scalability challenges arise when adapting models to diverse applications, as fine-tuning large pretrained models can be resource-intensive [117].

Transfer Knowledge from a Previously Explainable and Interpretable LLM to the Target Domain

Another way that seems promising is the transfer of knowledge from a previously explainable and interpretable LLM to the target domain. Adopting this approach may lead to a more explainable and interpretable model, but it cannot be guaranteed, and the resulting system's interpretability may be degraded. Despite some conducted research, this approach deserves further investigation, particularly through the development of transfer learning methods and techniques that preserve the existing pretrained model's explainability and interpretability as much as possible. In other words, this is not always the case that the interpretability or explainability of some source model is directly proportional to the interpretability or explainability of a target model. The challenges in preserving interpretability during transfer learning are multifaceted and arise from both the nature of transfer learning itself and the complexities of ensuring explainability in machine learning models. Based on the retrieved literature, the key challenges are mentioned as (A-G):

(A) **Concept Drift and Interpretability Drift**

In continual learning or incremental transfer learning, the rationale behind model predictions may change over time, leading to "interpretability concept drift." This drift complicates the ability to maintain consistent and user-friendly explanations for model decisions across tasks [118].

(B) **Weak Interpretability of the Transfer Process**

Many transfer learning methods lack mechanisms to explain how knowledge is transferred from the source to the target domain. This is particularly problematic in heterogeneous feature transfer learning, where the feature spaces of the source and target domains differ significantly [118].

(C) **Handling Uncertainty and Low-Quality Features**

In scenarios like multi-stream classification, where data streams may have redundant or low-quality features, ensuring interpretability while grappling with uncertainties becomes a challenge. This is compounded by the need to adapt to asynchronous concept drifts in the data [119].

(D) **Domain Shift, Dataset Bias, and Feature Transferability**

Transfer learning often involves adapting models to target domains with different data distributions or biases. This domain shift can degrade the interpretability of the model, as the features and patterns learned in the source domain may not align well with those in the target domain [120].

(E) **Feature Transformation and Representation Learning**

The process of transforming features to align the source and target domains often involves complex, non-intuitive methods (e.g., kernel-based mappings). These transformations can obscure the interpretability of the model, as the underlying relationships between features may become less transparent [121].

(F) **Integration of Explainability Tools**

Techniques like Shapley value analysis or fuzzy logic systems can enhance interpretability, but integrating these tools into transfer learning frameworks without compromising performance remains a challenge [121, 122].

(G) **Balancing Performance and Interpretability**

While transfer learning aims to improve performance in the target domain, this often comes at the cost of interpretability. For instance, models pre-trained on large datasets may excel in predictive accuracy but fail to provide clear explanations for their decisions in the target domain [122].

6.3 Approaches to Enhancing Interpretability and Transparency in Medical LLMs Through Transfer Learning

These challenges highlight the need for innovative methods that balance the trade-offs between performance and interpretability in transfer learning. Future research should focus on developing frameworks that explicitly incorporate interpretability as a core objective during the transfer process.

Fine-Tuning Pre-trained LLMs with Interpretability Constraints

One approach involves fine-tuning pre-trained LLMs with interpretability constraints. For instance, structured prompting techniques, such as Chain-of-Thought (CoT) reasoning, align model outputs with clinical logic, enhancing transparency without sacrificing accuracy [123]. Similarly, hybrid frameworks that combine deep learning with neuro-symbolic models offer inherent interpretability by embedding domain-specific knowledge into the model architecture [123, 124].

Ensemble Models

An ensemble can include a primary predictive model (e.g., a Transformer-based LLM) and an explainability module (e.g., SHAP or Grad-CAM). The explainability module interprets the predictions of the primary model, ensuring transparency; the point is that the predictive model can be transferred.

Federated Transfer Learning

Federated transfer learning further enhances privacy and explainability. Techniques like homomorphic encryption and differential privacy protect sensitive patient data while enabling decentralized training. Simultaneously, explainability tools such as Grad-CAM, counterfactual analysis, and SHAP provide localized insights into model decisions, ensuring transparency in distributed environments [125, 126]. These local insights are then aggregated into a global explanation of the model's behavior. The feasibility of applying FL to large-scale LLMs (FedLLM) relies on optimizing communication and protecting updates through specialized mechanisms:

- Parameter-Efficient Fine-tuning (PEFT/LoRA)

  Minimizes communication overhead by updating only a small, low-rank fraction of the model parameters [127, 128].
- Differential Privacy (DP)

  Adds calibrated noise to model updates to prevent inference attacks on individual patient data [129].
- Secure Aggregation

  Cryptographically ensures that the central server can only compute the average of updates, not inspect individual client contributions [130].

Dynamic Knowledge Updating for Transparent Evolution

Transparent evolution of medical LLMs requires shifting from parameter-based updates to externalized knowledge management, ensuring that changes are traceable and accountable. Medical LLMs must undergo Continual Learning (CL) to integrate new clinical guidelines and research, often through methods like Domain-Adaptive Pre-training or Continual Fine-Tuning [131]. However, this adaptation risks catastrophic forgetting and opaque shifts in model behavior, necessitating robust mechanisms for tracking knowledge evolution [132]. To maintain transparency, research focuses on separating evolving knowledge from static model parameters [133]. External memory systems, such as MemOS, treat knowledge as a manageable resource, utilizing "MemCubes" that store content alongside crucial metadata like provenance and versioning [134, 135]. This approach allows for efficient knowledge integration while enabling clear tracking of the source and timing of updates, reducing the risk of spurious forgetting [135].

Multimodal Integration for Richer Interpretability

Richer interpretability in medical LLMs requires the deep integration of multimodal data, such as images, clinical notes, and genomic information, to provide a contextually complete view of the model's reasoning process. Multimodal Large Language Models (MLLMs) enhance diagnostic support by simulating human physician collaboration and reasoning across diverse data types, moving toward trustworthy clinical assistance [136, 137]. Transfer learning, particularly through the fine-tuning of medical Vision-Language Pretrained (VLP) models, is essential for adapting these systems to specific clinical tasks efficiently [138]. These methods focus on achieving fine-grained alignment by disentangling visual and textual representations into anatomy-relevant and pathology-relevant streams, ensuring that the model captures specific concepts rather than rough correlations [139, 140].

Explaining the specific contribution of diverse modalities relies on visualization techniques embedded within the model architecture. These methods allow clinicians to trace the influence of each data type on the final prediction:

- Attention Masking

  Techniques like attention-masked image modeling identify image features that are highly responsive to specific textual tokens, linking visual evidence directly to report findings [141].
- Knowledge Alignment

  Injecting structured medical knowledge into the fusion module aligns vision and language representations, making the model's reasoning process more transparent and domain-aware [142].
- Modular Reasoning

Collaborative Chain-of-Thought (CoT) frameworks decompose complex tasks, allowing the model to explicitly show how visual observations and textual knowledge are integrated to conclude [143]. The successful deployment of medical LLMs relies on developing standardized, clinically relevant frameworks that rigorously assess the quality, fidelity, and usefulness of generated explanations. Current evaluations often lack clinical fidelity and comprehensive safety metrics, necessitating a shift toward robust, lifecycle-oriented assessment [144, 145]. Key XAI metrics must move beyond simple accuracy to quantify the trustworthiness of explanations, balancing plausibility with faithfulness—how accurately the explanation reflects the model's true reasoning [146, 147]. These evaluation aspects and metrics collectively move medical AI from opaque prediction engines toward transparent clinical partners. The final critical step might be the adoption of standardized evaluation frameworks, validated by clinicians, to rigorously assess the fidelity and clinical usefulness of explanations, ensuring safe and ethical integration into healthcare.

Finally, suppose we intend to conduct a comparative analysis of the different methods across various parameters, such as computational and memory complexity, fidelity, performance, and explainability. In that case, we can put it in this way: Full Fine-Tuning (FT) typically yields the highest task performance but is prohibitively expensive in terms of memory and compute, especially for large models [148, 149]. Parameter-Efficient Fine-Tuning (PEFT), such as LoRA or adapter-based methods, addresses this by achieving comparable performance while drastically reducing the number of trainable parameters and GPU memory consumption [150, 151]. Prompt-tuning offers superior flexibility and efficiency for few-shot scenarios and enhances cross-domain generalizability, particularly in decoder-only architectures [152]. For explainability, specialized transfer methods are necessary to ensure fidelity. XAI-driven knowledge distillation (KD), such as DiXtill, leverages local explanations (e.g., LIME/SHAP) to guide the transfer of knowledge from a large teacher LLM to a compact, self-explainable student model [153]. This approach ensures the student model maintains explanation consistency, which is often lost in traditional KD methods, paving the way for efficient edge AI applications [153]. The Table 2 summarizes what was mentioned above about different methods.

As discussed before, future directions in transfer learning for making medical LLMs more explainable include several promising areas. Federated learning is a key focus, allowing decentralized training of LLMs while preserving data privacy, and integrating explainability techniques like SHAP and LIME to ensure transparency in distributed environments [128, 154]. Multimodal integration is another critical avenue, combining textual and imaging data to enhance diagnostic accuracy and provide interpretable outputs, such as visual explanations alongside textual justifications [155, 156]. Dynamic knowledge updating is essential for keeping LLMs aligned with evolving medical knowledge, with future research exploring mechanisms for seamless updates while maintaining explainability [156]. Neuro-symbolic AI offers a hybrid approach by combining neural networks with symbolic reasoning, enabling models to provide structured, rule-based explanations that align with clinical logic [157, 158] Lastly, the development of standardized evaluation frameworks (not covered in this chapter) for explainability in medical LLMs is crucial to assess fidelity, interpretability, and clinical relevance, ensuring these models meet rigorous health-care standards [159] These directions aim to balance performance, transparency, and ethical considerations in high-stakes medical applications.

7 Conclusion

In this chapter, foundations and updates on Large Language Models (LLMs) and Explainable Artificial Intelligence (XAI), both developed through Transfer Learning, are critically explored with respect to medical applications. The discussion begins with the evolution of LLMs from early statistical methods to transformer technology and how they might transform medicine (clinical decision support, biomedical research, and patient data interpretation). It emphasizes the importance of transparency, interpretability, and explainability as prerequisites for the adoption of AI systems in the clinic, as well as for the associated trust and accountability. Principles and methodologies of transfer learning from instance-based, feature-based, and parameter-based to adversarial approaches are then discussed as essential strategies for less data and computing, but in return, highly adaptable to other domains. An analysis of medical LLMs, which include BioBERT, Med-PaLM, ClinicalBERT, and GPT-4, showed that even with their fantastic potential, they face significant challenges in terms of data privacy, bias and fairness issues, hallucination, computation needs, and a lack of transparency. The last part of the discussion presented emerging techniques on combining transfer learning with XAI, such as Attention-Based Cross-Modal frameworks, federated learning, parameter-efficient fine-tuning (PEFT), and multimodal interpretability approaches to improve the credibility, user-centered explainability, and interpretive clarity of clinical decision-making. Overall, this chapter concludes that medical LLMs will progress in a deliberate balance between performance, interpretability, and ethical responsibility. In this trajectory, transfer learning not only sustains economic efficiency but also acts as a link towards building transparent, accountable, and trustworthy AI systems in medicine.

References

1. Raza M, Jahangir Z, Riaz MB, Saeed MJ, Sattar MA. Industrial applications of large language models. Sci Rep [Internet]. Nature Research; 2025 [cited 2025 Oct 15];15:1–23. https://doi.org/10.1038/S41598-025-98483-1;SUBJMETA
2. Abuain T, Abuain WAK. Impact of Large Language Models in the Healthcare Industry: SOAR Analyses. 2024 3rd International Conference on Creative Communication and Innovative Technology, ICCIT 2024. Institute of Electrical and Electronics Engineers Inc.; 2024; https://doi.org/10.1109/ICCIT62134.2024.10701228
3. Antoshchuk SG, Kunup T V., Lytvynenko VI, Danchuk O V. Using large language models for video processing in the agricultural industry. Applied Aspects of Information Technology [Internet]. Odessa Polytechnic National University; 2025 [cited 2025 Oct 15];8:88–101. https://doi.org/10.15276/AAIT.08.2025.7
4. Zheng T, Deng Z, Tsang HT, Wang W, Bai J, Wang Z, et al. From Automation to Autonomy: A Survey on Large Language Models in Scientific Discovery. 2025 [cited 2025 Oct 15]; https://arxiv.org/pdf/2505.13259. Accessed 15 Oct 2025
5. Tom´ T, Mikolov T. Statistical Language Models Based on Neural Networks Overview Motivation.
6. Hochreiter S, Schmidhuber J. Long Short-Term Memory. Neural Comput. MIT Press Journals; 1997;9:1735–80. https://doi.org/10.1162/NECO.1997.9.8.1735
7. Burkov Andriy. The hundred-page language models book. True Positive Inc.; 2025 [cited 2025 Oct 16];140. https://dokumen.pub/the-hundred-page-language-models-book-hands-on-with-pytorch-1nbsped-1778042724-9781778042720.html. Accessed 16 Oct 2025
8. Vaswani A, Brain G, Shazeer N, Parmar N, Uszkoreit J, Jones L, et al. Attention is All you Need. Adv Neural Inf Process Syst. 2017;30.
9. Devlin J, Chang MW, Lee K, Toutanova K. BERT: Pre-training of Deep Bidirectional Transformers for Language Understanding. NAACL HLT 2019 - 2019 Conference of the North American Chapter of the Association for Computational Linguistics: Human Language Technologies - Proceedings of the Conference [Internet]. Association for Computational Linguistics (ACL); 2018 [cited 2025 Oct 15];1:4171–86. https://arxiv.org/pdf/1810.04805. Accessed 15 Oct 2025
10. Radford A, Wu J, Child R, Luan D, Amodei D, Sutskever I. Language Models are Unsupervised Multitask Learners. [cited 2025 Oct 15]; https://github.com/codelucas/newspaper. Accessed 15 Oct 2025

11. Brown TB, Mann B, Ryder N, Subbiah M, Kaplan J, Dhariwal P, et al. Language Models are Few-Shot Learners. Adv Neural Inf Process Syst [Internet]. 2020 [cited 2025 Oct 15];33:1877–901. https://commoncrawl.org/the-data/. Accessed 15 Oct 2025
12. Yin J, Bose A, Cong G, Lyngaas I, Anthony Q. Comparative Study of Large Language Model Architectures on Frontier. Proceedings - 2024 IEEE International Parallel and Distributed Processing Symposium, IPDPS 2024. Institute of Electrical and Electronics Engineers Inc.; 2024;556–69. https://doi.org/10.1109/IPDPS57955.2024.00056
13. Zayyanu M. REVOLUTIONISING TRANSLATION TECHNOLOGY: A COMPARATIVE STUDY OF VARIANT TRANSFORMER MODELS-BERT, GPT AND T5. Computer Science & Engineering: An International Journal (CSEIJ). 2024;14. https://doi.org/10.5121/cseij.2024.14302
14. Leiser F, Guse R, Sunyaev A. Large Language Model Architectures in Health Care: Scoping Review of Research Perspectives. J Med Internet Res [Internet]. JMIR Publications Inc.; 2025 [cited 2025 Oct 15];27:e70315. https://doi.org/10.2196/70315
15. Dong B, Zhang L, Yuan J, Chen Y, Li Q, Shen L. Large language models: game-changers in the healthcare industry. Sci Bull (Beijing) [Internet]. Elsevier B.V.; 2025 [cited 2025 Oct 15];70:283–6. https://doi.org/10.1016/j.scib.2024.11.031
16. Xu Z, Jain S, Kankanhalli M. Hallucination is Inevitable: An Innate Limitation of Large Language Models. 2024 [cited 2025 Oct 15]; https://arxiv.org/pdf/2401.11817. Accessed 15 Oct 2025
17. Chu Z, Jia H, Matricciani E, Feretzakis G, Papaspyridis K, Gkoulalas-Divanis A, et al. Privacy-Preserving Techniques in Generative AI and Large Language Models: A Narrative Review. Information 2024, Vol 15, Page 697 [Internet]. Multidisciplinary Digital Publishing Institute; 2024 [cited 2025 Oct 15];15:697. https://doi.org/10.3390/INFO15110697
18. Liu Y, Deng G, Li Y, Wang K, Wang Z, Wang X, et al. Prompt Injection attack against LLM-integrated Applications. 2023 [cited 2025 Oct 15]; https://arxiv.org/pdf/2306.05499. Accessed 15 Oct 2025
19. Zhu Z, Wang J, Cheng H, Liu Y. Unmasking and Improving Data Credibility: A Study with Datasets for Training Harmless Language Models. 12th International Conference on Learning Representations, ICLR 2024 [Internet]. International Conference on Learning Representations, ICLR; 2023 [cited 2025 Oct 15]; https://arxiv.org/pdf/2311.11202. Accessed 15 Oct 2025
20. Confalonieri R, Coba L, Wagner B, Besold TR. A historical perspective of explainable Artificial Intelligence. Wiley Interdiscip Rev Data Min Knowl Discov [Internet]. John Wiley & Sons, Ltd; 2021 [cited 2025 Oct 15];11:e1391. https://doi.org/10.1002/WIDM.1391
21. Liang B, Wang Y, Tong C. AI Reasoning in Deep Learning Era: From Symbolic AI to Neural–Symbolic AI. Mathematics 2025, Vol 13, Page 1707 [Internet]. Multidisciplinary Digital Publishing Institute; 2025 [cited 2025 Oct 15];13:1707. https://doi.org/10.3390/MATH13111707
22. Gill TG. Early expert systems: Where are they now? MIS Q. Management Information Systems Research Center; 1995;19:51–70. https://doi.org/10.2307/249711
23. Mohan Raja Pulicharla. Neurosymbolic AI: Bridging neural networks and symbolic reasoning. World Journal of Advanced Research and Reviews [Internet]. GSC Online Press; 2025 [cited 2025 Oct 15];25:2351–73. https://doi.org/10.30574/WJARR.2025.25.1.0287
24. Garcez AD, Gori M, Lamb LC, Serafini L, Spranger M, Tran SN. Neural-Symbolic Computing: An Effective Methodology for Principled Integration of Machine Learning and Reasoning. IfCoLoG Journal of Logics and their Applications [Internet]. College Publications; 2019 [cited 2025 Oct 15];6:611–31. https://arxiv.org/pdf/1905.06088. Accessed 15 Oct 2025
25. Murdoch WJ, Singh C, Kumbier K, Abbasi-Asl R, Yu B. Definitions, methods, and applications in interpretable machine learning. Proceedings of the National Academy of Sciences [Internet]. National Academy of Sciences; 2019 [cited 2025 Oct 15];116:22071–80. https://doi.org/10.1073/PNAS.1900654116

26. Carvalho D V., Pereira EM, Cardoso JS. Machine Learning Interpretability: A Survey on Methods and Metrics. Electronics 2019, Vol 8, Page 832 [Internet]. Multidisciplinary Digital Publishing Institute; 2019 [cited 2025 Oct 15];8:832. https://doi.org/10.3390/ELECTRONICS8080832
27. Linardatos P, Papastefanopoulos V, Kotsiantis S. Explainable AI: A Review of Machine Learning Interpretability Methods. Entropy 2021, Vol 23, Page 18 [Internet]. Multidisciplinary Digital Publishing Institute; 2020 [cited 2025 Oct 15];23:18. https://doi.org/10.3390/E23010018
28. Mi JX, Li A Di, Zhou LF. Review study of interpretation methods for future interpretable machine learning. IEEE Access. Institute of Electrical and Electronics Engineers Inc.; 2020;8:191969–85. https://doi.org/10.1109/ACCESS.2020.3032756
29. Ribeiro MT, Singh S, Guestrin C. "Why should i trust you?" Explaining the predictions of any classifier. Proceedings of the ACM SIGKDD International Conference on Knowledge Discovery and Data Mining [Internet]. Association for Computing Machinery; 2016 [cited 2025 Oct 15];13–17-August-2016:1135–44. https://doi.org/10.1145/2939672.2939778/SUPPL_FILE/KDD2016_RIBEIRO_ANY_CLASSIFIER_01-ACM.MP4
30. Lundberg SM, Allen PG, Lee S-I. A Unified Approach to Interpreting Model Predictions. Adv Neural Inf Process Syst [Internet]. 2017 [cited 2025 Oct 15];30. https://github.com/slundberg/shap. Accessed 15 Oct 2025
31. Muzellec S, Fel T, Boutin V, andéol L, VanRullen R, Serre T. Saliency strikes back: How filtering out high frequencies improves white-box explanations. 2023 [cited 2025 Oct 15]; https://arxiv.org/pdf/2307.09591. Accessed 15 Oct 2025
32. Zhuo Y, Ge Z. IG2: Integrated Gradient on Iterative Gradient Path for Feature Attribution. IEEE Trans Pattern Anal Mach Intell. IEEE Computer Society; 2024;46:7173–90. https://doi.org/10.1109/TPAMI.2024.3388092
33. Wang H, Joshi D, Wang S, Ji Q. Gradient-Based Uncertainty Attribution for Explainable Bayesian Deep Learning. 2023. p. 12044–53.
34. Kasmi G, Brunetto A, Fel T, Parekh J. One Wave To Explain Them All: A Unifying Perspective On Feature Attribution.
35. Kim J, Maathuis H, Sent D. Human-centered evaluation of explainable AI applications: a systematic review. Front Artif Intell. Frontiers Media SA; 2024;7:1456486. https://doi.org/10.3389/FRAI.2024.1456486/FULL
36. Schmitt V, Villa-Arenas LF, Feldhus N, Meyer J, Spang RP, Möller S. The Role of Explainability in Collaborative Human-AI Disinformation Detection. 2024 ACM Conference on Fairness, Accountability, and Transparency, FAccT 2024 [Internet]. Association for Computing Machinery, Inc; 2024 [cited 2025 Oct 15];2157–74. https://doi.org/10.1145/3630106.3659031
37. Rong Y, Leemann T, Nguyen TT, Fiedler L, Qian P, Unhelkar V, et al. Towards Human-Centered Explainable AI: A Survey of User Studies for Model Explanations. IEEE Trans Pattern Anal Mach Intell. IEEE Computer Society; 2024;46:2104–22. https://doi.org/10.1109/TPAMI.2023.3331846
38. Shulner-Tal A, Sheidin J. XAI4RE-Using Explainable AI for Responsible and Ethical AI. UMAP 2025 - Adjunct Proceedings of the 33rd ACM Conference on User Modeling, Adaptation and Personalization [Internet]. Association for Computing Machinery, Inc; 2025 [cited 2025 Oct 15];300–7. https://doi.org/10.1145/3708319.3733707
39. Zhao H, Chen H, Yang F, Liu N, Deng H, Cai H, et al. Explainability for Large Language Models: A Survey. ACM Trans Intell Syst Technol [Internet]. ACMPUB27New York, NY; 2024 [cited 2025 Oct 15];15:38. https://doi.org/10.1145/3639372
40. Kim J, Park S, Jeong K, Lee S, Han SH, Lee J, et al. Which is better? Exploring Prompting Strategy For LLM-based Metrics. Association for Computational Linguistics (ACL); 2023 [cited 2025 Oct 15];164–83. https://doi.org/10.18653/v1/2023.eval4nlp-1.14
41. Siegel NY, Camburu OM, Heess N, Perez-Ortiz M. The Probabilities Also Matter: A More Faithful Metric for Faithfulness of Free-Text Explanations in Large Language Models. Proceedings of the Annual Meeting of the Association for Computational Linguistics [Internet]. Association for Computational Linguistics (ACL); 2024 [cited 2025 Oct 15];2:530–46. https://doi.org/10.18653/v1/2024.acl-short.49

42. Admoni S, Amir O, Hallak A, Ziser Y. Towards Large Language Models with Self-Consistent Natural Language Explanations. 2025 [cited 2025 Oct 15]; https://arxiv.org/pdf/2506.07523. Accessed 15 Oct 2025
43. Ali AMO. Explainability in AI: Interpretable Models for Data Science. Int J Res Appl Sci Eng Technol. International Journal for Research in Applied Science and Engineering Technology (IJRASET); 2025;13:766–71. https://doi.org/10.22214/IJRASET.2025.66968
44. Pan SJ, Yang Q. A survey on transfer learning. IEEE Trans Knowl Data Eng. 2010;22:1345–59. https://doi.org/10.1109/TKDE.2009.191
45. Day O, Khoshgoftaar TM. A survey on heterogeneous transfer learning. J Big Data [Internet]. SpringerOpen; 2017 [cited 2025 Oct 15];4:1–42. https://doi.org/10.1186/S40537-017-0089-0/FIGURES/5
46. Zhang W, Deng L, Zhang L, Wu D. A Survey on Negative Transfer. IEEE/CAA Journal of Automatica Sinica. Institute of Electrical and Electronics Engineers Inc.; 2023;10:305–29. https://doi.org/10.1109/JAS.2022.106004
47. Chen Y, Qin X, Wang J, Yu C, Gao W. FedHealth: A Federated Transfer Learning Framework for Wearable Healthcare. IEEE Intell Syst. Institute of Electrical and Electronics Engineers Inc.; 2020;35:83–93. https://doi.org/10.1109/MIS.2020.2988604
48. Zhang W, Jiang N, Yang S, Li X. Federated transfer learning for remaining useful life prediction in prognostics with data privacy. Meas Sci Technol [Internet]. IOP Publishing; 2025 [cited 2025 Oct 15];36:076107. https://doi.org/10.1088/1361-6501/ADE552
49. Wu J, Zou D, Braverman V, Gu Q, Kakade SM. The Power and Limitation of Pretraining-Finetuning for Linear Regression under Covariate Shift. Adv Neural Inf Process Syst. 2022;35:33041–53.
50. Sugiyama M, Nakajima S, Kashima H, Buenau P, Kawanabe M. Direct Importance Estimation with Model Selection and Its Application to Covariate Shift Adaptation. Adv Neural Inf Process Syst. 2007;20.
51. Shimodaira H. Improving predictive inference under covariate shift by weighting the log-likelihood function. J Stat Plan Inference [Internet]. North-Holland; 2000 [cited 2025 Oct 15];90:227–44. https://doi.org/10.1016/S0378-3758(00)00115-4
52. Abbas S, Ahmed F, Khan WA, Ahmad M, Khan MA, Ghazal TM. Intelligent skin disease prediction system using transfer learning and explainable artificial intelligence. Sci Rep [Internet]. Nature Research; 2025 [cited 2025 Oct 15];15:1–13. https://doi.org/10.1038/S41598-024-83966-4;SUBJMETA
53. Zhong X, Guo S, Shan H, Gao L, Xue D, Zhao N. Feature-Based Transfer Learning Based on Distribution Similarity. IEEE Access. Institute of Electrical and Electronics Engineers Inc.; 2018;6:35551–7. https://doi.org/10.1109/ACCESS.2018.2843773
54. Zhou Y, Zhang X, Wang Y, Zhang B. Transfer learning and its application research. J Phys Conf Ser [Internet]. IOP Publishing; 2021 [cited 2025 Oct 15];1920:012058. https://doi.org/10.1088/1742-6596/1920/1/012058
55. Pan J. FEATURE-BASED TRANSFER LEARNING WITH REAL-WORLD APPLICATIONS. 2010;
56. Niu S, Hu Y, Wang J, Liu Y, Song H. Feature-based Distant Domain Transfer Learning. Proceedings - 2020 IEEE International Conference on Big Data, Big Data 2020. Institute of Electrical and Electronics Engineers Inc.; 2020;5164–71. https://doi.org/10.1109/BIGDATA50022.2020.9378493
57. Zheng W, Xue F, Chen Z, Chen D, Guo B, Shen C, et al. Disruption prediction for future tokamaks using parameter-based transfer learning. Commun Phys [Internet]. Nature Research; 2023 [cited 2025 Oct 15];6:1–11. https://doi.org/10.1038/S42005-023-01296-9;SUBJMETA
58. Xu H, Yang L, Long X. Underwater Sonar Image Classification with Small Samples Based on Parameter-based Transfer Learning and Deep Learning. Proceedings - 2022 Global Conference on Robotics, Artificial Intelligence and Information Technology, GCRAIT 2022. Institute of Electrical and Electronics Engineers Inc.; 2022;304–7. https://doi.org/10.1109/GCRAIT55928.2022.00071

59. Li Z, Su Y, Collier N. A Survey on Prompt Tuning. 2025 [cited 2025 Oct 15]; https://arxiv.org/pdf/2507.06085. Accessed 15 Oct 2025
60. Khan AM, Ugarte E, Kurniadi S, Al Shueili A, Wang WK. Physics-Informed Machine Learning for Hydraulic Fracturing—Part III: The Transfer Learning Validation. International Petroleum Technology Conference, IPTC 2025 [Internet]. OnePetro; 2025 [cited 2025 Oct 15]; https://doi.org/10.2523/IPTC-24850-MS
61. Csurka G, editor. Domain Adaptation in Computer Vision Applications. Cham: Springer International Publishing; 2017 [cited 2025 Oct 15]; https://doi.org/10.1007/978-3-319-58347-1
62. Tzeng E, Hoffman J, Saenko K, Darrell T. Adversarial Discriminative Domain Adaptation. 2017. p. 7167–76.
63. Guo Y, Zhang J, Sun B, Wang Y. Adversarial Deep Transfer Learning in Fault Diagnosis: Progress, Challenges, and Future Prospects. Sensors 2023, Vol 23, Page 7263 [Internet]. Multidisciplinary Digital Publishing Institute; 2023 [cited 2025 Oct 15];23:7263. https://doi.org/10.3390/S23167263
64. 65. Xiao J, Fan Y, Sun R, Wang J, Luo Z-Q. Stability Analysis and Generalization Bounds of Adversarial Training. Adv Neural Inf Process Syst. 2022;35:15446–59.
65. Zhao JC, Bagchi S. Are Fast Methods Stable in Adversarially Robust Transfer Learning? 2025 [cited 2025 Oct 15]; https://arxiv.org/pdf/2506.22602. Accessed 15 Oct 2025
66. Muhammad A, Bae SH. A Survey on Efficient Methods for Adversarial Robustness. IEEE Access. Institute of Electrical and Electronics Engineers Inc.; 2022;10:118815–30. https://doi.org/10.1109/ACCESS.2022.3216291
67. 68. Xing Y, Song Q, Cheng G. On the Algorithmic Stability of Adversarial Training. Adv Neural Inf Process Syst. 2021;34:26523–35.
68. Xiao J, Fan Y, Sun R, Luo Z-Q. Adversarial Rademacher Complexity of Deep Neural Networks. 2022 [cited 2025 Oct 15]; https://arxiv.org/pdf/2211.14966. Accessed 15 Oct 2025
69. Dong J, Chen J, Xie X, Lai J, Chen H. Survey on Adversarial Attack and Defense for Medical Image Analysis: Methods and Challenges. ACM Comput Surv [Internet]. ACMPUB27New York, NY; 2024 [cited 2025 Oct 15];57. https://doi.org/10.1145/3702638
70. Minagi A, Hirano H, Takemoto K. Natural Images Allow Universal Adversarial Attacks on Medical Image Classification Using Deep Neural Networks with Transfer Learning. J Imaging [Internet]. MDPI; 2022 [cited 2025 Oct 15];8:38. https://doi.org/10.3390/JIMAGING8020038/S1
71. Kamath U, Liu J, Whitaker J. Deep learning for NLP and speech recognition. Deep Learning for NLP and Speech Recognition. Springer International Publishing; 2019;1–621. https://doi.org/10.1007/978-3-030-14596-5/COVER
72. Kouw WM, Loog M. An introduction to domain adaptation and transfer learning. 2018 [cited 2025 Oct 15]; https://arxiv.org/pdf/1812.11806. Accessed 15 Oct 2025
73. Long M, Cao Y, Wang J, Jordan MI, Edu J. Learning Transferable Features with Deep Adaptation Networks [Internet]. PMLR; 2015 [cited 2025 Oct 15]. p. 97–105. https://proceedings.mlr.press/v37/long15.html. Accessed 15 Oct 2025
74. Kim YB, Stratos K, Kim D. Adversarial Adaptation of Synthetic or Stale Data. ACL 2017 - 55th Annual Meeting of the Association for Computational Linguistics, Proceedings of the Conference (Long Papers) [Internet]. Association for Computational Linguistics (ACL); 2017 [cited 2025 Oct 15];1:1297–307. https://doi.org/10.18653/V1/P17-1119
75. Xu B, Yin J, Lian C, Su Y, Zeng Z. Low-Rank Optimal Transport for Robust Domain Adaptation. IEEE/CAA Journal of Automatica Sinica. Institute of Electrical and Electronics Engineers Inc.; 2024;11:1667–80. https://doi.org/10.1109/JAS.2024.124344
76. Huang J, Smola AJ, Gretton A, Borgwardt KM, Schölkopf B. Correcting Sample Selection Bias by Unlabeled Data. Adv Neural Inf Process Syst. 2006;19.
77. Paul A, Rottensteiner F, Heipke C. ITERATIVE RE-WEIGHTED INSTANCE TRANSFER FOR DOMAIN ADAPTATION. ISPRS Annals of Photogrammetry, Remote Sensing and Spatial Information Sciences. Copernicus GmbH; 2016;III–3:339–46. https://doi.org/10.5194/ISPRSANNALS-III-3-339-2016

78. De Mathelin A, Richard G, Deheeger F, Mougeot M, Vayatis N. Adversarial Weighting for Domain Adaptation in Regression. Proceedings - International Conference on Tools with Artificial Intelligence, ICTAI. IEEE Computer Society; 2021;2021-November:49–56. https://doi.org/10.1109/ICTAI52525.2021.00015
79. Cao Z, You K, Long M, Wang J, Yang Q. Learning to Transfer Examples for Partial Domain Adaptation. 2019. p. 2985–94.
80. Li L, Wan Z, He H. Dual Alignment for Partial Domain Adaptation. IEEE Trans Cybern. Institute of Electrical and Electronics Engineers Inc.; 2021;51:3404–16. https://doi.org/10.1109/TCYB.2020.2983337
81. Ragab M, Eldele E, Tan WL, Foo CS, Chen Z, Wu M, et al. ADATIME: A Benchmarking Suite for Domain Adaptation on Time Series Data. ACM Trans Knowl Discov Data [Internet]. ACMPUB27New York, NY; 2023 [cited 2025 Oct 15];17. https://doi.org/10.1145/3587937
82. Lee J, Yoon W, Kim S, Kim D, Kim S, So CH, et al. BioBERT: a pre-trained biomedical language representation model for biomedical text mining. Bioinformatics [Internet]. Oxford Academic; 2020 [cited 2025 Oct 15];36:1234–40. https://doi.org/10.1093/BIOINFORMATICS/BTZ682
83. Singhal K, Azizi S, Tu T, Mahdavi SS, Wei J, Chung HW, et al. Large language models encode clinical knowledge. Nature [Internet]. Nature Research; 2023 [cited 2025 Oct 15];620:172–80. https://doi.org/10.1038/S41586-023-06291-2;SUBJMETA
84. Althobaiti FA. A Survey on Using Large Language Models in Healthcare. [cited 2025 Oct 15]; https://doi.org/10.7753/IJSEA1406.1001
85. Lataran PN, Farhadi A, Zamanifar A, Taheri A. Developing a BERT-Enhanced Blockchain Model for HealthInsurance Fraud Detection. Journal of Information Systems Research and Practice [Internet]. 2024 [cited 2025 Oct 13];2:56–61. https://ijps.um.edu.my/index.php/JISRP/article/view/54256. Accessed 13 Oct 2025
86. Rashidieranjbar F, Farhadi A, Zamanifar A. Revolutionizing Healthcare with Generative Artificial Intelligence Technologies. Springer, Cham; 2025 [cited 2025 Oct 16];189–221. https://doi.org/10.1007/978-3-031-76710-4_10
87. Yang X, Xiao Y, Liu D, Zhang Y, Deng H, Huang J, et al. Enhancing doctor-patient communication using large language models for pathology report interpretation. BMC Med Inform Decis Mak [Internet]. BioMed Central Ltd; 2025 [cited 2025 Oct 15];25:1–16. https://doi.org/10.1186/S12911-024-02838-Z/FIGURES/6
88. De Vito G, Ferrucci F, Angelakis A. HELIOT: LLM-Based CDSS for adverse drug reaction management. Knowl Based Syst [Internet]. Elsevier; 2025 [cited 2025 Oct 15];328:114184. https://doi.org/10.1016/J.KNOSYS.2025.114184
89. Sandmann S, Hegselmann S, Fujarski M, Bickmann L, Wild B, Eils R, et al. Benchmark evaluation of DeepSeek large language models in clinical decision-making. Nat Med [Internet]. Nature Research; 2025 [cited 2025 Oct 15];31:2546–9. https://doi.org/10.1038/S41591-025-03727-2;SUBJMETA
90. Yang X, Chen A, PourNejatian N, Shin HC, Smith KE, Parisien C, et al. A large language model for electronic health records. NPJ Digit Med [Internet]. Nature Research; 2022 [cited 2025 Oct 15];5:1–9. https://doi.org/10.1038/S41746-022-00742-2;SUBJMETA
91. Labrak Y, Bazoge A, Morin E, Gourraud PA, Rouvier M, Dufour R. BioMistral: A Collection of Open-Source Pretrained Large Language Models for Medical Domains. Proceedings of the Annual Meeting of the Association for Computational Linguistics [Internet]. Association for Computational Linguistics (ACL); 2024 [cited 2025 Oct 16];5848–64. https://doi.org/10.18653/v1/2024.findings-acl.348
92. Farhadi A, Taheri A. Application of GenAI in synthetic data generation in the healthcare system. Application of Generative AI in Healthcare Systems [Internet]. Springer Nature; 2025 [cited 2025 Oct 13];67–90. https://doi.org/10.1007/978-3-031-82963-5_3/FIGURES/4
93. Weber MT, Noll R, Marchl A, Facchinello C, Grünewaldt A, Hügel C, et al. MedBot vs RealDoc: efficacy of large language modeling in physician-patient communication for rare diseases. Journal of the American Medical Informatics Association [Internet]. Oxford Academic; 2025 [cited 2025 Oct 15];32:775–83. https://doi.org/10.1093/JAMIA/OCAF034

94. Dennstädt F, Hastings J, Putora PM, Schmerder M, Cihoric N. Implementing large language models in healthcare while balancing control, collaboration, costs and security. NPJ Digit Med [Internet]. Nature Research; 2025 [cited 2025 Oct 15];8:1–4. https://doi.org/10.1038/S41746-025-01476-7;SUBJMETA
95. Al-Garadi M, Mungle T, Ahmed A, Sarker A, Miao Z, Matheny ME. Large Language Models in Healthcare. 2025 [cited 2025 Oct 15]; https://arxiv.org/pdf/2503.04748. Accessed 15 Oct 2025
96. Yang Y, Jin Q, Zhu Q, Wang Z, Álvarez FE, Wan N, et al. Beyond Multiple-Choice Accuracy: Real-World Challenges of Implementing Large Language Models in Healthcare. Annu Rev Biomed Data Sci [Internet]. Annual Reviews Inc.; 2025 [cited 2025 Oct 15];8:305–16. https://doi.org/10.1146/ANNUREV-BIODATASCI-103123-094851/CITE/REFWORKS
97. Hussain W, Khoriba G, Maity S, Jyoti Saikia M. Large Language Models in Healthcare and Medical Applications: A Review. Bioengineering 2025, Vol 12, Page 631 [Internet]. Multidisciplinary Digital Publishing Institute; 2025 [cited 2025 Oct 15];12:631. https://doi.org/10.3390/BIOENGINEERING12060631
98. Kwan HY, Shell J, Fahy C, Yang S, Xing Y. Integrating Large Language Models into Medication Management in Remote Healthcare: Current Applications, Challenges, and Future Prospects. Systems 2025, Vol 13, Page 281 [Internet]. Multidisciplinary Digital Publishing Institute; 2025 [cited 2025 Oct 15];13:281. https://doi.org/10.3390/SYSTEMS13040281
99. Chaddad A, Lu Q, Li J, Katib Y, Kateb R, Tanougast C, et al. Explainable, Domain-Adaptive, and Federated Artificial Intelligence in Medicine. IEEE/CAA Journal of Automatica Sinica. Institute of Electrical and Electronics Engineers Inc.; 2023;10:859–76. https://doi.org/10.1109/JAS.2023.123123
100. Hao J. Deep learning-based medical image analysis with explainable transfer learning. Proceedings - 2023 International Conference on Computer Engineering and Distance Learning, CEDL 2023. Institute of Electrical and Electronics Engineers Inc.; 2023;106–9. https://doi.org/10.1109/CEDL60560.2023.00029
101. Karako K, Tang W. Applications of and issues with machine learning in medicine: Bridging the gap with explainable AI. Biosci Trends [Internet]. International Research and Cooperation Association for Bio & Socio-Sciences Advancement; 2024 [cited 2025 Oct 15];18:497–504. https://doi.org/10.5582/BST.2024.01342
102. Alsubai S, Ojo S, Nathaniel TI, Ayari M, Baili J, Almadhor A, et al. Transfer deep learning and explainable AI framework for brain tumor and Alzheimer's detection across multiple datasets. Front Med (Lausanne). Frontiers Media SA; 2025;12:1618550. https://doi.org/10.3389/FMED.2025.1618550/BIBTEX
103. Taiyeb Khosroshahi M, Morsali S, Gharakhanlou S, Motamedi A, Hassanbaghlou S, Vahedi H, et al. Explainable Artificial Intelligence in Neuroimaging of Alzheimer's Disease. Diagnostics [Internet]. Multidisciplinary Digital Publishing Institute (MDPI); 2025 [cited 2025 Oct 15];15:612. https://doi.org/10.3390/DIAGNOSTICS15050612/S1
104. Bergomi L, Nicora G, Orlowska MA, Podrecca C, Bellazzi R, Fregosi C, et al. Which explanations do clinicians prefer? A comparative evaluation of XAI understandability and actionability in predicting the need for hospitalization. BMC Med Inform Decis Mak [Internet]. BioMed Central Ltd; 2025 [cited 2025 Oct 15];25:1–13. https://doi.org/10.1186/S12911-025-03045-0/FIGURES/7
105. Räz T, Pahud De Mortanges A, Reyes M. Explainable AI in medicine: challenges of integrating XAI into the future clinical routine. Frontiers in Radiology. Frontiers Media SA; 2025;5:1627169. https://doi.org/10.3389/FRADI.2025.1627169/BIBTEX
106. Corti L, Oltmans R, Jung J, Balayn A, Wijsenbeek M, Yang J. "It Is a Moving Process": Understanding the Evolution of Explainability Needs of Clinicians in Pulmonary Medicine. Conference on Human Factors in Computing Systems - Proceedings [Internet]. Association for Computing Machinery; 2024 [cited 2025 Oct 15]; https://doi.org/10.1145/3613904.3642551/SUPPL_FILE/3613904.3642551-TALK-VIDEO.VTT
107. Zhen L, Hu P, Peng X, Goh RSM, Zhou JT. Deep Multimodal Transfer Learning for Cross-Modal Retrieval. IEEE Trans Neural Netw Learn Syst. Institute of Electrical and Electronics Engineers Inc.; 2022;33:798–810. https://doi.org/10.1109/TNNLS.2020.3029181

108. Shen Y. Geometric Optimal Transport for Cross-Modal Medical Manifold Alignment: A Differential Approach to Multimodal Diagnosis. IEEE Access. Institute of Electrical and Electronics Engineers Inc.; 2025;13:147666–78. https://doi.org/10.1109/ACCESS.2025.3587298
109. V. JP, S. AAV, P. GK, Karthikeyan NK. A novel attention-based cross-modal transfer learning framework for predicting cardiovascular disease. Comput Biol Med [Internet]. Pergamon; 2024 [cited 2025 Oct 15];170:107977. https://doi.org/10.1016/J.COMPBIOMED.2024.107977
110. Zhang Z, Zhou Z, Wang H, Zhang Z, Huang H, Cao Q. Grasp Stability Assessment Through Attention-Guided Cross-Modality Fusion and Transfer Learning. IEEE International Conference on Intelligent Robots and Systems. Institute of Electrical and Electronics Engineers Inc.; 2023;9472–9. https://doi.org/10.1109/IROS55552.2023.10342411
111. Liu M, Liu Y, Cui H, Li C, Ma J. MGCT: Mutual-Guided Cross-Modality Transformer for Survival Outcome Prediction using Integrative Histopathology-Genomic Features. Proceedings - 2023 2023 IEEE International Conference on Bioinformatics and Biomedicine, BIBM 2023. Institute of Electrical and Electronics Engineers Inc.; 2023;1306–12. https://doi.org/10.1109/BIBM58861.2023.10385897
112. Xu H, Sun Y, Mei X, Tian X, Ma J. Attention-Guided Polarization Image Fusion Using Salient Information Distribution. IEEE Trans Comput Imaging. Institute of Electrical and Electronics Engineers Inc.; 2022;8:1117–30. https://doi.org/10.1109/TCI.2022.3228633
113. Li S, Xie Y, Wang G, Zhang L, Zhou W. Adaptive Multimodal Fusion With Attention Guided Deep Supervision Net for Grading Hepatocellular Carcinoma. IEEE J Biomed Health Inform. Institute of Electrical and Electronics Engineers Inc.; 2022;26:4123–31. https://doi.org/10.1109/JBHI.2022.3161466
114. Zhou F, Chen H. Cross-Modal Translation and Alignment for Survival Analysis [Internet]. 2023 [cited 2025 Oct 15]. p. 21485–94. https://github.com/FT-ZHOU-ZZZ/CMTA. Accessed 15 Oct 2025
115. Zhang J, Wang Q, Wang Q. A Sign Language Recognition Framework Based on Cross-Modal Complementary Information Fusion. IEEE Trans Multimedia. Institute of Electrical and Electronics Engineers Inc.; 2024;26:8131–44. https://doi.org/10.1109/TMM.2024.3377095
116. Kashyap R, Billore R, Kumar P, Gupta R. Cross-Modal Transfer Learning for Multimodal Enhancement in Human-Computer Interaction. Proceedings of the 6th International Conference on Inventive Research in Computing Applications, ICIRCA 2025. Institute of Electrical and Electronics Engineers Inc.; 2025;1938–44. https://doi.org/10.1109/ICIRCA65293.2025.11089658
117. Agrawal T, Guermal M, Balazia M, Bremond F. CM3T: Framework for Efficient Multimodal Learning for Inhomogeneous Interaction Datasets. Proceedings - 2025 IEEE Winter Conference on Applications of Computer Vision, WACV 2025. Institute of Electrical and Electronics Engineers Inc.; 2025;7379–88. https://doi.org/10.1109/WACV61041.2025.00717
118. Rymarczyk D, Van De Weijer J, Zieli´nski BZ, Twardowski B. ICICLE: Interpretable Class Incremental Continual Learning. 2023. p. 1887–98.
119. Yu E, Lu J, Zhang G. Fuzzy Shared Representation Learning for Multistream Classification. IEEE Transactions on Fuzzy Systems. Institute of Electrical and Electronics Engineers Inc.; 2024;32:5625–37. https://doi.org/10.1109/TFUZZ.2024.3423024
120. Anwar RW, Abrar M, Ullah F. Transfer Learning in Brain Tumor Classification: Challenges, Opportunities, and Future Prospects. International Conference on ICT Convergence. IEEE Computer Society; 2023;24–9. https://doi.org/10.1109/ICTC58733.2023.10392830
121. Xu P, Deng Z, Wang J, Zhang Q, Choi KS, Wang S. Transfer Representation Learning with TSK Fuzzy System. IEEE Transactions on Fuzzy Systems. Institute of Electrical and Electronics Engineers Inc.; 2021;29:649–63. https://doi.org/10.1109/TFUZZ.2019.2958299
122. Miccio LA. Understanding Polymers Through Transfer Learning and Explainable AI. Applied Sciences (Switzerland) [Internet]. Multidisciplinary Digital Publishing Institute (MDPI); 2024 [cited 2025 Oct 15];14:10413. https://doi.org/10.3390/APP142210413/S1

123. Wankhade K, Jain D, Suryavanshi S. Developing Interpretable Large Language Models for High-Stakes Decision-Making in Healthcare: Insights from an XAI Review Perspective. International Journal of Research Publication and Reviews. Genesis Global Publication; 2025;6:405–15. https://doi.org/10.55248/GENGPI.6.0525.1615
124. Kamal MS, Nimmy SF, Islam MR, Naseem U. Explainable Medical Image Captioning. WWW Companion 2025 - Companion Proceedings of the ACM Web Conference 2025 [Internet]. Association for Computing Machinery, Inc; 2025 [cited 2025 Oct 15];2253–61. https://doi.org/10.1145/3701716.3717549
125. Sumariya S, Rami S, Revadekar S, Bhadane C. Federated Transfer Learning for Lung Disease Detection. Int J Imaging Syst Technol [Internet]. John Wiley & Sons, Ltd; 2025 [cited 2025 Oct 15];35:e70080. https://doi.org/10.1002/IMA.70080
126. Naidu UG, Lakkshmanan A, Krishna JSVG, Elamathi E, Reddy TS, V. T. Ram Pavan Kumar M. Federated AI Framework for Privacy-Preserving Differential Diagnosis Across Distributed Medical Networks. Proceedings of the 6th International Conference on Inventive Research in Computing Applications, ICIRCA 2025. Institute of Electrical and Electronics Engineers Inc.; 2025;932–40. https://doi.org/10.1109/ICIRCA65293.2025.11089748
127. Liu XY, Zhu R, Zha D, Gao J, Zhong S, White M, et al. Differentially Private Low-Rank Adaptation of Large Language Model Using Federated Learning. ACM Trans Manag Inf Syst [Internet]. ACMPUB27New York, NY; 2025 [cited 2025 Oct 15];16:24. https://doi.org/10.1145/3682068
128. Song I, Lee K. Optimizing Communication and Performance in Federated Learning for Large Language Models. Institute of Electrical and Electronics Engineers (IEEE); 2025;0964–7. https://doi.org/10.1109/ICAIIC64266.2025.10920742
129. Pan L, Huang M, Wang L, Qin P, Peng S. FedDP: Privacy-preserving method based on federated learning for histopathology image segmentation. Proceedings - 2024 IEEE International Conference on Bioinformatics and Biomedicine, BIBM 2024. Institute of Electrical and Electronics Engineers Inc.; 2024;2325–31. https://doi.org/10.1109/BIBM62325.2024.10822021
130. Dodevski Z, Pavleska T, Trajkovik V. Leveraging Federated Learning for Secure Transfer and Deployment of ML Models in Healthcare *. 2024 [cited 2025 Oct 15]; https://doi.org/10.70314/is.2024.chtm.5
131. Shi H, Xu Z, Wang H, Qin W, Wang W, Wang Y, et al. Continual Learning of Large Language Models: A Comprehensive Survey. ACM Comput Surv [Internet]. Association for Computing Machinery (ACM); 2024 [cited 2025 Oct 15];1. https://doi.org/10.1145/3735633
132. Wu T, Luo L, Li Y-F, Pan S, Vu T-T, Haffari G. Continual Learning for Large Language Models: A Survey. 2024 [cited 2025 Oct 15]; https://arxiv.org/pdf/2402.01364. Accessed 15 Oct 2025
133. Zheng J, Qiu S, Shi C, Ma Q. Towards Lifelong Learning of Large Language Models: A Survey. ACM Comput Surv [Internet]. ACMPUB27New York, NY; 2025 [cited 2025 Oct 15];57. https://doi.org/10.1145/3716629
134. Li Z, Song S, Xi C, Wang H, Tang C, Niu S, et al. MemOS: A Memory OS for AI System. 2025 [cited 2025 Oct 15]; https://arxiv.org/pdf/2507.03724. Accessed 15 Oct 2025
135. Zheng J, Cai X, Qiu S, Ma Q. Spurious Forgetting in Continual Learning of Language Models. 13th International Conference on Learning Representations, ICLR 2025 [Internet]. International Conference on Learning Representations, ICLR; 2025 [cited 2025 Oct 15];76881–946. https://arxiv.org/pdf/2501.13453. Accessed 15 Oct 2025
136. Wang W, Ma Z, Wang Z, Wu C, Ji J, Chen W, et al. A Survey of LLM-based Agents in Medicine: How far are we from Baymax? Association for Computational Linguistics (ACL); 2025 [cited 2025 Oct 15];10345–59. https://doi.org/10.18653/v1/2025.findings-acl.539
137. Gu Z, Liu F, Chen J, Yin C, Zhang P. A Proactive Agent Collaborative Framework for Zero-Shot Multimodal Medical Reasoning. Advanced Intelligent Systems [Internet]. John Wiley & Sons, Ltd; 2025 [cited 2025 Oct 15];7:2400840. https://doi.org/10.1002/AISY.202400840
138. Chen J, Jiang Y, Yang D, Li M, Wei J, Qian Z, et al. Can LLMs' Tuning Methods Work in Medical Multimodal Domain? Lecture Notes in Computer Science (including subseries

Lecture Notes in Artificial Intelligence and Lecture Notes in Bioinformatics) [Internet]. Springer Science and Business Media Deutschland GmbH; 2024 [cited 2025 Oct 15];15005 LNCS:112–22. https://doi.org/10.1007/978-3-031-72086-4_11
139. Fan W, Suvon MNI, Zhou S, Liu X, Alabed S, Osmani V, et al. MeDSLIP: Medical Dual-Stream Language-Image Pre-training with Pathology-Anatomy Semantic Alignment. 2024 [cited 2025 Oct 15]; https://arxiv.org/pdf/2403.10635. Accessed 15 Oct 2025
140. Ni X, Wu L, Zhuang J, Wang Q, Wu M, Vardhanabhuti V, et al. MG-3D: Multi-Grained Knowledge-Enhanced 3D Medical Vision-Language Pre-training. 2024 [cited 2025 Oct 15]; https://arxiv.org/pdf/2412.05876. Accessed 15 Oct 2025
141. Wu B, Xie Y, Zhang Z, Minh V, Phan H, Chen Q, et al. MMCLIP: Cross-modal Attention Masked Modelling for Medical Language-Image Pre-Training. 2024 [cited 2025 Oct 15];1. https://arxiv.org/pdf/2407.19546. Accessed 15 Oct 2025
142. Chen Z, Li G, Wan X. Align, Reason and Learn: Enhancing Medical Vision-and-Language Pre-training with Knowledge. MM 2022 - Proceedings of the 30th ACM International Conference on Multimedia [Internet]. Association for Computing Machinery, Inc; 2022 [cited 2025 Oct 15];5152–61. https://doi.org/10.1145/3503161.3547948/SUPPL_FILE/MM22-FP0814.MP4
143. Wei L, Wang W, Shen X, Xie Y, Fan Z, Zhang X, et al. MC-CoT: A Modular Collaborative CoT Framework for Zero-shot Medical-VQA with LLM and MLLM Integration. 2024 [cited 2025 Oct 15]; https://arxiv.org/pdf/2410.04521. Accessed 15 Oct 2025
144. Ma Z, Wang W, Yu G, Cheung Y-F, Ding M, Liu J, et al. Beyond the Leaderboard: Rethinking Medical Benchmarks for Large Language Models. 2025 [cited 2025 Oct 15]; https://arxiv.org/pdf/2508.04325. Accessed 15 Oct 2025
145. Budler LC, Chen H, Chen A, Topaz M, Tam W, Bian J, et al. A Brief Review on Benchmarking for Large Language Models Evaluation in Healthcare. Wiley Interdiscip Rev Data Min Knowl Discov [Internet]. John Wiley & Sons, Ltd; 2025 [cited 2025 Oct 15];15:e70010. https://doi.org/10.1002/WIDM.70010
146. Zaman K, Srivastava S. A Causal Lens for Evaluating Faithfulness Metrics. 2025 [cited 2025 Oct 15]; https://arxiv.org/pdf/2502.18848. Accessed 15 Oct 2025
147. Shailya K, Rajpal S, Krishnan GS, Ravindran B. LExT: Towards Evaluating Trustworthiness of Natural Language Explanations. ACMF AccT 2025 - Proceedings of the 2025 ACM Conference on Fairness, Accountability,and Transparency [Internet]. Association for Computing Machinery, Inc; 2025 [cited 2025 Oct 15];1565–87. https://doi.org/10.1145/3715275.3732104
148. Shin J, Tang C, Mohati T, Nayebi M, Wang S, Hemmati H. Prompt Engineering or Fine-Tuning: An Empirical Assessment of LLMs for Code. Proceedings - 2025 IEEE/ACM 22nd International Conference on Mining Software Repositories, MSR 2025. Institute of Electrical and Electronics Engineers Inc.; 2025;490–502. https://doi.org/10.1109/MSR66628.2025.00082
149. Prasad K, Srinivasan V, Gumpena P, Yattapu M, Brahmbhatt VH. Comparative Analysis of Different Efficient Fine Tuning Methods of Large Language Models (LLMs) in Low-Resource Setting. 2024 [cited 2025 Oct 15]; https://arxiv.org/pdf/2405.13181. Accessed 15 Oct 2025
150. Hu Z, Wang L, Lan Y, Xu W, Lim EP, Bing L, et al. LLM-Adapters: An Adapter Family for Parameter-Efficient Fine-Tuning of Large Language Models. EMNLP 2023 - 2023 Conference on Empirical Methods in Natural Language Processing, Proceedings [Internet]. Association for Computational Linguistics (ACL); 2023 [cited 2025 Oct 15];5254–76. https://doi.org/10.18653/v1/2023.emnlp-main.319
151. Damana AA, Hitesh P. Innovative Adapter-Based Fine-Tuning and Streamlined Training Strategies for Text Summarization. International Conference on Electrical and Computer Engineering Researches, ICECER 2024. Institute of Electrical and Electronics Engineers Inc.; 2024; https://doi.org/10.1109/ICECER62944.2024.10920340
152. Peng C, Yu Z, Smith KE, Lo-Ciganic W-H, Bian J, Wu Y. Enhancing Cross-Domain Generalizability in Social Determinants of Health Extraction with Prompt-Tuning Large Language Models. AMIA Summits on Translational Science Proceedings [Internet]. 2025 [cited 2025 Oct 15];2025:432. https://pmc.ncbi.nlm.nih.gov/articles/PMC12150740/. Accessed 15 Oct 2025

153. Cantini R, Orsino A, Talia D. Xai-driven knowledge distillation of large language models for efficient deployment on low-resource devices. J Big Data [Internet]. Springer Nature; 2024 [cited 2025 Oct 15];11:1–17. https://doi.org/10.1186/S40537-024-00928-3/FIGURES/5
154. Wang C, Li M, He J, Wang Z, Darzi E, Chen Z, et al. A survey for large language models in biomedicine. Artif Intell Med [Internet]. Elsevier; 2025 [cited 2025 Oct 15];170:103268. https://doi.org/10.1016/J.ARTMED.2025.103268
155. AlSaad R, Abd-Alrazaq A, Boughorbel S, Ahmed A, Renault MA, Damseh R, et al. Multimodal Large Language Models in Health Care: Applications, Challenges, and Future Outlook. J Med Internet Res 2024;26:e59505 https://www.jmir.org/2024/1/e59505 [Internet]. Journal of Medical Internet Research; 2024 [cited 2025 Oct 15];26:e59505. https://doi.org/10.2196/59505
156. Urooj B, Fayaz M, Ali S, Dang LM, Kim KW. Large Language Models in Medical Image Analysis: A Systematic Survey and Future Directions. Bioengineering 2025, Vol 12, Page 818 [Internet]. Multidisciplinary Digital Publishing Institute; 2025 [cited 2025 Oct 15];12:818. https://doi.org/10.3390/BIOENGINEERING12080818
157. Gaur M, Sheth A. Building trustworthy NeuroSymbolic AI Systems: Consistency, reliability, explainability, and safety. AI Mag [Internet]. John Wiley & Sons, Ltd; 2024 [cited 2025 Oct 15];45:139–55. https://doi.org/10.1002/AAAI.12149
158. Sheth A, Roy K. Neurosymbolic Value-Inspired Artificial Intelligence (Why, What, and How). IEEE Intell Syst. Institute of Electrical and Electronics Engineers Inc.; 2024;39:5–11. https://doi.org/10.1109/MIS.2023.3344353
159. Xiao H, Zhou F, Liu X, Liu T, Li Z, Liu X, et al. A comprehensive survey of large language models and multimodal large language models in medicine. Information Fusion [Internet]. Elsevier; 2025 [cited 2025 Oct 15];117:102888. https://doi.org/10.1016/J.INFFUS.2024.102888

Explainable NLP in Healthcare: Enhancing Clinical Documentation and Information Extraction

Azadeh Zamanifar, Alireza Taheri, Amirfarhad Farhadi, and Fouad Bahrpeyma

List of Abbreviations

AI	Artificial intelligence
BoW	Bag-of-Words
CDSS	Clinical Decision Support System
CNN	Convolutional neural network
DL	Deep learning
EHR	Electronic Health Record
GDPR	General Data Protection Regulation
HCE	Human-centered evaluations
HIPAA	Health Insurance Portability and Accountability Act
IAI	Interpretable AI
LIME	Local Interpretable Model-agnostic Explanations
LLM	Large language model
LSTM	Long short-term memory
NER	Named Entity Recognition
NLEs	Natural language explanations
NLP	Natural Language Processing

A. Zamanifar · A. Taheri
Department of Computer Engineering, SR.C, Islamic Azad University, Tehran, Iran
e-mail: azamanifar@iau.ac.ir

A. Taheri
e-mail: alirezataheri@srbiau.ac.ir

A. Farhadi
School of Computer Engineering, Iran University of Science and Technology, Tehran, Iran
e-mail: am_farhadi@mail.iust.ac.ir

F. Bahrpeyma (✉)
University of Applied Sciences Dresden (HTW Dresden), Dresden, Germany
e-mail: Bahrpeyma@ieee.org

A. Zamanifar et al. (eds.), *Explainable Large Language Models in Healthcare Applications*, Bio-IT and AI, https://doi.org/10.1007/978-3-032-15088-2_7

OB-CIE	Ontology-based Clinical Information Extraction System
OCR	Optical character recognition
PCD	Patient Clinical Data
PIPL	Personal Information Protective Law
RAI	Rational AI
RE	Relation Extraction
RNN	Recurrent Neural Network
SHAP	SHapley Additive exPlanations
XAI	Explainable Artificial Intelligence
XNLP	Explainable NLP

1 Introduction

The proliferation of Artificial Intelligence (AI), particularly through deep learning (DL) and Natural Language Processing (NLP), is driving a massive shift in healthcare. Every day, healthcare systems generate large volumes of data, a significant portion of which is unstructured, residing in formats like Electronic Health Records (EHRs), free-text clinical notes, discharge summaries, and the vast body of medical literature [1, 2]. This unstructured data represents a rich, yet historically underutilized, repository of critical clinical information and nuanced patient context [3]. NLP technologies are proving indispensable for unlocking the value contained within this data, offering powerful capabilities to automate clinical documentation, streamline complex workflows, extract actionable insights, and ultimately enhance the quality of patient care [4]. The evolution of NLP from traditional methods, such as rule-based systems or Bag-of-Words (BoW) models, to sophisticated neural architectures like Recurrent Neural Networks (RNNs) and, more recently, transformer-based models like BERT, has dramatically improved performance on a wide array of healthcare tasks. These modern models can analyze medical texts with unprecedented accuracy, aiding in applications ranging from clinical decision support and disease prediction to drug discovery and population health management [5].

However, these remarkable gains in predictive power have come at a significant cost: transparency. The very architectural complexity that makes models like BERT and GPT-4o so effective, also renders their internal decision-making processes opaque to human understanding [2, 6]. These systems often function as inscrutable black boxes, accepting inputs and producing outputs without revealing the underlying logic or reasoning that led to a particular conclusion [7]. This lack of interpretability poses a formidable barrier to their widespread and responsible adoption, especially in high-stakes domains such as healthcare, where decisions can have profound and irreversible consequences for human lives. For clinicians to confidently integrate AI-driven recommendations into their practice, they must be able to comprehend, verify, and ultimately trust the system's reasoning. A model that suggests a diagnosis or a treatment plan without justification is of limited utility, as it forces the medical professional to blindly follow its guidance without the ability to scrutinize the evidence or rationale, a practice that is untenable in modern medicine [8]. This fundamental need for clarity and accountability has catalyzed the rise of a crucial subfield in AI dedicated to demystifying these complex models.

In response to this critical challenge, the field of Explainable Artificial Intelligence (XAI) has emerged, providing a collection of methods and frameworks designed to render the operations of AI systems transparent, understandable, and trustworthy to human users. Within this broader movement, Explainable Natural Language Processing (XNLP) specifically focuses on illuminating the decision-making pathways of language-based models. Figure 1 illustrates the difference in using XNLP in contrast to NLP in a hypothetical scenario. In this case, the clinician can easily verify the prediction using the extra explanation. The primary objective of XNLP is to move beyond mere prediction and provide clear, human-comprehensible justifications for why a model arrived at a particular output. By doing so, XNLP aims to bolster the credibility and reliability of NLP systems, allowing developers to debug their models, helping clinicians to validate AI-generated insights, and empowering patients and other stakeholders to understand the logic behind automated decisions that affect them [9]. An explainable system is one that can articulate the factors that influenced its outcome, transforming it from an opaque oracle into a transparent and collaborative partner in the decision-making process.

A foundational step in navigating the landscape of XNLP is clarifying its core terminology, as concepts like interpretability and explainability are often used interchangeably in the literature, leading to ambiguity. Some researchers draw a distinction, defining interpretability as the capacity to design models that are inherently transparent by nature, so-called white-box or Interpretable AI (IAI) models like linear regression or decision trees, where the causal links between inputs and outputs are

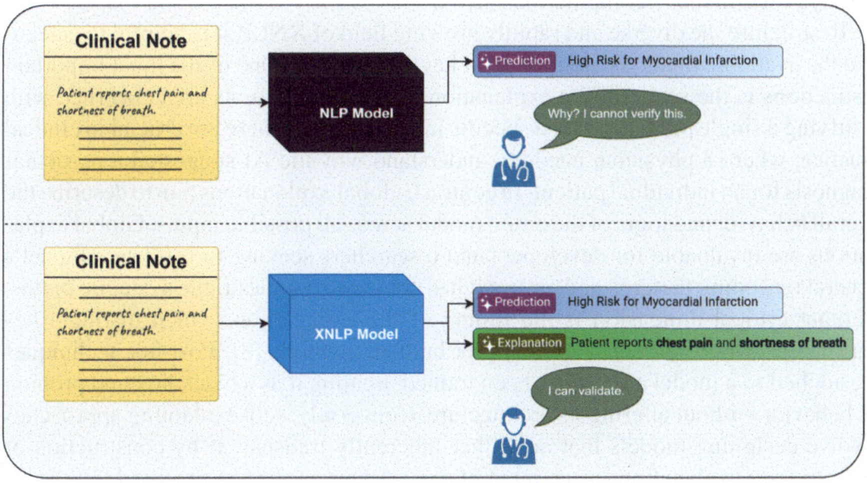

Fig. 1 XNLP versus NLP in the analysis of medical documents

clear. In this context, explainability refers to the post-hoc application of techniques to make an existing black-box model's predictions understandable [1, 5]. Other perspectives treat explainability as a specialization of interpretability, where an explanation must be not only understandable but also complete, meaning it accurately describes the system's operations to a degree that allows a human to anticipate its behavior. Yet another view uses the terms synonymously to refer to the broader goal of presenting a model's internal mechanics in understandable terms to a human. For the purposes of this chapter, we will often use the umbrella term XIAI, as introduced in some sources, to encompass both inherently interpretable models and post-hoc explanation methods, acknowledging that both approaches work toward the common goal of enhancing transparency in AI systems [6, 7].

Central to any discussion of explanation quality is the principle of faithfulness, which is arguably the most fundamental desideratum for an explanation method. Faithfulness dictates that an explanation must accurately reflect the actual reasoning process of the model it is describing. It ensures that the justification provided is not merely a post-hoc rationalization that sounds convincing but is, in fact, a true representation of the model's internal logic [10]. This is a critical distinction from plausibility, which refers to how convincing or understandable an explanation appears to a human user. An explanation can be highly plausible, which is aligning perfectly with human intuition, while being completely unfaithful, meaning it has little to no connection to the model's actual decision-making mechanism. Such plausible yet unfaithful explanations are particularly pernicious, as they can engender a false sense of security, leading users to over-trust a model that may be relying on spurious correlations or undesirable biases. The ongoing debate over whether attention weights in transformer models constitute a faithful explanation of their predictions serves as a prominent example of this tension, underscoring the necessity of rigorously evaluating explanation methods not just for their intuitive appeal but for their causal fidelity to the model's computations [7, 9].

To structure the diverse and rapidly growing field of XNLP, it is helpful to categorize the available methods along several key dimensions. One of the most important distinctions is the scope of the explanation. Local explanations are concerned with justifying a single prediction for a specific input instance. These are crucial in clinical practice, where a physician needs to understand why the AI suggested a particular diagnosis for an individual patient. In contrast, global explanations aim to describe the overall behavior and logic of the entire model across all possible inputs. Global explanations are invaluable for developers and researchers seeking to validate a model's general reasoning patterns and ensure it has not learned undesirable systemic biases. Another critical dimension is the timing of the explanation, which distinguishes between post-hoc and self-explaining or built-in methods [8]. Post-hoc techniques are applied to a model after it has been trained, treating it as a black box and probing its behavior without altering its architecture. Conversely, self-explaining approaches involve designing models that are either inherently transparent by construction or that generate explanations as an integral part of their prediction process [1].

XNLP methods can be classified by their generality, separating model-agnostic techniques from model-specific ones. Model-agnostic methods, as their name suggests, can be applied to any type of machine learning model, regardless of its internal architecture. This flexibility makes them powerful and widely applicable tools. Model-specific techniques, on the other hand, are tailored to a particular class of models, leveraging knowledge of their specific architecture to provide deeper and often more efficient explanations. For example, methods designed to interpret the attention mechanism are specific to transformer-based models. These categorizations, which are scope, timing, and generality provide a foundational taxonomy for navigating the diverse landscape of XNLP and understanding the unique strengths and applications of different approaches [4, 8].

Within this taxonomic framework, several distinct forms of explanation have become prominent in the field of XNLP. One of the most common is feature attribution, which seeks to explain a prediction by assigning an importance score to each feature in the input text, such as individual words or tokens. These scores indicate the degree to which each feature contributed, either positively or negatively, to the final outcome. Techniques that produce feature attributions are numerous and varied, including gradient-based methods like Saliency Maps, Integrated Gradients, and DeepLIFT, which use backpropagation to trace the output back to the input features, as well as perturbation-based methods that systematically alter parts of the input to observe the impact on the model's prediction [11]. Another powerful approach is the use of surrogate models, which explain a complex black-box model by approximating its behavior locally with a simpler, inherently interpretable model [12]. Local Interpretable Model-agnostic Explanations (LIME) is a well-known example of this technique; it generates a small, linear model to explain an individual prediction, making the reasoning behind that specific decision transparent. A related and highly influential method is SHapley Additive exPlanations (SHAP), which is grounded in cooperative game theory. SHAP calculates Shapley values to fairly distribute the payout or the model's prediction among the players or the input features thereby providing a robust and theoretically sound measure of each feature's contribution [3, 8].

A particularly intuitive and increasingly popular form of explanation is rationalization, which involves justifying a model's prediction by providing an explanation in natural language. These natural language explanations or rationales are especially valuable in healthcare because they are accessible and comprehensible to non-technical stakeholders, such as clinicians and patients, who may not have expertise in machine learning [2]. Rationalization techniques can be broadly divided into two categories: extractive rationalization and abstractive rationalization. Extractive methods produce a rationale by identifying and selecting the most salient snippets of text directly from the original input that are sufficient to justify the model's prediction. Abstractive methods, in contrast, are generative; they create a novel, free-text explanation that summarizes the model's reasoning, often paraphrasing or synthesizing information rather than simply extracting it. The remarkable recent advances in large language models (LLMs) like GPT and T5 have significantly enhanced the capabilities of abstractive rationalization, enabling models to generate highly fluent and coherent justifications that can almost seem as if the models are talking for themselves. This powerful approach has led to the proposal of a new subfield within XAI, termed Rational AI (RAI), dedicated to advancing research into these human-centric, natural language-based explanation methods.

The practical application of these XNLP techniques spans a wide spectrum of tasks and domains within the healthcare ecosystem. In clinical settings, explainable models are being developed for crucial tasks like medical diagnosis, disease prediction from clinical notes, automated medical coding, and providing decision support for treatment planning. For instance, an XNLP system might not only classify a pathology report as indicative of cancer but also highlight the specific phrases or sentences in the report that led to this conclusion, allowing a pathologist to quickly verify the finding [1, 3]. The application of XNLP extends to more specialized areas as well, including mental health, where models analyze patient-generated text from social media or therapy notes to identify signs of depression or suicidal ideation, with explanations providing clinicians with the context needed for intervention. Beyond direct clinical care, XNLP is also being leveraged in pharmaceutical research for tasks like drug discovery and analyzing clinical trial data, in public health for monitoring disease outbreaks from news reports, and in financial contexts within healthcare for fraud detection. The methods are being applied to increasingly diverse data sources, including not only structured EHR data and free-text clinical notes but also scanned documents, which require a complex pipeline of optical character recognition (OCR) before NLP can even be applied [5].

To ensure that the explanations generated by these diverse methods are genuinely useful and reliable, the field of XNLP places a strong emphasis on evaluation. Unlike traditional NLP models that can be judged primarily on predictive accuracy, the quality of an explanation is a more multifaceted concept that cannot be captured by a single metric. Consequently, the evaluation of XNLP systems involves a combination of quantitative and qualitative approaches. Quantitative metrics aim to objectively measure properties of an explanation, such as its faithfulness (how accurately it reflects the model's process), sufficiency (whether the explanation alone is enough to produce the same prediction), and comprehensiveness (the degree to which removing the explanatory features changes the prediction). In contrast, qualitative evaluation often takes the form of human-centered evaluations (HCE), where end-users such as clinicians are asked to assess the clarity, usefulness, and trustworthiness of the explanations provided by a system. This human-in-the-loop approach is critical for ensuring that XNLP solutions are not just technically sound but also practically valuable and aligned with the needs of the stakeholders who will ultimately rely on them in real-world settings. The development of robust, standardized evaluation benchmarks is an active and vital area of research, as it is essential for systematically tracking progress and fostering the development of more effective and trustworthy XNLP technologies [8, 11].

This chapter aims to provide a comprehensive overview of the burgeoning field of explainable NLP in healthcare. We will begin by reviewing the foundational modeling techniques that underpin XNLP, from traditional interpretable models to the complex transformer architectures that dominate modern NLP. Subsequently, we will focus on the applications of XNLP in healthcare, illustrating through case studies how these technologies are being deployed in practice. Another portion of the chapter will be dedicated to the critical aspects of XNLP, where we will delve into the nuances of evaluation metrics, the trade-offs between performance and interpretability, the different approaches to rationalization, and the indispensable role of human evaluation. Finally, we will look toward the horizon, discussing promising future directions and research opportunities that are poised to shape the next generation of transparent, trustworthy, and impactful NLP solutions for healthcare.

2 What is Explainable NLP

XNLP represents a critical subfield within the broader domain of XAI, specifically designed to address the inherent opacity of advanced language-based models. At its core, explainability, often used interchangeably with interpretability, refers to the degree to which a model's internal mechanisms can be presented in understandable terms to a human user. This includes both how a model makes specific predictions and what knowledge it encodes [7]. XAI, in general, aims to provide transparent and interpretable decision-making processes for AI systems. Within this context, XNLP focuses on making these language-based models comprehensible to both experts and non-experts, moving beyond mere yes or no answers to provide rich rationales that answer Why, When, and Where questions regarding a model's outputs.

The necessity for XNLP arises directly from the widespread adoption of highly complex DL models, including RNNs, long short-term memory (LSTM) networks, convolutional neural networks (CNNs), and particularly modern transformer-based models like BERT, GPT-4o, and Gemini. While these models have significantly advanced NLP performance across diverse applications, such as medical diagnosis, patient monitoring, drug discovery, and financial decision-making, their intricate architectures render them largely black-box systems. This opaqueness means their internal workings and reasoning processes are difficult for humans to understand, hindering their credibility, reliability, and widespread adoption, especially in high-stakes environments like healthcare where trustworthy and transparent decision-making is paramount for patient safety and clinical efficacy. For instance, understanding how complex NLP embeddings translate into deep learning decision-making mechanisms remains a significant challenge.

XNLP employs various techniques to shed light on these black-box models. One common approach involves feature importance or feature attribution methods, also known as saliency maps, which aim to quantify the relative importance of each input token (word or phrase) to a model's prediction. The underlying assumption is that tokens causing a greater change in the output when removed or altered are more important. Methods like SHAP are considered a gold standard for providing local explanations based on feature importance [11]. Another widely used technique involves attention mechanisms, which can highlight specific parts of the input text that the model attends to most when making a prediction, often visualized as heatmaps. However, the faithfulness of attention weights as direct explanations for a model's causal reasoning is a subject of ongoing debate [7].

Beyond simply highlighting features, XNLP also leverages NLEs, a technique that generates free-form text to explain a model's predictions. This approach, sometimes called rationalization, can be either extractive, where relevant fragments are highlighted directly from the input text, or abstractive, where new textual justifications are generated. NLEs are particularly attractive because they are intuitive and human-comprehensible, making model reasoning more accessible to non-expert users. Additionally, some XAI methods are designed to produce rule-based explanations, offering clear and direct insights that are highly suitable for applications demanding significant transparency [1].

Explanations can be categorized based on their scope and generation method. Local explanations provide insights for individual predictions, detailing why a specific output was generated for a particular input. In contrast, global explanations offer a broader understanding of the model's overall predictive mechanisms [9]. Most current XAI methods tend to be local in scope. Furthermore, explanations can be post-hoc, meaning they are generated after a pre-trained black-box model makes a prediction, or they can be derived from inherently interpretable models, which are designed from scratch to be transparent and explainable by design. While post-hoc methods are widely applied to existing complex models, developing inherently interpretable models often involves significant architectural modifications and retraining, which can be computationally expensive. Some explainability methods are also model-agnostic, meaning they can be applied to any machine learning model, while others are model-specific, tailored for particular architectures [4].

Key principles guide the design and evaluation of XNLP. Faithfulness, also known as fidelity or reliability, is paramount, requiring that an explanation accurately reflects the true reasoning process behind the model's prediction. An unfaithful explanation, even if plausible, has little to do with how the model actually operates. Plausibility, or human comprehensibility, is also crucial, referring to how understandable an explanation is to a human user. There can be an inherent tension between faithfulness and plausibility, as a perfectly faithful explanation might be too complex for a human to understand, and vice versa. Stability ensures that similar inputs yield similar explanations, which is vital for trust. Finally, transparency refers to the degree to which a person can grasp the underlying mechanisms of a model [7].

The application of XNLP to textual data presents unique challenges compared to other data modalities. Text is inherently unstructured and context-dependent, unlike the more structured nature of image or tabular data. The use of complex word embeddings further adds to the opacity of models, making it difficult to visualize their inner workings when trained on textual data. Explaining decisions in terms of input features is particularly challenging in text, where syntactic and semantic features are deeply fused and interdependent. Additionally, explaining long-term dependencies within text, where relevant words may not be consecutive, and providing multilingual support, introduces further complexity in deciphering encoded language semantics and context. Despite these challenges, XNLP is crucial for enabling medical professionals and patients to understand the rationale behind AI-generated diagnoses and

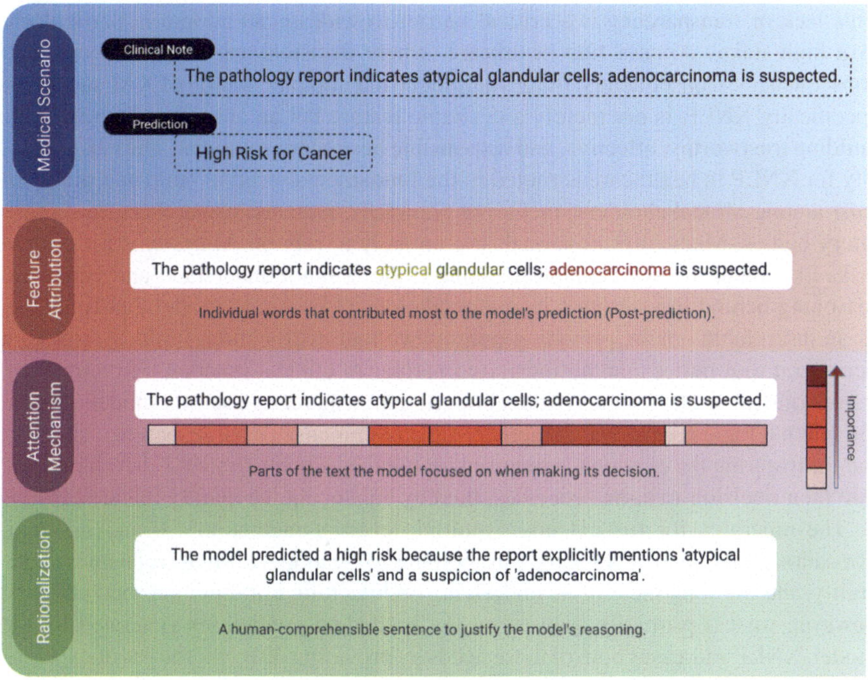

Fig. 2 Three methods for explaining XNLP predictions: feature attribution, attention mechanism, and rationalization

recommendations, fostering trust, and leading to more accurate and informed treatment decisions. It also assists developers in optimizing models for improved performance and stability [6]. Figure 2 represent different explanation formats. Word-level Feature Attribution and Attention Mechanism in addition to full Rationalization that are used to justify an NLP model's prediction of a *High Risk for Cancer* based on a medical note.

3 Imperative of XNLP in Healthcare

The integration of AI and NLP into the healthcare sector is catalyzing a profound transformation, offering unprecedented capabilities for analyzing vast quantities of unstructured data and streamlining clinical workflows. While these technologies hold immense promise for enhancing patient outcomes and operational efficiency, their increasing complexity has given rise to a significant challenge: the black-box nature of many advanced models [13, 14]. Deep learning architectures, such as transformers, can achieve remarkable predictive accuracy, but their internal decision-making processes are often opaque and uninterpretable to human users.

This lack of transparency is a critical barrier to widespread adoption, particularly in a high-stakes domain like healthcare, where decisions can carry life-or-death consequences. Consequently, the development and implementation of XAI, and more specifically XNLP, is not merely a desirable feature but an absolute imperative for building trustworthy, effective, and responsible healthcare solutions. The core necessity for XNLP in healthcare is rooted in the fundamental need to build and maintain trust among all stakeholders, including clinicians, patients, researchers, and regulatory bodies. Medical professionals are unlikely to rely on AI systems for critical tasks like diagnosis or treatment planning if they cannot understand and verify the reasoning behind the system's recommendations [15]. An AI model that functions as an inscrutable oracle, providing outputs without justification, limits its role to a peripheral tool rather than an integrated partner in clinical practice. Explainability transforms these black-box systems into more transparent glass-box models, where the internal logic is rendered comprehensible to humans. This transparency is essential for fostering the confidence required for clinicians to integrate AI-driven insights into their decision-making processes, thereby enhancing the quality of care [14].

The imperative for trust extends to multiple levels of interaction with an AI system. For data scientists and developers, trust pertains to the model itself, its robustness, reliability, and the soundness of its underlying architecture. For clinicians and patients, however, trust is primarily focused on the individual predictions generated by the model. XNLP addresses both of these needs by providing different scopes of explanation [16]. By offering justifications for individual predictions, XNLP methods allow a healthcare professional to scrutinize the evidence the model used, compare it with their own domain knowledge, and ultimately make a more informed and defensible clinical judgment. This ability to probe and understand individual outputs is what makes AI a truly supportive tool in patient care.

A central tenet of modern healthcare is the synergy between human expertise and technological support, a concept often referred to as AI-assisted decision-making. The most effective clinical outcomes are typically achieved not by replacing human clinicians with algorithms, but by combining the strengths of both. Humans bring broad domain experience, creative thinking, and an intuitive understanding of complex, nuanced patient contexts, whereas AI algorithms excel at specific, data-intensive tasks with high accuracy. XNLP is the critical bridge that facilitates this collaboration. When an AI system provides an explanation for its output, it allows the clinician to integrate that information with their own expertise. For instance, in the context of a Clinical Decision Support System (CDSS), an explainable model not only suggests a potential diagnosis but also highlights the key features from the patient's record, such as specific symptoms, lab results, or phrases in clinical notes that led to its conclusion [17]. This empowers the clinician to validate the AI's reasoning, identify potential shortcomings in the system, and maintain ultimate control and responsibility over the patient's care. Without this explanatory layer, the risk is that opaque algorithms could inadvertently lead to a situation where patients become passive spectators in a decision-making process they cannot understand [15].

The sheer volume and complexity of healthcare data further underscore the necessity of XNLP. It is estimated that up to 80% of all healthcare data is unstructured, existing in formats such as free-text clinical notes, EHRs, discharge summaries, and medical literature. This data is a rich reservoir of valuable clinical information, but its unstructured nature makes it difficult for traditional computational methods to process and analyze. NLP technologies are essential for unlocking the potential of this data by extracting, structuring, and interpreting it on a massive scale [14, 18]. However, as NLP models have evolved from simpler rule-based systems to highly complex deep learning architectures to handle the nuances of medical language, the need for explainability has grown in parallel. When a deep learning model analyzes a clinical note to predict a diagnosis or recommend a treatment, XNLP techniques are required to reveal how the model is interpreting the text, which words or phrases it deems most significant, and what patterns it has identified. This is crucial not only for validating the model's predictions but also for ensuring that the subtle, context-dependent meanings inherent in clinical language are being correctly interpreted [3].

The imperative for XNLP is also driven by its direct contributions to improving and refining the AI models themselves. By making the internal workings of a model transparent, developers and researchers can more easily debug the system, identify areas for improvement, and discover potential biases learned from the training data. For example, an explanation might reveal that a model is relying on spurious correlations or that it was trained on a dataset with limited diversity, which could affect its performance on different patient populations. This insight allows for targeted interventions to refine the model and enhance its fairness and robustness. Furthermore, explainability is a prerequisite for advancing scientific knowledge [19]. When an XNLP system reveals the connections it has learned between different medical concepts, such as a drug and a previously unknown adverse reaction it can generate new, testable hypotheses for biomedical researchers. This transforms the AI from a simple prediction tool into an engine for discovery, accelerating pharmaceutical research and expanding our understanding of diseases. Knowledge graphs, for instance, structure information in a way that allows AI to perform reasoning and expose novel links between entities like genes, diseases, and treatments, with the paths through the graph serving as a form of explanation.

Specific applications within healthcare highlight the indispensable role of XNLP in achieving tangible improvements in both clinical practice and research. In automating clinical documentation and medical coding, NLP systems can parse unstructured physician notes to extract key information and assign standardized codes, such as ICD-10, for billing and record-keeping. XNLP is imperative in this context to ensure accuracy and provide an auditable trail for why a particular code was assigned. For example, Relation Extraction (RE) models can identify the semantic link between a procedure like a CT Scan and a body part like the chest mentioned in a clinical note. This enriched entity, CT Scan Chest, can then be mapped to a more specific and accurate medical code [18]. The extracted relation itself serves as the explanation, making the automated coding process both efficient and transparent. Similarly, in the development of CDSS, explainability allows developers to identify shortcomings in a system and gives clinicians the confidence to use AI-powered recommendations for tasks like sepsis detection or suggesting personalized cancer treatments.

The field of mental health presents a particularly compelling case for the imperative of XNLP. Unlike many other medical conditions that can be diagnosed with objective laboratory tests, mental health disorders are often identified through subjective self-reports and the interpretation of language and behavior. The relationships between symptoms and underlying disorders are often probabilistic and highly nuanced. Research in Mental Illness Detection and Analysis on Social Media (MIDAS) has shown the potential of NLP to analyze patient-generated text for signs of conditions like depression, anxiety, and psychological stress. However, this field has historically prioritized predictive accuracy over model interpretability. Given the sensitive nature of mental health data and the life-altering consequences of an incorrect assessment, it is crucial that models used in this domain are explainable. An XNLP model can provide clinicians with insights into which linguistic features, such as the use of certain word categories, syntactic complexity, or expressions of sentiment are driving a prediction, allowing for a more thorough and human-centered evaluation [20].

A powerful driver for the adoption of XNLP in healthcare comes from the evolving legal and regulatory landscape. Global data privacy regulations, such as the European Union's General Data Protection Regulation (GDPR), the American Health Insurance Portability and Accountability Act (HIPAA), and China's Personal Information Protective Law (PIPL), all contain principles that mandate transparency in how personal and sensitive health information is processed [21]. These legal frameworks implicitly create a demand for explainable AI, as organizations must be able to account for how their algorithms use patient data to arrive at decisions. The forthcoming EU Artificial Intelligence Act further solidifies this imperative by proposing strict transparency and documentation obligations for AI systems classified as high-risk, a category that will undoubtedly include many healthcare applications. Compliance with these regulations is not optional, making the development of explainable systems a legal and operational necessity for healthcare organizations seeking to leverage AI [15].

The imperative for XNLP also arises from its potential to enable more effective and personalized patient care. By analyzing patient-specific data from sources like EHRs, clinical notes, and even genetic reports, NLP can help tailor treatments to an individual's unique circumstances. For instance, in pharmacogenomics, an NLP system can analyze a patient's genetic profile alongside their clinical history to recommend medications that are most likely to be effective and least likely to cause adverse reactions. Explainability is crucial in this process because it allows clinicians to understand why a particular treatment is being recommended, based on specific genetic markers or clinical factors identified by the model [14]. This justification is essential for verifying the recommendation and communicating the treatment plan to the patient. In this way, XNLP supports a shift away from one-size-fits-all medicine toward a more precise, data-driven, and patient-centric approach to healthcare. Furthermore, by analyzing patient feedback and sentiment from sources like surveys or telehealth communications, NLP can provide healthcare organizations with valuable insights into the patient experience, identifying areas for improvement and enabling more responsive and empathetic care. The ability to explain how these sentiment analyses are derived ensures that the insights are credible and actionable [3].

To meet these multifaceted demands, a variety of XNLP techniques have been developed and applied in healthcare contexts. Attention mechanisms, which are often inherent to transformer-based models, have become a dominant approach for achieving interpretability [5]. These mechanisms provide a form of local explanation by highlighting which words or tokens in an input text were given the most weight by the model when making a prediction. This allows a user to see, for example, which specific symptoms mentioned in a clinical note most influenced a diagnostic prediction. Knowledge graphs offer another powerful, inherently explainable approach by structuring data in a way that makes relationships between entities explicit, allowing the system to provide reasoning by tracing paths between concepts [17]. For more opaque, black-box models, post-hoc explanation methods like LIME and SHAP can be applied [22]. These techniques work by creating simpler, interpretable surrogate models that approximate the behavior of the complex model for a specific prediction, thereby providing a justification without needing to alter the original model [3]. The availability of both model-intrinsic (interpretable by design) and post-hoc (explainable via an external layer) methods ensures that even the most advanced and complex NLP architectures can be made more transparent, reinforcing the feasibility of meeting the imperative for explainability across a wide range of healthcare applications [19].

A comprehensive investigation into XAI reveals that knowledge graphs are also a pivotal technology for demystifying the black box of artificial intelligence in healthcare by structuring complex medical information into intuitive, interconnected networks. This research found that such graphs significantly enhance the transparency of AI models across numerous critical applications, including the detection of healthcare misinformation, the prediction of adverse drug reactions and interactions, and the complex process of drug repurposing. The core achievement of this approach lies in its ability to represent entities and their relationships in a human-understandable format, which allows AI systems to perform logical reasoning, extract key features from clinical notes, and provide clear justifications for their outputs, such as explaining an automated medical code assignment [19]. Ultimately, the findings demonstrate that by providing this semantic structure and enabling evidence-based reasoning, knowledge graphs effectively bridge the knowledge gap between opaque AI systems and medical experts, fostering greater trust and facilitating more informed clinical decision-making. In Fig. 3 XAI techniques are categorized by their explanatory scope (Local vs. Global) and timing (Self-Explaining vs. Post-Hoc).

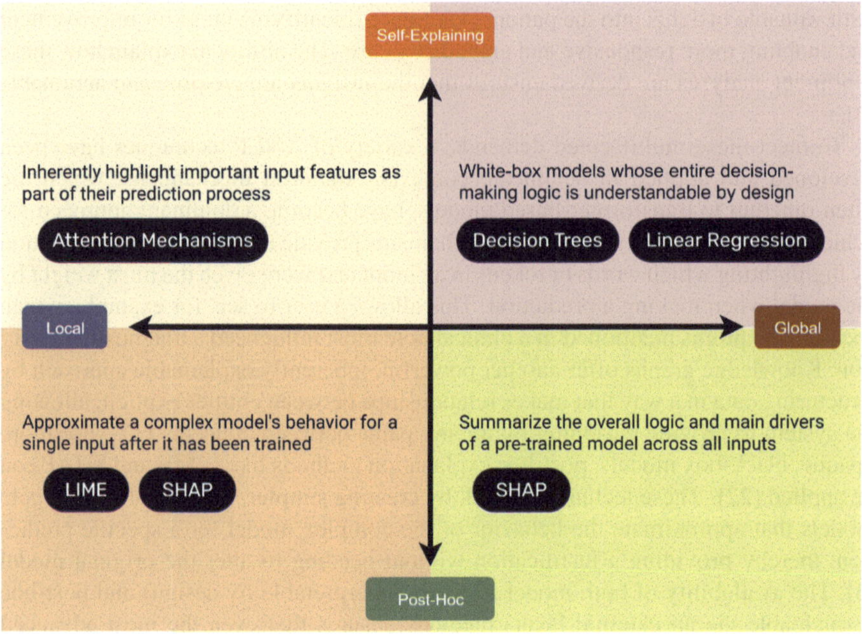

Fig. 3 XAI methods categorized by scope and timing

4 Applications and Use Cases

XNLP plays a critical role in enhancing clinical documentation and information extraction within healthcare by addressing the inherent challenges of unstructured textual data. Clinical records, such as EHRs, frequently contain vast amounts of free-text notes, making systematic extraction of crucial patient information, like smoking status, incredibly difficult and time-consuming for manual review. NLP systems are specifically designed to transform this unstructured clinical free text into structured data that machines can readily process and analyze [23]. In this section, we will take a look at several studies that focused on medical documentation and information extraction which may or may not employ XAI techniques. These studies are reviewed to better understand the role of NLP in this specific domain of healthcare.

The integration of XNLP is crucial because it provides transparency to these information extraction processes, which is paramount in a high-stakes domain like healthcare. By offering explanations for its decisions, XNLP allows clinicians to understand why the model classified a patient in a certain way, highlighting specific words or phrases that were most influential in the prediction. For example, a model might identify rygeophør (smoking cessation) as a key feature for a Former-Smoker or fortsat (continued) for an Active-Smoker. This interpretability fosters trust, enables medical professionals to validate AI-generated insights against their domain expertise, and helps in identifying and mitigating potential systematic errors or biases in the model, thereby improving overall performance and reliability [24]. Ultimately, XNLP ensures that the information extracted from clinical documentation is not only accurate but also understandable and justifiable, leading to more informed clinical decision-making and better patient outcomes.

Through a meticulous investigation into how AI systems articulate their reasoning from textual data in healthcare, a significant body of research uncovered that attention mechanisms stand as the overwhelmingly dominant approach for introducing explainability into AI systems that support clinical decision-making. This approach essentially allows the AI to highlight the specific words, phrases, or sentences within a document that most profoundly influence its predictions or classifications, mirroring how a human expert would focus on critical information. The process often involves converting textual data into embeddings, which then pass through convolutional layers, with an attention layer subsequently pinpointing the influential segments. Explanations are typically communicated through intuitive visualizations like highlighted text and heatmaps, making the model's focus clearer to users. A key achievement of this inquiry was revealing a critical gap in the field: a widespread absence of standardized, objective, and comparable evaluation metrics for these explainability components across various studies [17]. This finding underscores an urgent necessity to develop robust methods for assessing the correctness and user-friendliness of explanations, thereby fostering the creation of more trustworthy and comprehensible AI models for healthcare applications.

A comprehensive exploration [25] delves into the transformative capacity of NLP within healthcare AI, focusing on its power to unlock crucial clinical insights from the vast amounts of unstructured patient data. This work meticulously details key NLP tasks such as Named Entity Recognition (NER), which identifies and categorizes specific medical entities like diseases, symptoms, and medications, and RE, which uncovers semantic links between these identified entities. The methodology encompasses various underlying technologies, ranging from traditional machine learning models to advanced deep learning architectures like Transformers (e.g., ClinicalBERT [26]), highlighting their ability to effectively process and interpret the complex, nuanced clinical language found in physician's notes and radiology reports. A significant achievement lies in demonstrating how NLP can bridge the insight gap, enabling the conversion of raw textual data into structured, analyzable information that would otherwise remain underutilized. The findings underscore NLP's pivotal role in enhancing diagnostic accuracy, personalizing treatment strategies by extracting crucial details like patient preferences and social determinants of health, advancing medical research through efficient cohort identification, and optimizing healthcare operations via automated coding and quality reporting. Ultimately, this exploration reveals NLP's profound potential to create a more data-driven, efficient, and patient-centered healthcare system, while also emphasizing the imperative of explainable AI to address algorithmic bias and ensure trust in these powerful technologies.

Hossain et al. [27] introduced an innovative and comprehensive integrated NLP framework designed to significantly enhance healthcare decision-making by thoroughly analyzing both structured and unstructured data within EHRs. Their methodology began by leveraging spaCy for fundamental preprocessing, which included tasks such as tokenization and NER to meticulously identify crucial clinical entities within the textual data. For more profound contextual understanding of complex clinical narratives, they incorporated and fine-tuned BERT, a deep learning model featuring a transformer-based architecture, through custom NLP pipelines specifically adapted to recognize nuanced medical terminology and linguistic patterns inherent in clinical notes. Model training and refinement were meticulously conducted using PyTorch, employing a supervised learning approach, expert annotation, rigorous cross-validation, and hyperparameter tuning to ensure robust and accurate performance. The achievements of this research are evident in its qualitative outcomes. Their NLP models effectively identified and classified critical clinical entities, uncovered broad medical themes, and accurately differentiated between high- and low-risk patients, which is vital for providing personalized care. Moreover, their framework substantially reduced the time required for processing vast amounts of data compared to traditional manual methods, provided a more comprehensive understanding of intricate clinical narratives, and brought to light previously unrecorded comorbidities, potential treatment disparities across patient demographics, and emerging health concerns. These accomplishments collectively demonstrate the framework's profound capacity to streamline clinical workflows and elevate the quality of patient care through data-driven insights. Thatoi et al. [28] conducted a review to comprehensively analyze the role of NLP in extracting clinical information from EHRs specifically for cancer prognosis. Their methodology involved a rigorous process of collecting and evaluating peer-reviewed publications, adhering to PRISMA-P statements, utilizing broad search terms across multiple reputable databases, and applying strict inclusion criteria to focus solely on studies detailing NLP techniques for clinical data extraction relevant to cancer prognosis. A key achievement of this extensive review was to qualitatively illuminate the significant developments in NLP techniques such as NER, Sentiment Analysis, and Text Classification, underscoring their potential to automate the extraction of critical insights from vast, unstructured clinical data. The authors revealed how NLP can streamline data analysis, uncover hidden patterns in clinical notes to enhance prognostic models, and identify novel biomarkers that might otherwise be missed through manual review, thereby ensuring more accurate and efficient information collection in oncology. Furthermore, their work brought to light substantial challenges, including the pervasive variability in clinical language, the critical need for high-quality annotated data, and paramount privacy and security concerns inherent in handling sensitive patient information. The review's final achievement was to propose crucial future research directions, advocating for continued advancements toward more accurate, individualized cancer prognosis and emphasizing the growing imperative for greater explainability and transparency in NLP models deployed in clinical settings.

Ebrahimi et al. [23] developed an innovative and efficient NLP pipeline, a pioneering effort to classify patients' smoking status directly from unstructured Danish EHRs and provide explanations for these classifications. Their approach involved a multi-stage methodology: first, comprehensive preprocessing techniques were applied to address the unique linguistic characteristics of Danish text, including abbreviations and textual irregularities. Next, four advanced feature extraction methods such as Embedding, BERT, Word2Vec, and Count Vectorizer were employed to convert the raw text into numerical representations. For classification, they utilized a Stacking-based Ensemble (SE) model, which combines multiple base machine learning algorithms using a meta-learner, and a Convolutional Long Short-Term Memory Neural Network (CNN-LSTM), designed to capture both local features and long-term dependencies within the text. Crucially, to ensure transparency, they integrated LIME, a post-hoc Explainable AI (XAI) technique, to clarify the models' decisions by highlighting the significance of individual linguistic features. The research achieved a high level of classification performance, with the BERT-CNN-LSTM architecture demonstrating strong capabilities in both binary and multiclass distinctions of smoking status. Furthermore, the Embedding-CNN-LSTM architecture showed particularly consistent detection rates across different smoking categories. The LIME explanations successfully unveiled clinically relevant features, such as rygeophør (smoking cessation) for former smokers and fortsat (continued) for active smokers, thereby significantly enhancing the interpretability and trustworthiness of their smoking status identification system in the complex EHR environment.

An innovative framework [29], termed the Neurosymbolic Explainable AI-Augmented EHR Disease Prediction Framework (NXAI—AEHRDP), was designed to significantly enhance disease prediction by integrating neurosymbolic AI with XAI using multimodal EHRs. This pioneering approach operates in three stages: initially, it meticulously integrates and preprocesses diverse multimodal data, encompassing structured details like lab results, unstructured clinical notes, and imaging data, leveraging NLP for text and specialized techniques for images. Subsequently, the processed data feeds into a neurosymbolic AI model, where a neural network identifies complex patterns across these varied data types, while a symbolic AI component applies logical reasoning based on encoded medical knowledge and guidelines to ensure predictions are consistent and medically sound. Finally, the framework incorporates XAI techniques, such as best feature analysis and rule extraction, to provide clear, human-understandable insights into the model's decision-making process, highlighting precisely which parts of the patient's EHR contributed most significantly to a prediction. The achievements of this work include developing a model that not only predicts diseases with high accuracy but also explains its reasoning in a way comprehensible to clinicians, fostering trust and enabling earlier, more accurate diagnoses. This comprehensive, transparent approach aims to lead to personalized care, reduce misdiagnoses, and improve overall patient outcomes by accounting for the full spectrum of a patient's health data.

Another approach [30], the Ontology-based Clinical Information Extraction System (OB-CIE), was developed to revolutionize how clinical concepts are extracted from physicians' free-text notes, specifically targeting handwritten entries, and seamlessly converted into structured information within EHRs. This pioneering approach was designed to facilitate physician workflow by eliminating the need for manual data entry into predefined forms. The methodology begins with meticulous data pre-processing, employing OCR to transform handwritten notes into editable text, followed by advanced spell-checking customized with medical lexicons to ensure accuracy. Subsequently, an NLP module refines the text through stages like sentence detection, tokenization, and Part-of-Speech tagging to identify critical noun phrases that represent medical concepts. The core innovation lies in the Information Extraction module, which is built upon a custom-designed Patient Clinical Data (PCD) Ontology. This ontology, crafted from EHR analysis, expert knowledge, and integrated UMLS concepts, acts as the domain knowledge base for NER, accurately categorizing clinical noun phrases (e.g., diagnoses, medications, symptoms) and their attributes. Further, a rule-based approach classifies entire clinical sentences into predefined categories and extracts relationships between identified concepts, mapping them directly to the EHR database structure. The achievements of the OB-CIE system are significant: it demonstrated robust performance in classifying clinical concepts and sentences, qualitatively surpassing traditional data mining techniques. It effectively enhanced the usability of extracted information, solving a critical problem in current IE systems. By providing a structured representation of clinical notes directly from handwriting, the system streamlines documentation, saves physicians considerable time, and fosters better patient interaction, thereby addressing physician resistance to EHR adoption and making patient data more valuable for decision-making and knowledge discovery.

To tackle the persistent challenge of extracting vital clinical information from unstructured, scanned documents within EHRs, an end-to-end data pipeline was successfully engineered and validated [31]. The methodology uniquely integrated tailored image preprocessing, OCR, and a sophisticated neural network architecture that combined both text sequence data and structured features representing the document's layout, such as word positions and page numbers. The results demonstrated that advanced deep learning models, especially ClinicalBERT, substantially outperformed traditional machine learning approaches in accurately identifying key clinical values. A significant achievement was the discovery that incorporating the document's structural and layout information as inputs to the model significantly boosted its predictive performance, a novel contribution to the field. Furthermore, the study achieved a critical milestone by showing that these high-performing models could attain impressive accuracy with a remarkably small amount of training data, underscoring the pipeline's efficiency and practical applicability in real-world clinical settings where extensive data annotation is often infeasible.

In an effort to improve the prediction of hospital readmission for patients following a heart attack, a study explored whether social risk factors extracted from clinical notes using NLP could enhance machine learning models [32]. The methodology involved using an NLP tool, Moonstone, to identify seven distinct social risks within patient records and then incorporating these variables into various prediction models alongside standard structured clinical data. The primary and most significant achievement of this research was the clear demonstration of a null result: contrary to initial hypotheses, the inclusion of these NLP-derived social risk factors did not provide any statistically significant improvement in predicting 30-day readmission. While the social factors showed some association with readmission on their own, their predictive value was attenuated once combined with the comprehensive clinical data. The investigation concluded that this lack of improvement was likely due to information overlap, suggesting the structured clinical variables were already acting as surrogates for the social factors, a finding supported by multicollinearity analysis. This outcome is a crucial contribution, highlighting that simply applying NLP to extract social data from highly variable clinician documentation may be insufficient and that alternative, more standardized methods are needed to effectively capture these vital determinants of health.

To combat the overwhelming burden of clinical documentation in modern healthcare, Yalçın and Korkmaz engineered a transformative system for automating EHR management [33]. The cornerstone of their work is a novel hybrid architecture that synergistically combines the power of transformer-based attention mechanisms with domain-specific knowledge embeddings. This approach was designed to intelligently process unstructured clinical narratives, such as physician's notes and convert them into structured, actionable data. Their methodology was exceptionally comprehensive, encompassing the development of a sophisticated mathematical framework to optimize the balance between computational performance and clinical utility, as well as a detailed, layered system architecture for practical deployment in complex hospital IT environments. The achievements of this framework, validated through implementation case studies across multiple hospital systems, were profound. The system delivered significant improvements in documentation accuracy and overall quality, drastically accelerated data processing speeds, and provided substantial daily time savings for clinicians, thereby helping to reduce burnout and unlock the potential of healthcare data for improved decision support (Table 1).

5 Ethical Considerations

The integration of XNLP into healthcare represents a significant technological leap, offering the potential to revolutionize clinical documentation, decision support, and patient engagement. However, this promise is accompanied by a complex and critical landscape of ethical considerations that must be carefully navigated to ensure these powerful tools are deployed responsibly, equitably, and safely. The high-stakes nature of medicine, where decisions can have life-altering consequences, demands a level of scrutiny for NLP systems that goes far beyond mere technical performance [34]. The core ethical challenge stems from the inherent opacity of many advanced models, often referred to as black boxes, which conflicts with the fundamental requirements of trust, accountability, and transparency in clinical practice. Therefore, a proactive and integrated approach to ethics is not an optional add-on but a foundational requirement for the successful and beneficial implementation of XNLP in healthcare. These considerations span the entire machine learning development lifecycle, from initial study design and data collection to algorithm development, deployment, and

Table 1 Summary of the selected applications and use cases of NLP in Healthcare

Application	Key methods	Findings
Smoking status classification	An NLP pipeline using BERT, CNN-LSTM, and LIME was developed to classify smoking status from unstructured Danish EHRs	The system achieved high classification accuracy, and the use of LIME for explainability successfully identified clinically relevant terms, enhancing the model's trustworthiness
Disease prediction from multimodal EHRs	The NXAI—AEHRDP framework was designed, integrating neurosymbolic AI (neural networks and symbolic reasoning) with XAI techniques	The framework produces highly accurate disease predictions while also providing clear, human-understandable explanations for its reasoning, fostering clinical trust
Information extraction from handwritten notes	The OB-CIE system used OCR, an NLP module, and a custom medical ontology to convert handwritten notes into structured EHR data	This approach successfully automated clinical documentation, saving physicians significant time and making patient data more valuable for decision-making
Information extraction from scanned documents	An end-to-end data pipeline was engineered using ClinicalBERT that combined text data with structured features representing the document's layout	Incorporating the document's structural information was found to significantly boost predictive performance, and the model achieved high accuracy with a small amount of training data
Improving clinical documentation	A hybrid architecture was engineered that combines transformer-based attention mechanisms with domain-specific knowledge embeddings to automate EHR management	The system was validated in multiple hospitals and shown to significantly improve documentation accuracy, accelerate data processing, and save clinicians time, helping to reduce burnout
Predicting hospital readmission	An NLP tool was used to extract social risk factors from clinical notes to see if they could improve machine learning models for predicting 30-day readmission after a heart attack	A significant null result was found: adding the NLP-derived social factors did not improve prediction, likely because the information was already captured by other clinical variables

ongoing critique [35]. Table 2 illustrates the content of this section in a structured and digestible format.

A primary and paramount ethical concern is the privacy and security of sensitive patient data, which serves as the lifeblood of healthcare NLP systems. These models necessitate access to vast quantities of personal health information, including

Table 2 Ethical considerations for utilizing XNLP in healthcare

Ethical consideration	Description
Privacy and security of patient data	Protecting confidential patient data through regulatory compliance (e.g., HIPAA, GDPR) and robust technical safeguards
Informed consent and patient autonomy	Requiring clear, transparent patient consent for data use, respecting the right to be informed and to withdraw participation
Algorithmic bias and fairness	Mitigating the risk of models amplifying societal biases from training data, which can cause health disparities
Transparency and explainability	Making "black box" models understandable so clinicians can trust, validate, and take responsibility for AI-assisted decisions
Medical liability and accountability	Defining legal responsibility for AI-driven errors and ensuring ultimate accountability remains with human professionals
Patient safety and human oversight	Mandating essential human oversight to safeguard against AI errors (e.g., "hallucinations") and ensure patient safety

unstructured clinical notes, lab results, and patient communications, all of which are highly confidential. Consequently, robust measures must be implemented to protect this information from unauthorized access, misuse, or data breaches. Adherence to stringent data protection regulations, such as HIPAA in the United States and GDPR in Europe, is an absolute necessity [36]. Safeguarding data involves a multi-layered approach, including the use of strong encryption for data both at rest and in transit, the implementation of rigorous access control mechanisms like role-based access control (RBAC) and multi-factor authentication, and conducting regular security audits to identify and mitigate vulnerabilities. Furthermore, principles of data minimization, which is collecting only the data essential for the intended purpose and data anonymization are crucial for reducing privacy risks. Even with these measures, the risk of data leakage or re-identification persists, particularly as LLMs may inadvertently memorize and reproduce portions of their training data. This necessitates a culture of security within healthcare organizations, reinforced by ongoing staff training on the proper handling of sensitive patient information [37].

Closely linked to data privacy is the principle of informed consent and patient autonomy. Often, patient data is used to train and fine-tune NLP models without the explicit and specific consent of the individuals involved, raising significant ethical questions. A truly ethical framework requires that patients are comprehensively informed about how their data will be used in NLP systems, including the potential benefits, risks, and implications for their care. This communication must be clear and transparent, allowing patients to ask questions and make genuinely informed decisions about granting or withholding consent. Patients have the right to be aware of the involvement of AI in their healthcare and should, where feasible, have the option to choose alternatives or withdraw their consent at any time without compromising the quality of their care. The ethical considerations of autonomy, however, extend beyond individual consent, especially when data is used to build models that affect entire communities or subgroups. Traditional bioethical frameworks, which often focus on the individual, may be insufficient to address potential group-level harms. This calls for a reframing of autonomy to include community engagement and concepts like informed refusal, which empower historically marginalized groups to guide, modify, or refuse research involving their data, thereby ensuring that technology does not perpetuate structural injustices [35].

Perhaps one of the most pervasive and discussed ethical challenges in NLP is the issue of algorithmic bias and fairness. NLP models are not created in a vacuum; they learn from the data they are trained on, and if this data reflects existing societal inequities and biases, the models will inevitably inherit, perpetuate, and even amplify them. In healthcare, this can lead to discriminatory or unfair outcomes that exacerbate health disparities for underrepresented racial, ethnic, gender, socioeconomic, or linguistic groups. Biases can originate from multiple sources, including historical inequities in healthcare access, subjective and biased language in clinical documentation practices, and the underrepresentation of certain populations in training datasets [35]. For example, a model trained primarily on data from one demographic group may perform poorly for others, leading to inequitable quality of care. Mitigating bias is a complex, ongoing process that must be integrated throughout the model development lifecycle. Key strategies include the intentional curation of diverse and representative training data, regular auditing and testing of models for biased behavior across different demographic groups, and the development of algorithmic debiasing techniques. It is also crucial to involve diverse stakeholders, including ethicists, sociologists, and representatives from marginalized communities, in the design and evaluation process to achieve a more comprehensive understanding of potential biases. The very act of measuring fairness is also a challenge, as there is no single, universally accepted metric, and different fairness criteria can be mutually exclusive, requiring careful consideration of the specific context and potential harms [35].

The principles of transparency and explainability are at the very heart of ethical NLP in healthcare, as they are prerequisites for building trust and ensuring accountability. The black box nature of many sophisticated NLP models, such as deep neural networks and LLMs, makes it difficult to understand the reasoning behind their predictions or recommendations [38]. This opacity is fundamentally at odds with the needs of clinical practice, where healthcare providers must be able to critically appraise, validate, and take responsibility for their decisions. Clinicians and patients need to understand how an NLP system arrives at a conclusion to assess its reliability, identify potential errors or biases, and ultimately, trust its output. Explainability is not merely a technical feature but a key enabler of other core bioethical principles, including beneficence (promoting well-being), non-maleficence (avoiding harm), and justice. Efforts to enhance transparency include developing XAI methods, visualizing attention mechanisms in transformer models, and generating human-readable rationales for model outputs. However, a significant challenge remains in balancing interpretability with predictive accuracy, as the most powerful models are often the least transparent. Furthermore, explanations themselves can be misleading if they do not faithfully reflect the model's true causal reasoning, underscoring the need for rigorous evaluation of the explainability methods themselves.

The introduction of NLP applications into clinical decision-making raises complex questions of medical liability and accountability. When an NLP system provides an erroneous or harmful recommendation that contributes to an adverse patient outcome, determining who is responsible becomes a critical and unresolved issue. Liability could potentially rest with the software developers who created the model, the healthcare institution that deployed it, the clinician who acted on its recommendation, or some combination of all three. This ambiguity highlights the urgent need for robust legal and regulatory frameworks that clearly define the roles, responsibilities, and accountability of all stakeholders involved. Without such clarity, the fear of legal uncertainty could impede the adoption of beneficial technologies. Establishing these frameworks requires a collaborative effort among developers, healthcare providers, regulatory bodies, and legal experts to create standards for the responsible development and use of these tools. A key principle emerging from this discussion is that accountability must ultimately remain with human professionals, reinforcing the need for continuous and meaningful human oversight [37].

Ultimately, all these ethical considerations converge on the imperative of patient safety and the necessity of human oversight. The potential for NLP models to produce factually incorrect information, a phenomenon known as hallucination in LLMs, poses a direct and serious threat to patient safety if relied upon for clinical decisions. Similarly, omissions of critical details or biased recommendations can lead to delayed care, incorrect diagnoses, or inappropriate treatments. Given these risks, human oversight is not just a recommendation but an essential safeguard. NLP outputs should be utilized as decision-making aids or supplements to augment the capabilities of healthcare professionals, rather than as autonomous substitutes for human expertise and judgment. Clinicians must retain the final say and ultimate responsibility for patient care, using their contextual understanding, empathy, and ethical reasoning to interpret and contextualize the outputs generated by NLP systems. The integration of NLP also has the potential to impact the clinician-patient relationship; while it could depersonalize care if used improperly, it could also free clinicians from administrative burdens, allowing for more meaningful and empathetic interaction with patients [34]. Therefore, thoughtful integration into clinical workflows is essential to ensure that technology enhances, rather than detracts from, the therapeutic alliance and the delivery of responsible, accountable, and patient-centered care. Addressing these multifaceted ethical challenges requires a continuous, collaborative, and interdisciplinary effort to ensure that the deployment of XNLP in healthcare genuinely promotes well-being, upholds justice, and respects human dignity.

6 Conclusion

The journey through the landscape of XNLP in healthcare culminates in a clear and compelling realization: explainability is not a peripheral feature or a technical luxury, but the foundational bedrock upon which the future of trustworthy and effective medical AI must be built. We began by establishing the central paradox of modern AI: the very complexity that grants sophisticated models their unprecedented power to unlock insights from vast reserves of unstructured clinical data also shrouds their reasoning in an opaque closed black box. In a high-stakes domain like healthcare, where every decision can have profound human consequences, this opacity represents a formidable barrier to adoption, clashing with the fundamental principles of trust, verification, and accountability that govern clinical practice. XNLP has emerged as the essential discipline dedicated to dismantling this barrier, providing the methods and frameworks necessary to render these powerful systems transparent, understandable, and ultimately reliable to the clinicians, patients, and researchers who depend on them.

Throughout this chapter, we have systematically navigated the core tenets, techniques, and applications of XNLP. We have clarified key terminology and emphasized the critical distinction between a merely plausible explanation and one that is truly faithful to a model's internal logic, a crucial requirement for preventing the over-trust of systems that may rely on spurious correlations. A foundational taxonomy was presented, categorizing an array of explanation methods to provide a structured framework for understanding the field. We explored a spectrum of specific techniques, from feature attribution methods that assign importance to input tokens, to the increasingly powerful and intuitive approach of rationalization, where models generate human-readable, natural language justifications for their outputs, demonstrating their value across a multitude of healthcare applications.

The imperative for this technological shift is driven by a convergence of practical, ethical, and regulatory forces. Clinicians require explanations to validate AI-generated insights, facilitate effective human-AI collaboration, and maintain ultimate responsibility for patient care. Developers need transparency to debug models, identify and mitigate harmful biases, and improve system robustness. Moreover, the evolving legal landscape increasingly mandates a right to explanation, making transparency a matter of compliance. This imperative is deeply intertwined with a complex web of ethical considerations, from safeguarding patient privacy and ensuring fairness to establishing clear lines of accountability. A proactive and integrated approach to ethics, coupled with a steadfast commitment to human oversight, is indispensable for the responsible deployment of these technologies. However, the path forward is not without significant challenges. This chapter has highlighted the critical need for more robust and standardized methods for evaluating the quality of explanations, a task that requires a sophisticated blend of quantitative metrics and HCE to assess real-world utility and clarity. The principles of transparency and explainability are at the very heart of ethical NLP, as they are prerequisites for building trust and ensuring accountability. Yet, a significant challenge remains in balancing interpretability with predictive accuracy, as the most powerful models are often the least transparent. Furthermore, the explanations themselves can be misleading if they do not faithfully reflect the model's true causal reasoning, underscoring the need for rigorous evaluation of the explainability methods themselves.

Looking ahead, the future of XNLP in healthcare will be shaped by progress on several key fronts that include the development of novel techniques that can provide more causally sound explanations, the advancement of abstractive rationalization to enable models to articulate their reasoning more fluently, and the creation of comprehensive benchmarks to rigorously assess and compare different methods. The ultimate vision is a healthcare ecosystem where AI is not an inscrutable oracle but a transparent and collaborative partner, augmenting the expertise of medical professionals and empowering patients with understandable insights. By continuing to champion and advance the principles of explainability, we can ensure that the remarkable potential of NLP is harnessed not only to innovate, but to do so responsibly, equitably, and in the unwavering service of human health and well-being.

References

1. Luo, S., et al., *Local interpretations for explainable natural language processing: A survey.* ACM Computing Surveys, 2024. **56**(9): p. 1–36.
2. Gurrapu, S., et al., *Rationalization for explainable NLP: a survey.* Frontiers in artificial intelligence, 2023. **6**: p. 1225093.
3. Mohammadi, H., et al., *Explainability in practice: A survey of explainable nlp across various domains.* arXiv preprint arXiv:2502.00837, 2025.
4. Zini, J.E. and M. Awad, *On the explainability of natural language processing deep models.* ACM Computing Surveys, 2022. **55**(5): p. 1–31.
5. Huang, G., et al., *From explainable to interpretable deep learning for natural language processing in healthcare: How far from reality?* Computational and structural biotechnology journal, 2024. **24**: p. 362–373.
6. Xu, Q., et al., *Applications of explainable AI in natural language processing.* Global Academic Frontiers, 2024. **2**(3): p. 51–64.
7. Lyu, Q., M. Apidianaki, and C. Callison-Burch, *Towards faithful model explanation in nlp: A survey.* Computational Linguistics, 2024. **50**(2): p. 657–723.
8. Danilevsky, M., et al., *A survey of the state of explainable AI for natural language processing.* arXiv preprint arXiv:2010.00711, 2020.
9. Balkir, E., et al., *Challenges in applying explainability methods to improve the fairness of NLP models.* arXiv preprint arXiv:2206.03945, 2022.
10. Farhadi, A. and A. Taheri, *Application of GenAI in Synthetic Data Generation in the Healthcare System*, in *Application of Generative AI in Healthcare Systems.* 2025, Springer. p. 67–89.
11. Dhaini, M., et al., *Evalxnlp: A framework for benchmarking post-hoc explainability methods on nlp models.* arXiv preprint arXiv:2505.01238, 2025.
12. Mosca, E., et al. *SHAP-based explanation methods: a review for NLP interpretability.* in *Proceedings of the 29th international conference on computational linguistics.* 2022.
13. Moradi, M. and M. Samwald, *Deep learning, natural language processing, and explainable artificial intelligence in the biomedical domain.* arXiv preprint arXiv:2202.12678, 2022.
14. Upadhyaya, N., H. Joshi, and C. Agrawal, *Examining NLP for Smarter, Data-Driven Healthcare Solutions*, in *Intelligent Systems and IoT Applications in Clinical Health.* 2025, IGI Global. p. 393–420.

15. Hulsen, T., *Explainable artificial intelligence (XAI): concepts and challenges in healthcare.* Ai, 2023. **4**(3): p. 652–666.
16. Saraswat, D., et al., *Explainable AI for healthcare 5.0: opportunities and challenges.* IEEe Access, 2022. **10**: p. 84486–84517.
17. Amjad, H., et al., *Attention-Based Explainability Approaches in Healthcare Natural Language Processing.* HEALTHINF, 2023: p. 689–696.
18. Haq, H.U., V. Kocaman, and D. Talby, *Deeper clinical document understanding using relation extraction.* arXiv preprint arXiv:2112.13259, 2021.
19. Rajabi, E. and S. Kafaie, *Knowledge graphs and explainable AI in healthcare.* Information, 2022. **13**(10): p. 459.
20. Kerz, E., et al., *Toward explainable AI (XAI) for mental health detection based on language behavior.* Frontiers in psychiatry, 2023. **14**: p. 1219479.
21. Taheri, A., A. Farhadi, and A. Zamanifar, *Application of GenAI in Clinical Administration Support*, in *Application of Generative AI in Healthcare Systems.* 2025, Springer. p. 91–117.
22. Dhaini, M., et al., *When Explainability Meets Privacy: An Investigation at the Intersection of Post-hoc Explainability and Differential Privacy in the Context of Natural Language Processing.* arXiv preprint arXiv:2508.10482, 2025.
23. Ebrahimi, A., et al., *Identification of patients' smoking status using an explainable AI approach: a Danish electronic health records case study.* BMC Medical Research Methodology, 2024. **24**(1): p. 114.
24. Bennetot, A., et al., *A practical guide on explainable ai techniques applied on biomedical use case applications.* arXiv preprint arXiv:2111.14260, 2021.
25. Derek, V. and P. Collings, *Natural Language Processing (NLP) in Healthcare AI: Enhancing Clinical Insight Extraction from Unstructured Patient Data.* 2025.
26. Huang, K., J. Altosaar, and R. Ranganath, *Clinicalbert: Modeling clinical notes and predicting hospital readmission.* arXiv preprint arXiv:1904.05342, 2019.
27. Hossain, M.R., et al., *Natural Language Processing (NLP) in Analyzing Electronic Health Records for Better Decision Making.* Journal of Computer Science and Technology Studies, 2024. **6**(5): p. 216–228.
28. Thatoi, P., et al., *Natural language processing (NLP) in the extraction of clinical information from EHRs for cancer prognosis.* International Journal, 2023. **10**(4): p. 2676–2694.
29. Vidya, R., *LEVERAGING NEUROSYMBOLIC AI AND EXPLAINABLE AI FOR ENHANCED DISEASE PREDICTION IN ELECTRONIC HEALTH RECORDS USING MULTIMODAL DATASETS.* AI AND ITS APPLICATIONS. **145**.
30. Yehia, E., et al., *Ontology-based clinical information extraction from physician's free-text notes.* Journal of biomedical informatics, 2019. **98**: p. 103276.
31. Hsu, E., et al., *Deep learning-based NLP data pipeline for EHR-scanned document information extraction.* JAMIA open, 2022. **5**(2): p. ooac045.
32. Brown, J.R., et al., *Information extraction from electronic health records to predict readmission following acute myocardial infarction: does natural language processing using clinical notes improve prediction of readmission?* Journal of the American Heart Association, 2022. **11**(7): p. e024198.
33. Yalçın, S. and M. Korkmaz, *The Application of Natural Language Processing Technologies to Automate Documentation and Improve the Efficiency of Electronic Health Record Systems in Hospital Administration.* Applied Science, Engineering, and Technology Review: Innovations, Applications, and Directions, 2024. **14**(12): p. 1–23.
34. Jung, K.-H., *Large language models in medicine: Clinical applications, technical challenges, and ethical considerations.* Healthcare Informatics Research, 2025. **31**(2): p. 114–124.
35. Bear Don't Walk IV, O.J., et al., *A scoping review of ethics considerations in clinical natural language processing.* JAMIA open, 2022. **5**(2): p. ooac039.
36. Alam, M.N., M. Kaur, and M.S. Kabir, *Explainable AI in healthcare: enhancing transparency and trust upon legal and ethical consideration.* Int Res J Eng Technol, 2023. **10**(6): p. 1–9.
37. Khattak, W.A. and F. Rabbi, *Ethical considerations and challenges in the deployment of natural language processing systems in healthcare.* International Journal of Applied Health Care Analytics, 2023. **8**(5): p. 17–36.

38. Dhanda, S.S., et al., *Advancement in public health through machine learning: a narrative review of opportunities and ethical considerations.* Journal of Big Data, 2025. **12**(1): p. 1–58.

Case Studies of Explainable LLMs in Diagnosis, Treatment Planning, and Patient Interaction

Azadeh Tabatabaei, Mohammad Gorji, and Abolfazl Rahimi

1 Introduction

The integration of large language models (LLMs) into medicine heralds a transformative era across the clinical spectrum, from diagnosis and treatment planning to direct patient interaction. However, the inherent "black-box" nature of these models presents a fundamental challenge to their clinical adoption, creating a critical tension between their immense potential and the non-negotiable requirements for transparency, safety, and trust. Clinicians are rightly cautious of AI systems that generate recommendations without a transparent, verifiable rationale, especially in high-stakes scenarios where patient safety is paramount [1–3]. An LLM that is highly accurate but inexplicable is ultimately less useful in practice than a slightly less precise model that offers a coherent, evidence-based justification that a clinician can employ to make a final, informed decision. This represents a crucial evolution of the field: a transition beyond a sole emphasis on predictive accuracy towards a more comprehensive goal of credibility, which includes coherence, verifiability, and alignment with established medical knowledge.

Explainable LLMs (XLLMs) aim to bridge this gap by not only achieving high performance but also providing clear justifications for their outputs. The form and function of this explainability are customized to the specific clinical domain. In

A. Tabatabaei (✉)
Departmet of Computer Engineering, University of Science and Culture, Tehran, Iran
e-mail: a.tabatabaei@usc.ac.ir

M. Gorji
School of Electrical and Computer Engineering, College of Engineering, University of Tehran, Tehran, Iran

A. Rahimi
Ophthalmology Department, TeMS.C., Islamic Azad University, Tehran, Iran

A. Zamanifar et al. (eds.), *Explainable Large Language Models in Healthcare Applications*, Bio-IT and AI, https://doi.org/10.1007/978-3-032-15088-2_8

diagnosis, explainability techniques—such as chain-of-thought reasoning, uncertainty quantification, and evidence retrieval from knowledge graphs or literature—are designed to emulate a clinician's differential diagnostic process. This enables a physician to follow the model's logic and validate its conclusions against their own expertise and available data [4–6].

The stakes are substantially elevated in treatment planning, where decisions have direct, immediate, and frequently irreversible consequences. In this context, transparency is not merely an auxiliary feature but a fundamental prerequisite for safety. XLLMs address this necessity through the utilization of techniques such as retrieval-augmented generation (RAG) derived from clinical guidelines and trial data, structured reasoning that emulates a physician's workflow, and multimodal analysis. These methodologies serve as cognitive safeguards, limiting the model's generative scope and compelling it to adhere to logical, verifiable pathways, thereby mitigating the risk of hazardous hallucinations [7–9].

Furthermore, in patient interactions, large language model (LLM)-powered conversational agents are employed for symptom monitoring, chronic disease management, and health education. Their effectiveness depends not only on accuracy but also on their capacity to establish trust with patients. Explainability in this context refers to providing clear, personalized explanations that correspond with a patient's level of health literacy, thereby enabling understanding and adherence to recommendations, while also offering the assurance that the patient's clinician can verify these explanations [10, 11].

A compelling insight emerging from recent research is that explainability is not merely a mechanism for fostering trust; it acts as a catalyst for improved performance. Techniques that enforce structured, evidence-based reasoning—such as iterative hypothesis verification and adherence to knowledge pathways—serve as a form of cognitive scaffolding. This compels the Large Language Model (LLM) to engage in more rigorous and constrained reasoning, thereby reducing errors, minimizing dependence on spurious correlations, and ultimately enhancing the model's overall effectiveness and reliability [12, 13], see Fig. 1.

This chapter presents a series of advanced case studies that examine the application of explainable large language models (LLMs) across three vital domains: diagnosis, treatment planning, and patient engagement. The cases encompass a wide variety of medical specialties and methodologies, including interactive diagnostic assistants, models that incorporate uncertainty, hybrid systems integrating external knowledge sources, virtual participants in multidisciplinary tumor boards, and empathetic chatbots designed for patient interaction. Each case study is meticulously analyzed, detailing the clinical context, the specific explainability technique utilized, a concrete application example, evaluation results, and a critical discussion of strengths and limitations. A comparative analysis is subsequently conducted to synthesize overarching themes and methodological differences. The chapter concludes with a synthesis of best practices for development and evaluation, outlining future research directions that maintain a focus on scientific rigor, real-world datasets, and ethical, user-centric considerations. Through these exemplars, we delineate a pathway toward LLM-based

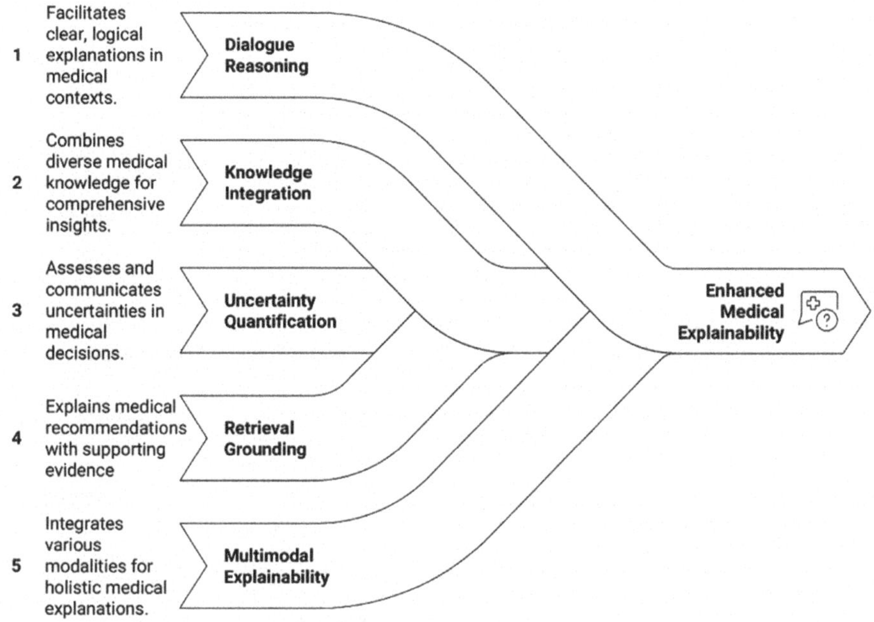

Fig. 1 Unified approaches to medical explainability

tools that are not only accurate but also interpretable, accountable, and collaborative partners in clinical practice.

2 Explainable Large Language Models in Diagnosis

The case studies presented in this section examine recent efforts to operationalize explainability in large language models (LLMs) applied to clinical diagnosis. While differing in scope and methodology, these approaches share the objective of making model reasoning more transparent, verifiable, and clinically actionable. Some frameworks emphasize dialogue as a means of externalizing reasoning processes, others implement structured inference pipelines to systematically evaluate diagnostic hypotheses, and still others integrate external medical knowledge or explicitly represent diagnostic uncertainty. Taken together, these case studies illustrate the range of strategies being explored to reconcile predictive performance with interpretability, offering complementary perspectives on how explainable LLMs can support clinical decision-making. For a comprehensive overview of the surveyed systems, see Table 1.

Table 1 Overview of explainable LLM case studies in diagnosis (compact)

System	Primary clinical task	Input modality	Core explainability technique	Knowledge source
AMIE [1, 10]	Differential diagnosis	Conversational text	Dialogue as in-context CoT	Internal LLM + UpToDate
Dual-Inf [13]	Differential diagnosis	Clinical notes	Forward/ backward inference with self-check	Internal LLM
ConfiDx [5]	Diagnosis and uncertainty	Clinical notes	Criterion-based reasoning; flags unmet criteria	Internal LLM (guideline-tuned)
DR.KNOWS [4]	Diagnosis from EHR	EHR notes (Text)	Retrieve KG paths to guide reasoning	External KG (UMLS)
RetCare [14]	Outcome/risk prediction	EHR data	RAG over medical literature	External (PubMed)
DrHouse [16]	Differential diagnosis	Text + sensor data	Multi-turn dialog with live DDx probabilities	Internal LLM + UpToDate/ PubMed

2.1 Articulate Medical Intelligence Explorer (AMIE): The Dialogue-Driven Diagnostician

Articulate Medical Intelligence Explorer (AMIE) is a large language model developed by researchers at Google to assist clinicians with differential diagnosis and diagnostic reasoning [1, 10]. The motivation for AMIE stems from the recognition that forming a comprehensive differential diagnosis is an iterative, dialogue-driven process in clinical care. Prior AI systems for diagnosis showed promise in specific tasks, such as classifying images or suggesting likely diseases, but lacked interactive capabilities and often failed to explain their reasoning. AMIE was designed to fill this gap by engaging in a conversational diagnostic dialogue with physicians or even patients, mimicking the process of history-taking and clinical reasoning in plain language. This system was trained and optimized specifically for medical dialogue across a wide range of conditions, using a novel self-play simulation environment to generate realistic patient-doctor conversations for training data. By doing so, AMIE aims to provide not just a single answer, but a thought process—asking follow-up questions, summarizing relevant findings, and generating a differential list of possible diagnoses along with reasoning. AMIE's innovation lies in reframing the diagnostic process from a human-queries-AI "oracle" to a collaborative dialogue, where the diagnostic hypothesis is built jointly, enhancing user trust and agency.

Explainability Technique AMIE's explainability emerges organically from its interactive and iterative approach to diagnosis. Instead of producing a one-shot answer,

AMIE conducts a multi-turn exchange that naturally exposes its line of reasoning. For example, the model may ask the patient or clinician clarifying questions about symptom details or risk factors, thereby revealing which information it deems important and why. This process is akin to a physician explaining their thought process by gathering data; it inherently provides a form of chain-of-thought explanation. Additionally, AMIE was trained with automated feedback mechanisms that encourage it to adhere to clinically meaningful dialogue behaviors, including summarizing findings and considering multiple hypotheses, which improves transparency. In effect, the explainability technique is an in-context explanation where the reasoning is woven into the dialogue. AMIE does not rely on a separate post-hoc explanation module; rather, the dialogue itself serves as the explanation of the model's thought process. Another aspect of AMIE's design is the incorporation of up-to-date medical knowledge. The system uses a continuously updated knowledge base, including medical literature and resources like UpToDate, to ground its responses. By citing or referring to medical knowledge in its answers, AMIE can provide justifications for its conclusions, which enhances interpretability. In summary, AMIE's explainability is achieved through conversational reasoning and explicit knowledge grounding, allowing users to follow how the model moves from symptoms to diagnoses in a step-by-step manner.

Evaluation and Outcomes AMIE's impact on diagnostic accuracy and thoroughness has been evaluated in both standalone and clinician-assisted settings. In one study, twenty physicians were given difficult cases and assigned to use either traditional resources (e.g., web search, textbooks) or AMIE for assistance. The results were telling: AMIE on its own achieved a top-10 diagnostic accuracy of 59.1%, outperforming the unaided clinicians who managed 33.6%. More importantly, when clinicians used AMIE as an assistive tool, their diagnostic lists became significantly more comprehensive and accurate. The clinicians assisted by AMIE attained a top-10 accuracy of 51.7% in their differential diagnoses, compared to 36.1% when the same clinicians worked without AMIE. This improvement was statistically significant, indicating that AMIE's suggestions meaningfully improved human diagnostic performance. Clinicians with AMIE support also outperformed those with just internet search support (44.4% accuracy). Another evaluation focused on conversational performance using a simulated clinical exam setting (remote Objective Structured Clinical Examinations, OSCE) with standardized patient actors. AMIE was pitted against real primary care physicians in managing 149 mock patient cases via text-based chat. According to specialist physician evaluators, AMIE demonstrated greater diagnostic accuracy and outperformed the human doctors on 28 of 32 evaluation axes, including history-taking thoroughness, management reasoning, and communication clarity. Patient actors also rated AMIE highly, with the system scoring better on 24 of 26 patient-experience measures such as empathy and information delivery. These outcomes underscore that AMIE can not only match or exceed human diagnostic reasoning in challenging cases but does so while maintaining strong communication—a proxy for explainability and user acceptability. AMIE's underlying medical knowledge is robust, leveraging both a general LLM backbone and domain specialization (e.g., a version of Med-PaLM 2 was used in development), which has shown expert-level accuracy on medical exams.

Strengths and Limitations AMIE's primary strength lies in its interactive diagnostic reasoning. The multi-turn dialogue approach naturally provides transparency, as the model's questions and explanations mirror a clinician's reasoning process, enhancing trust. This is complemented by demonstrated improvements in clinician performance, where studies showed that AMIE's assistance led to more accurate and comprehensive differentials by physicians, indicating a tangible benefit in clinical decision support. Furthermore, its broad clinical scope, achieved through self-play training on diverse simulated cases, makes it a versatile diagnostic aide. A particularly notable strength is its user-centric communication; the system was evaluated on communication skills and empathy, scoring on par with or better than human doctors, suggesting AMIE can explain diagnoses in a patient-friendly manner. Finally, its integration of up-to-date medical knowledge ensures its explanations and suggestions align with current clinical guidelines, increasing credibility.

However, the system has several limitations. The evaluations of AMIE occurred in controlled, experimental settings, such as case report reviews and simulated text chats, which are not typical of real clinical practice, limiting the generalizability of the results to live patient care. Despite its extensive training, AMIE could still occasionally hallucinate incorrect medical facts or miss subtleties not present in its training data, and there is no guarantee it will recognize its own knowledge gaps. The system's performance is also dependent on the quality of user input; poor or misleading symptom descriptions can lead it down the wrong path, a challenge it shares with human clinicians. As with any AI providing medical guidance, there are significant ethical and oversight concerns regarding liability and the potential for its suggestions to bias a clinician's judgment. While AMIE's dialogue is understandable, the model's internal decision-making remains a complex neural network, and its reasoning is not fully captured by the conversational output. It does not, for instance, provide quantified confidence scores for each diagnosis unless explicitly prompted, which could be a valuable addition for enhancing interpretability.

2.2 Dual-Inf: Bidirectional Inference for High-Fidelity Explanations

Dual-Inf is a novel framework proposed to improve explainability in automatic differential diagnosis (DDx) generation by LLMs [13]. Developed by Zhou et al., Dual-Inf was motivated by two key challenges: the lack of datasets annotated with expert diagnostic explanations and the difficulty LLMs have with complex multi-step clinical reasoning. Prior to this work, most automatic DDx systems could list probable diseases but offered little insight into why those diseases were chosen. To address this, the researchers first created Open-XDDx, the first public dataset of clinical cases annotated with differential diagnoses and detailed expert-written explanations for each potential diagnosis. This dataset spans 570 clinical notes across nine specialties. With this resource, they developed Dual-Inf, a pipeline that elicits an LLM to reason in a bidirectional manner, akin to how doctors verify a diagnosis by checking if expected symptoms are present or absent. The overall goal of Dual-Inf is to produce precise, high-quality differential diagnosis explanations alongside the list of possible conditions, moving beyond simple prediction to generate a cogent narrative explaining each candidate diagnosis.

Explainability Technique The hallmark of Dual-Inf is its bidirectional inference strategy combined with iterative self-refinement. This method acts as an internal "sanity check" or cognitive checksum for the LLM's reasoning process. The framework consists of three LLM-based components and a feedback loop. First, in Forward Inference, the LLM acts as a diagnostician, reading the patient's note and producing an initial set of possible diagnoses with an explanation for each. Second, in Backward Inference, a separate LLM process takes each proposed diagnosis and works backward to generate the expected symptoms and findings typical for that condition, drawn from the LLM's internal medical knowledge. This backward pass serves as a verification mechanism. Third, an Examination & Synthesis component compares the backward-generated list with the actual patient case to evaluate each diagnosis. If a proposed diagnosis is missing key supportive evidence or has contradictions, this module can discard it, mark it as low confidence, or enrich the forward explanation by merging patient-specific findings with the general knowledge from the backward inference. This process ensures that each diagnosis is supported by a sufficient set of findings. Finally, Dual-Inf employs an Iterative Self-Reflection loop where diagnoses deemed low-confidence are fed back to the forward inference module with a prompt to "think again" or consider alternatives. This self-correction cycle can run multiple iterations until the diagnoses and explanations reach an acceptable confidence threshold. Through these steps, Dual-Inf's explainability technique forces the LLM to articulate why a diagnosis fits (forward reasoning) and verify its consistency with known manifestations (backward reasoning), reducing hallucinations and yielding a more reliable and complete reasoning chain.

Evaluation and Outcomes The Dual-Inf framework was rigorously evaluated on the curated Open-XDDx dataset, using both automated metrics and expert human review with LLMs such as GPT-4 and LLaMA-3 70B. Metrics for explanation quality included correctness, completeness, and usefulness. Dual-Inf showed clear improvements over baseline prompting strategies. Compared to a standard Chain-of-Thought (CoT) prompting baseline, Dual-Inf's explanations omitted significantly fewer important details, with a reported 13.4% reduction in missing content errors in a 100-case sample using GPT-4. Additionally, it achieved higher correctness, as the backward checking helped filter out the forward module's mistakes. The diagnostic accuracy of the model's final answers was also improved; Dual-Inf achieved a higher median accuracy per case than the baseline CoT approach. Perhaps most telling was the human evaluation: experts rated Dual-Inf's explanations as more useful and aligned with clinical reasoning best practices, especially noting its superior ability to justify why a diagnosis was included or excluded. The iterative self-refinement was found to contribute meaningfully, with performance improving up to a point with each iteration, confirming that the "think twice" feedback loop helped the model correct itself. These gains were consistent across different model sizes and types, indicating that Dual-Inf's method is model-agnostic to an extent.

Strengths and Limitations The structured reasoning approach of Dual-Inf is a significant strength, as it explicitly breaks the diagnostic task into forward and backward passes, improving the coherence and trustworthiness of explanations by reducing unsupported claims and hallucinations. Evaluations confirmed that this leads to improved explanation quality, providing more complete and correct rationales than standard prompting. This enhanced reasoning also translates to enhanced diagnostic accuracy, with Dual-Inf's differential lists more often including and correctly prioritizing the true condition. The framework demonstrated generalizability across multiple LLMs and nine medical specialties. A major contribution is the creation of the Open-XDDx dataset, which provides a valuable public benchmark for explainable clinical reasoning research. Finally, the iterative refinement loop allows the system to perform quality control on its own answers, an innovative approach for tasks requiring high reliability.

However, the framework has notable limitations. Its backward inference relies on the LLM's internal medical knowledge, which, if incorrect or incomplete, could introduce hallucinated evidence and lead to convincingly wrong explanations. Unlike some systems, Dual-Inf does not integrate external knowledge bases in real-time, potentially missing uncommon diseases or the latest research findings. The output is a qualitative differential diagnosis without explicit probability scoring, which can be a drawback in clinical practice where prioritizing the most likely or dangerous condition is critical. The multi-step inference process is computationally complex and resource-intensive, which could be a bottleneck in time-sensitive scenarios. While tested across nine specialties, the dataset size of 570 notes is relatively small, and its performance on more free-form or noisy clinical narratives outside this curated set is unproven. Finally, integrating this standalone reasoning engine into a real clinical workflow would require significant engineering and usability testing with practitioners.

2.3 ConfiDx: Articulating Uncertainty in Clinical Reasoning

ConfiDx is an approach focusing on a critical aspect of clinical AI explainability: diagnostic uncertainty. In practice, a patient's data often does not point conclusively to a single diagnosis. This uncertainty is something clinicians routinely recognize and communicate. However, most computational diagnostic models either do not express uncertainty or are over-confident even when evidence is incomplete. ConfiDx, introduced by Zhou et al., addresses this gap by training an LLM-based diagnostic system to explicitly identify when a case is ambiguous and to explain the source of that ambiguity [5]. This work represents a shift in the fundamental question being asked of the AI, from "What is the diagnosis?" to "How certain are we, and why?". The developers formalized a new task of uncertainty-aware disease diagnosis, creating richly annotated datasets that label not just the diagnosis for each clinical note, but also the degree of diagnostic ambiguity and the reasons for it. By fine-tuning on data that emphasizes adhering to formal diagnostic criteria and noting when they are unmet, ConfiDx seeks to align the LLM with the cautious reasoning style of clinicians, making its output more trustworthy and clinically realistic.

Explainability Technique ConfiDx's explainability technique centers on criterion-based reasoning and uncertainty labeling. The model was fine-tuned on open-source LLMs using diagnostic criteria from clinical guidelines as part of its training data. When confronted with a new case, ConfiDx internally checks which criteria are satisfied by the patient's data and which are not. The output is typically twofold: a diagnostic prediction or differential diagnosis, and an explanation that includes identification of any gaps or uncertainties in the evidence. For instance, ConfiDx might state, "This patient likely has Condition X, however the absence of [a certain definitive test result] creates uncertainty, so Condition Y is also possible." This is achieved by structuring the model's prompt and instruction format during fine-tuning to elicit such behavior, a form of attribution-based prompting where the model attributes its uncertainty to specific missing or conflicting pieces of evidence. Another element is the use of multiple prediction types during training. The model can output a class indicating "uncertain" or "unknown" if criteria are insufficient, teaching it that "I don't know" is an acceptable and sometimes correct answer. In sum, ConfiDx's explainability is delivered via transparent uncertainty quantification: by design, the model's explanation will highlight what information is lacking or ambiguous and how that affects the confidence in each diagnostic conclusion.

Evaluation and Outcomes ConfiDx was evaluated on several real-world clinical note datasets annotated for diagnostic uncertainty, benchmarked against off-the-shelf LLMs and other baselines. In terms of diagnostic accuracy, ConfiDx showed notable gains. For example, when fine-tuning a large 70B model, the area under the ROC curve (AUC) for diagnosis prediction increased from 0.594 to 0.658. More impressively, ConfiDx excelled at uncertainty recognition, significantly outperforming base models in detecting uncertain cases, achieving an average AUC around 0.50 for uncertainty identification and an F1-score of 0.57 on that task, with improvements being statistically significant ($p < 0.001$). Across all explanation quality metrics evaluated, ConfiDx showed notable gains over baseline LLMs, with fine-tuned models consistently scoring higher on metrics like BERTScore for explanation similarity to the gold-standard ($p < 0.001$). In one analysis, a fine-tuned ConfiDx (70B) model outperformed even other large models including GPT-4 on the combined task of diagnosis, uncertainty detection, and explanation. The model also demonstrated generalizability, maintaining robust performance on cases from institutions not seen in training, suggesting it learned general medical reasoning patterns rather than dataset-specific quirks.

Strengths and Limitations A primary strength of ConfiDx is its explicit communication of diagnostic uncertainty, a rare and valuable feature that provides a safety check against overconfident diagnoses and aligns its outputs with clinical thinking. This is achieved through improved diagnostic alignment, as fine-tuning on clinical criteria and guidelines made the model more accurate and its reasoning more medically sound. The system provides rich, comprehensive narrative explanations covering both why a diagnosis is considered and why uncertainty exists, which can help educate users and guide next steps. Its robustness and generalization across datasets from different hospitals are crucial for real-world deployment. By addressing the often-neglected area of diagnostic ambiguity, ConfiDx meets a realistic clinical need, encouraging caution and potentially improving patient outcomes by avoiding premature closure on a single diagnosis.

The approach is not without limitations. Training ConfiDx requires richly annotated data, which is costly and time-consuming to produce, and its performance is tied to the quality and scope of these annotations. The complex, multi-part output format may be difficult to integrate into clinical workflows without overwhelming busy clinicians. While calibration is improved, the model is not perfect and may still exhibit residual overconfidence or underconfidence. Its understanding of diagnostic criteria is learned from text and may lack the nuance and flexibility of an experienced clinician who knows when to deviate from formal guidelines. Finally, while technical metrics are strong, prospective user studies are needed to confirm that clinicians find these uncertainty-aware explanations beneficial in practice.

2.4 *DR.KNOWS: Grounding Diagnosis in Structured Medical Knowledge*

DR.KNOWS (Diagnostic Reasoning Knowledge Graph System) is a system that augments LLM-based diagnosis with structured medical knowledge, addressing the challenge of information overload and omission in electronic health records (EHRs) [4]. In modern healthcare, clinicians must sift through lengthy EHR notes where critical information can be buried, leading to diagnostic errors. DR.KNOWS was created to tackle this issue by integrating a medical knowledge graph (KG), built from the Unified Medical Language System (UMLS) ontology, into the diagnostic reasoning process. This approach demonstrates the value of using structured knowledge to impose beneficial constraints on an LLM's generative capabilities. By leveraging this structured knowledge, DR.KNOWS aims to improve diagnostic accuracy and produce explanations grounded in known medical relationships. Instead of an LLM free-associating diagnoses, it is guided by a "map" of medical knowledge, making its suggestions more interpretable and verifiable by clinicians, as the knowledge graph paths can serve as an audit trail.

Explainability Technique DR.KNOWS employs a knowledge retrieval and reasoning technique to generate explainable diagnoses. First, given a patient's EHR notes, the system uses a graph neural network to encode the medical KG and learn vector representations of medical concepts. Then, it uses an attention-based path ranking algorithm to retrieve the most relevant knowledge subgraphs, or "paths," that connect the patient's findings to potential diagnoses. For example, if a patient has fever and cough, the system might retrieve a path like: "fever"—"cough"—"pneumonia." These knowledge paths are then integrated into an LLM's prompt. The LLM's job is to generate the final diagnostic prediction and a rationale that incorporates the knowledge path. Because the model is prompted with a valid, pre-existing medical relationship, its explanation is constrained to follow established medical science. This is a form of retrieval-augmented generation using structured data. The output might be phrased as: "Based on the patient's symptom A and B, which are connected to diagnosis D in medical knowledge, D is a likely diagnosis." The use of the UMLS-based KG ensures that the concepts are standardized medical terms, making the explanations formal and precise. In summary, the technique constrains and guides the LLM with a knowledge graph so that diagnostic suggestions come with a traceable chain of reasoning.

Evaluation and Outcomes The DR.KNOWS system was evaluated on two real-world EHR datasets from different geographic locations to test generalizability. Its performance was compared with baseline models, including LLMs without knowledge graph support. The results showed notable improvements in multiple areas. First, in extracting relevant diagnostic concepts from notes, DR.KNOWS had higher accuracy than baselines, meaning it better recognized the key features that mattered for diagnosis. For the diagnostic prediction task, DR.KNOWS achieved enhanced metrics, including higher ROUGE-L and concept-F1 scores, suggesting its diagnoses and explanations more closely matched the expected answers and covered the necessary medical concepts more completely than other models. In a human evaluation, clinicians found that DR.KNOWS's rationales were strongly aligned with correct clinical reasoning, indicating the model was not just accurate but also explained itself in a logical and acceptable manner. The system also demonstrated improved abstraction and reasoning, inferring higher-level concepts from raw data. While acknowledging the potential for bias within the KG data, the authors proposed strategies to mitigate this by focusing on case-specific path selection. Overall, DR.KNOWS demonstrated that integrating a knowledge graph yields a more accurate and interpretable diagnostic system.

Strengths and Limitations The foremost strength of DR.KNOWS is its provision of knowledge-grounded explanations. The ability to trace an output back to known medical relationships via a UMLS concept path makes the reasoning transparent and verifiable, reducing the "black-box" nature of predictions. This knowledge guidance also led to improved diagnostic accuracy and recall of relevant findings from patient records, reducing critical diagnostic omissions. The system's attention-based path ranker enhances its focus on pertinent data, allowing it to ignore noise in EHR notes, a significant challenge in clinical data analysis. DR.KNOWS demonstrated generalizability across different hospital datasets, suggesting that its reliance on universal medical knowledge enhances robustness. Furthermore, the inclusion of diagnostic safety metrics in its human evaluation reflects a commendable focus on aligning with clinical standards.

However, the system's reliance on its underlying KG introduces limitations. The KG's completeness and currency are critical; if it is missing a link for a newly discovered association or a rare disease, DR.KNOWS may fail to make that connection. The KG might also encode biases, such as over-representing common conditions, which could skew the model's predictions. The implementation is complex, combining GNNs, KGs, and LLMs, which could hinder reproducibility and deployment. Explanations, while grounded, might be lengthy or use technical jargon that is difficult for non-experts to comprehend. The evaluation, while positive, was retrospective, and there is no evidence yet of its performance in a prospective trial or its impact on clinicians' diagnostic accuracy in practice. Finally, practical integration into EHR workflows would pose significant technical and training challenges for clinical staff.

2.5 RetCare: Evidence-Based Reasoning via Retrieval-Augmented Generation

RetCare is a workflow developed to enhance the interpretability of clinical decision-making models by leveraging retrieval-augmented generation (RAG) with medical literature [14]. The impetus for RetCare comes from the observation that many high-performing but opaque machine learning models for tasks like predicting patient outcomes are rarely used by clinicians due to a lack of trust. RetCare addresses this by inserting an intermediate step: whenever a model makes a prediction, the system retrieves relevant information from authoritative medical sources like PubMed to support or challenge that prediction. This approach mirrors a physician's practice of consulting a textbook or journal when uncertain. RetCare operationalizes the principle that medical knowledge is not a static entity but a constantly evolving body of literature. By bridging the gap between data-driven predictions and evidence-based medicine, it aims to build clinician trust, as advice is more likely to be accepted if it is backed by references to established clinical guidelines or research.

Explainability Technique The explainability technique in RetCare is the retrieval of authoritative knowledge followed by context-specific explanation generation. The pipeline begins after a predictive model analyzes a patient's EHR data and produces an outcome (e.g., "30% risk of mortality") along with the features that contributed most. RetCare takes these outputs and formulates a query for an external knowledge base of over two million PubMed entries. For example, if the model identifies "age $ > 80$" and "high troponin levels" as key factors for high mortality risk, RetCare will query PubMed for articles linking these factors. The system then retrieves a set of relevant literature. Finally, a large language model is prompted with the patient context, the model's prediction, and the retrieved evidence snippets. The LLM's task is to compose a final explanation that validates the model's prediction by weaving in the retrieved knowledge. The explanation reads like an evidence-based justification, anchoring the model's claims in authoritative sources and aligning AI reasoning with the norms of human medical knowledge. This is a form of attribution-based explanation where the importance of a feature is justified not by its internal model weight, but by external clinical evidence.

Evaluation and Outcomes RetCare was evaluated on two real-world datasets to assess its impact on the accuracy and reliability of model predictions. The key finding was that RetCare's approach significantly improved model performance metrics compared to the base predictive model alone. This suggests that processing a model's prediction through an LLM augmented with external knowledge can actually correct errors or filter out unsupported conclusions, leading to more accurate final decisions. The reliability improvements likely refer to enhanced consistency and trustworthiness, possibly measured by the higher precision of predictions that are supported by strong literature evidence. Another important result area is trust and interpretability. The workflow facilitates more informed and trustworthy decision-making by aligning AI outputs with real-world clinical practices, where clinicians often consult guidelines or literature to double-check recommendations. The public release of the code on GitHub further suggests the development of a reproducible and generalizable method. In summary, the evaluations indicate that combining LLM reasoning with retrieval of medical literature resulted in quantitatively better predictive performance and qualitatively better explanations, building a crucial bridge between black-box model outputs and the evidence-based reasoning clinicians expect.

Strengths and Limitations RetCare's greatest strength is its delivery of evidence-based explanations, which ground model predictions in published medical research and dramatically increase their credibility for clinicians trained in evidence-based medicine. A rare and powerful feature is that this interpretability method also improves predictive performance by validating and filtering model outputs with literature, catching unsupported predictions. The hybrid approach elegantly combines data-driven insights with human-curated knowledge, leveraging the strengths of both. The framework is general and not tied to a specific predictive model or clinical condition, making it widely applicable. The detailed explanations with references can also serve as an educational tool for practitioners and can enhance transparency by highlighting when no supporting literature is found for a model's reasoning, preventing blind trust.

However, the approach is only as good as the literature it retrieves; there is a risk of retrieving irrelevant or low-quality studies, which could lead to misleading explanations. The explanatory output can become lengthy or technical, potentially causing information overload for busy clinicians. For novel scenarios or rare conditions, there may be no published evidence to retrieve, limiting the system's utility in such cases. The process of searching millions of PubMed entries and running an LLM is computationally expensive and may not be feasible for real-time clinical settings. Seamless integration into EHR systems would be necessary for practical use, raising issues of data privacy and regulatory compliance [15]. Finally, while accuracy improvements were shown, prospective clinical trials are needed to determine how RetCare affects actual clinician behavior and decision-making in practice.

2.6 DrHouse: Multi-Turn Multimodal Diagnostic Reasoning

DrHouse is an LLM-powered virtual doctor system for diagnostic reasoning, distinguished by its incorporation of continuous sensor data and external knowledge into a conversational framework [16]. The system's name reflects its emphasis on using objective data to solve complex cases, rather than relying solely on patients' subjective symptom descriptions. DrHouse was motivated by the limitations of existing single-turn, text-only medical chatbots, which depend heavily on potentially incomplete or biased patient descriptions and operate with static, outdated medical knowledge. DrHouse overcomes these limitations by enabling a multi-turn consultation where the AI can actively ask for more information and by integrating data from wearable sensors (e.g., heart rate, physical activity) into the diagnostic process. Additionally, it continuously refreshes its medical knowledge by accessing resources like UpToDate and PubMed, ensuring its knowledge base remains current. This combination of interactivity and multimodality is expected to produce more accurate and context-aware diagnoses, closing the loop between what a patient feels and what their body is objectively doing.

Explainability Technique DrHouse's explainability is rooted in its dialogue-driven reasoning and explicit probability estimates for diagnoses. The system conducts a multi-turn interview, asking questions that both gather missing information and externalize its reasoning process. Each question provides insight into what diagnoses the model is considering. A major component is a novel diagnostic algorithm that assigns likelihoods to potential diseases concurrently during the conversation. DrHouse maintains and updates a differential diagnosis list with associated probabilities, providing the user with a transparent view of its diagnostic certainty. This is effectively a form of Bayesian-like reasoning, where each new piece of information updates the model's beliefs. The integration of sensor data enhances explainability by grounding the reasoning in concrete, objective figures. For instance, DrHouse might use a high resting heart rate from a patient's smartwatch as direct evidence in its explanation. By tapping into continuously updated medical knowledge, the system can bolster its explanations with up-to-date facts. In summary, DrHouse's explainability technique is interactive and iterative, delivered through questioning and explicit, probabilistic reasoning outputs at each turn.

Evaluation and Outcomes DrHouse was evaluated on both technical metrics and user acceptance. On the technical side, tests on three public datasets and a self-collected dataset showed that DrHouse achieved up to an 18.8% increase in diagnosis accuracy over state-of-the-art baselines. This substantial jump indicates that the inclusion of sensor data and the multi-turn approach led to more correct diagnoses. A user study with 32 participants, including medical experts and patients, yielded very positive results: 75% of medical experts and 91.7% of patients were willing to use DrHouse. This high acceptance rate suggests that both groups found the system's diagnostic process and interaction credible and useful. For medical experts, this level of willingness is notable and implies that the system's transparent, data-driven design addressed many of their concerns. For patients, the high acceptance likely reflects an appreciation for the thorough, interactive consultation. Combining improved accuracy with strong user willingness to adopt, DrHouse appears to succeed both as a diagnostic engine and as a communicative, explainable agent, demonstrating that explainability and performance can go hand in hand.

Strengths and Limitations DrHouse's foremost strength is its multimodal data integration, which leverages objective information from wearables for more accurate and personalized diagnoses, providing concrete reference points for explanations. Its interactive, iterative reasoning mirrors a normal clinical encounter, making it intuitive for users while reducing the risk of misdiagnosis from incomplete information. The explicit quantification of uncertainty through probabilistic outputs is an advanced form of explanation that manages user expectations and highlights subtle possibilities. By accessing up-to-date medical databases, the system's knowledge base remains current, a critical feature in the rapidly evolving field of medicine. The high user satisfaction reported in its evaluation indicates a human-centric design that addresses end-user needs effectively. Finally, the validated accuracy improvement over baselines confirms that its novel combination of techniques produces tangible results.

However, the system's complexity and resource requirements are a significant limitation, potentially hindering real-time performance and accessibility. Its dependency on the connectivity and accuracy of sensor data introduces a potential point of failure. The multi-turn dialogue, while thorough, could lead to information overload or fatigue for some users. The user study, though positive, was small, and larger, more diverse population studies are needed to validate its effectiveness and safety, particularly regarding cultural sensitivity and handling of atypical presentations. As with any powerful AI, there are concerns about medical oversight and the risk of patient over-trust, especially given the precise-sounding probabilities. Lastly, the use of personal sensor data raises significant privacy concerns that must be addressed with robust security and clear consent processes.

3 Comparative Analysis of Explainability Techniques for Diagnostic LLMs

Across these diverse case studies, several common themes and comparative insights emerge regarding how explainable LLMs are being used for diagnosis. There is no single "best" technique; rather, the optimal approach depends on the specific clinical task, data modality, and user needs. This analysis compares and contrasts the different frameworks along several key dimensions.

3.1 Techniques for Explainability

Chain-of-Thought and Dialogue: Systems like Dual-Inf explicitly prompt LLMs to articulate their reasoning step-by-step via a structured two-step infer-and-verify process [13]. Similarly, AMIE and DrHouse rely on interactive dialogue; their multi-turn Q&A format naturally externalizes reasoning as the models ask clarifying questions and summarize intermediate conclusions. This iterative questioning is essentially a human-interpretable chain-of-thought, as seen when DrHouse queries sensor readings and updates probabilities based on the answers [16]. This represents a significant departure from classical diagnostic AI, which often kept its reasoning hidden.

Knowledge Integration: A number of systems incorporate external knowledge to ground the model's reasoning. DR.KNOWS pulls in structured knowledge from UMLS-based graphs to guide its diagnostic suggestions [4], while RetCare uses unstructured knowledge from PubMed literature to support or contest a model's prediction with citations [14]. Both improve explainability by providing an evidence trail. In contrast, systems like ConfiDx and Dual-Inf rely more on learned internal knowledge, ensuring consistency with medical criteria and reasoning patterns but not explicitly citing external sources. The knowledge-integrated systems arguably provide stronger verifiability, whereas the internally learned approaches provide a coherent narrative but require more trust in the model's internal knowledge.

Quantitative Uncertainty and Probabilities: Explainable diagnostic LLMs increasingly embrace uncertainty communication. ConfiDx explicitly identifies diagnostic ambiguity and states when criteria are unmet [5]. DrHouse goes a step further by assigning probabilities to each considered diagnosis in real-time [16]. This trend underscores that knowing what the model does not know is as important as knowing what it suggests. Communicating uncertainty is a key aspect of trust; ConfiDx and DrHouse both showed that acknowledging uncertainty did not impede accuracy and resonated well with users [5, 16].

3.2 Domain-Specific Use Cases

Differential Diagnosis from Symptoms: AMIE, Dual-Inf, and DrHouse all focus on diagnostic reasoning from patient symptoms and signs. For these tasks, where complex reasoning under uncertainty is the main challenge, techniques like CoT prompting (Dual-Inf) and interactive Q &A (AMIE, DrHouse) excel because they can break down multi-symptom reasoning into manageable steps [1, 10, 13, 16]. In such domains, explainability is crucial for safety, ensuring that rare but critical diagnoses are not overlooked.

Outcome/Risk Prediction from EHR Data: RetCare addresses predictive modeling based on structured or large-scale data. In this use case, the primary explanation need is feature attribution and evidence support: RetCare's RAG approach connects model outputs to authoritative evidence about risk factors and provides citations to the literature [14]. This alignment with evidence highlights that interpretability goes hand-in-hand with ethical AI and clinician trust.

Knowledge-intensive Diagnosis: DR.KNOWS's knowledge graph and AMIE's conversational knowledge integration hint at solving cases that require broad or deep medical knowledge. The explainability approach here tends to be knowledge-based reasoning, where the system explicitly maps patient information onto a knowledge base to surface a rationale [1, 4]. The advantage is twofold: more accurate diagnoses on tough cases and more credible explanations.

3.3 Evaluation Methods and Outcomes

Accuracy versus Explainability: Encouragingly, many systems showed that adding explainability features either maintained or improved predictive performance. Dual-Inf's prompting scheme improved diagnostic accuracy [13], ConfiDx's uncertainty-awareness led to superior accuracy [5], and DrHouse's interactive approach achieved a significant accuracy boost [16]. This dispels the myth that interpretability comes at the cost of accuracy.

User-Centric Evaluation: Many projects involved human experts or patients in evaluation. AMIE was tested with physicians who achieved better diagnoses with its assistance [1], DrHouse underwent a user study showing high willingness to use among doctors and patients [16], and DR.KNOWS had human evaluators judge its rationales as strongly aligned with correct reasoning [4]. This emphasis reflects a broader understanding that explainability is ultimately about the end-user. The positive reception suggests that when AI clearly explains itself, clinicians feel more comfortable using it as a collaborator.

Comparative Performance: Several studies allow for direct or indirect comparisons. Dual-Inf outperformed a simpler CoT baseline in explanation quality [13]. DR.KNOWS outperformed standard LLMs without its KG [4]. ConfiDx, after fine-tuning, beat off-the-shelf GPT-4 in accuracy and explanation metrics [5]. This pattern suggests a synergy: combining general LLM intelligence with specialized frameworks produces the best outcomes.

4 Explainable Large Language Models in Treatment Planning

The application of explainable large language models (LLMs) in treatment planning represents an emerging frontier in clinical decision support. Whereas diagnostic reasoning emphasizes the identification of potential conditions, treatment planning requires integrating patient-specific factors with evolving medical evidence to recommend appropriate management strategies. This task poses unique challenges for explainability, as clinicians must not only understand the rationale behind suggested interventions but also evaluate their alignment with guidelines, safety considerations, and patient preferences. Recent frameworks have explored diverse techniques to address these challenges, ranging from systems that ground recommendations in evidence-based literature to models that generate stepwise justifications for clinical management decisions, as shown in Table 2. The case studies in this section highlight such approaches, illustrating how explainable LLMs are being developed to support transparent, trustworthy, and context-sensitive treatment planning.

4.1 *GPT-4 Assistance in Clinical Management Tasks*

While LLMs have shown proficiency in diagnostic reasoning, their utility in management reasoning—which involves complex trade-offs, risk assessment, and patient-specific adjustments—has been less clear. To address this, Goh et al. conducted a prospective, randomized, controlled trial to assess whether assistance from GPT-4 could improve physician performance on complex, open-ended management

Table 2 Overview of explainable LLM case studies in treatment planning (compact)

System	Primary clinical task	Input modality	Core explainability technique	Knowledge source
GPT-4 Assist [11]	General management	Text vignettes	Narrative CoT	Internal LLM
ChatGPT-CARE [7]	COVID-19 Therapy	Text descriptions	Guideline-graph CoT	Encoded guidelines
ChatGPT-4 [17]	Oncology planning	Text summaries	Comparative concordance	Internal LLM
MEREDITH [18]	Precision oncology	Text + genomics	RAG + expert feedback	PubMed, trials, guidelines
GPT-RadPlan [19]	Radiotherapy planning	Images + DVHs	Multimodal iterative critique	Internal LLM + Exemplars
DeepDR-LLM [20]	Diabetes management	Retinal images + Text	Multimodal grounding	Guidelines (LLM) + Fundus evidence

reasoning tasks [11]. The study was motivated by the understanding that real-world clinical care is not a single-point classification problem but an iterative process of decision-making under uncertainty. The researchers designed five clinical vignettes based on real patient encounters, where information was revealed sequentially to mimic the evolving nature of a clinical case. Ninety-two practicing physicians were randomized to either use conventional resources (e.g., UpToDate, Google) or to have access to GPT-4 in addition to those resources. The goal was to quantify the value added by the LLM in scenarios that demand nuanced management scripts rather than simple fact recall.

Explainability Technique The explainability in this study was not derived from a specialized post-hoc module but was inherent in the natural language output of GPT-4 itself. The model was used "out of the box," and its ability to generate a structured, reasoned narrative served as the explanation. When prompted with a clinical scenario, GPT-4 produces a response that typically includes a summary of key findings, a ranked differential diagnosis, a management plan ordered by priority, and a rationale for its recommendations. This output mirrors the structure of a clinical note or a consultation summary, making it an operational form of explanation that is immediately legible to a clinician. The model's "chain-of-thought" ability to articulate why certain actions are indicated (e.g., "given the risk of hemodynamic compromise...") or contraindicated allows the physician to scrutinize its logic directly. The study's design leveraged this inherent narrative transparency to see if it would influence and improve human decision-making.

Evaluation and Outcomes The study's primary outcome was physician performance measured on expert-developed scoring rubrics. The results were significant: physicians assisted by GPT-4 scored substantially higher than those using only conventional resources (mean score of 43.0% vs. 35.7%, a mean difference of 6.5%) [11]. This improvement was consistent across different domains, including management decisions, diagnostic reasoning, and context-specific questions. Interestingly, the performance of the LLM-augmented physicians was statistically indistinguishable from the performance of GPT-4 alone (43.0% vs. 43.7%), suggesting that the LLM was able to elevate human performance to its own level in this setting. A notable secondary finding was that physicians using the LLM spent more time on each case, which may reflect deeper engagement and deliberation prompted by the AI's detailed output.

Strengths and Limitations The primary strength of this study is its rigorous randomized controlled trial design, which provides strong evidence for the LLM's utility in improving management reasoning. The use of sequentially revealed information in the vignettes created a more realistic simulation of clinical practice. The inherent, narrative-based explainability of GPT-4 proved effective and aligned well with physicians' cognitive workflows. However, the study has limitations inherent to simulated environments. The clinical vignettes, while complex, do not capture the full spectrum of real-world pressures, such as patient preferences, communication challenges, and systemic constraints. The explainability, while clear, is not grounded in specific evidence (i.e., it does not provide citations), which limits its verifiability in novel or controversial cases. Furthermore, while the study measured rubric-based scores, it could not assess real patient outcomes.

4.2 ChatGPT-CARE: Guideline-Grounded COVID-19 Treatment

The COVID-19 pandemic created a challenging clinical environment where treatment guidelines for outpatients evolved rapidly in response to new evidence on antivirals, risk factors, and viral variants. For busy clinicians, keeping up with these frequent updates was a significant burden. This context motivated Wang et al. to develop ChatGPT-CARE, a tool designed to enhance ChatGPT with up-to-date clinical practice guidelines for COVID-19 outpatient management [7]. The primary goal was to create a more reliable and accurate decision support tool that could reduce the risk of hallucinations or outdated recommendations from a general-purpose LLM by anchoring its reasoning in authoritative clinical documents from sources like the CDC and IDSA.

Explainability Technique ChatGPT-CARE's explainability technique is a form of **reasoning scaffold** implemented via in-context learning and chain-of-thought prompting. The developers transformed the official COVID-19 outpatient treatment guideline into a Python-coded directed knowledge graph, where nodes represent clinical decision points (e.g., "Patient is at high risk," "GFR < 30 mL/min") and edges represent the logical flow between them [7]. This entire graph structure is provided to the LLM within the system prompt. When a user inputs a patient description, the LLM is instructed to use a "chain-of-thought" process: first, it maps the patient's details to the corresponding nodes in the graph, and second, it traverses the graph to arrive at a treatment recommendation. The explanation is the explicit output of this traversal, stating which patient facts map to which guideline criteria at each step, making the reasoning process transparent and directly traceable to the underlying guideline.

Evaluation and Outcomes ChatGPT-CARE (using both GPT-3.5 and GPT-4) was evaluated against the original ChatGPT models on 45 patient descriptions categorized by difficulty (easy, medium, hard). Two physicians evaluated the responses for accuracy, clarity, and the presence of hallucinations. The results showed that ChatGPT-CARE models were consistently judged to be more accurate and to provide clearer responses than their original counterparts across all difficulty levels [7]. ChatGPT-CARE (GPT-4) was the top performer, generating the most accurate responses. While both versions of ChatGPT-CARE had a low rate of hallucination, they occasionally occurred when the model had to make inferences from specific details, whereas the original ChatGPT often avoided hallucinations by providing vague, non-committal answers.

Strengths and Limitations The primary strength of ChatGPT-CARE is its explicit grounding in clinical guidelines, which makes its recommendations both more reliable and highly explainable. The chain-of-thought technique provides a transparent, step-by-step rationale that clinicians can easily follow and verify. However, the system has limitations. Its knowledge is confined to the specific guideline encoded in the prompt; it cannot reason about conditions outside of that scope. The approach requires significant manual effort to translate guidelines into a coded graph and must be diligently maintained as guidelines evolve. Furthermore, when faced with incomplete information, the model tended to make arbitrary assumptions rather than asking for clarification, a critical gap for a real-world clinical tool.

4.3 ChatGPT-4 in Multidisciplinary Tumor Boards (General Oncology)

Multidisciplinary Tumor Boards (MDTs), or tumor councils, are the standard of care for cancer treatment planning. These meetings bring together experts from surgery, medical oncology, radiation oncology, radiology, and pathology to form a consensus on the best treatment plan for each patient. This process is knowledge-intensive and relies on integrating complex patient data with rapidly changing clinical guidelines. Dogan et al. conducted a prospective study to evaluate whether a general-purpose LLM, ChatGPT-4, could function as a decision-support tool in this setting [17]. The study presented 100 real, consecutive cancer cases to ChatGPT-4 after they had been discussed by the hospital's MDT and compared the AI's recommendations to the board's final decisions.

Explainability Technique The explainability in this study is primarily **comparative and observational**. No special explainability framework was engineered. Instead, ChatGPT-4 was given a detailed, anonymized clinical summary and asked to propose a treatment plan. The transparency comes from analyzing the concordance and discordance between the AI's output and the human expert consensus. By examining the cases where the AI's recommendation differed from the MDT's, the researchers could infer the model's reasoning gaps. For example, if the AI consistently recommended standard-of-care chemotherapy while the MDT opted for a less aggressive approach due to patient frailty, it explains that the model is not adequately weighing performance status. The AI's own natural language rationale, when provided, adds a layer of direct explanation.

Evaluation and Outcomes The study found a high rate of concordance between ChatGPT-4's recommendations and the MDT's decisions. The overall agreement rate was 76.4%, with a Cohen's Kappa of 0.764, indicating substantial agreement [17]. The AI's most frequent recommendations were surgery (39%) and neoadjuvant treatment (37%), closely mirroring the MDT's preferences (36 and 45%, respectively). The inconsistencies (23.6% of cases) were most common in scenarios requiring highly individualized decisions, where factors like patient comorbidities, surgical risk, or quality of life—not fully detailed in the text summary—heavily influenced the human experts' choices.

Strengths and Limitations The study's prospective design is a major strength, as it evaluates the AI on real-time, consecutive cases, providing a realistic assessment of its capabilities. The high concordance rate demonstrates that a general LLM has successfully internalized standard oncological guidelines. However, the study also highlights significant limitations. The AI's inability to process "contextual clinical judgment" (e.g., patient preferences, psychosocial factors) is a critical gap. Its reasoning is opaque (a "black box"), as it does not cite the specific evidence or guidelines it uses. The study was single-center, and its findings may not generalize to other institutions with different practice patterns.

4.4 MEREDITH: Precision Oncology with Gemini Pro and RAG

Precision oncology relies on translating complex genomic data from tumor profiling into actionable treatment plans. This task is challenging due to the sheer volume of medical literature, the constant emergence of new clinical trial data, and the need to consider drug availability and approval status. General-purpose LLMs often fail in this domain because they lack the specialized, up-to-date knowledge required. To address this, Lammert et al. developed MEREDITH (Medical Evidence Retrieval and Data Integration for Tailored Healthcare), an expert-guided LLM system for precision oncology [18]. Built on Google's Gemini Pro, MEREDITH was designed to replicate the evidence-gathering and reasoning process of a molecular tumor board (MTB).

Explainability Technique MEREDITH's core explainability technique is retrieval-augmented generation (RAG) combined with an expert-in-the-loop design. The system's knowledge base is not limited to its pre-trained data; it actively retrieves information from a curated set of external sources, including PubMed, clinical trial databases (ClinicalTrials.gov), drug approval lists (FDA/EMA), and oncology guidelines [18]. It uses a "chain of thought" prompting structure that guides it to: (1) summarize relevant literature, (2) identify applicable guidelines and drug approvals, (3) find recruiting trials, and (4) synthesize these findings into ranked recommendations. The explanation is the evidence itself—every treatment option is presented with its supporting data (e.g., trial name, evidence level). Furthermore, human MTB experts review the AI's output, and their feedback is used to refine the system, creating an auditable, collaborative reasoning process.

Evaluation and Outcomes MEREDITH was evaluated on 10 fictional but complex oncology cases. Its recommendations were compared to those of a human MTB. The enhanced version of MEREDITH, which incorporated the full range of data sources, achieved a very high concordance of 94.7% with the expert panel's recommendations [18]. Quantitative analysis using semantic cosine similarity showed a significant improvement from the draft system (PubMed only) to the enhanced system (from 0.71 to 0.76, $p = 0.01$), confirming that the enriched data sources improved the quality and relevance of the AI's suggestions. Notably, MEREDITH often identified a broader range of options than the human experts (median of 4 vs. 2), including preclinical and combination therapies, productively expanding the MTB's discussion.

Strengths and Limitations The system's primary strength is its high degree of explainability, achieved through evidence-grounded RAG and a transparent reasoning chain. The expert-in-the-loop design ensures continuous alignment with clinical best practices. It successfully broadened the treatment search space without generating irrelevant options. A key limitation is its reliance on curated data sources; its performance depends on their quality and comprehensiveness. While it performed well on fictional cases, validation on real, anonymized patient data is a necessary next step. Its sophisticated architecture may also present deployment challenges in resource-limited settings.

4.5 GPT-RadPlan: Automated Radiotherapy Planning with GPT-4Vision

Radiotherapy treatment planning is a highly technical process where medical physicists iteratively adjust parameters (e.g., beam angles, weights) to create a plan that delivers a lethal dose of radiation to a tumor while sparing surrounding healthy organs-at-risk (OARs). This trial-and-error process is time-consuming and requires significant expertise. Liu et al. developed GPT-RadPlan, an innovative framework that uses a large multi-modal model, GPT-4Vision, to automate this process [19]. The system acts as an AI agent that can "see" and interpret the outputs of a treatment planning system (TPS) and provide human-like feedback to guide the optimization.

Explainability Technique GPT-RadPlan's explainability is **multi-modal and iterative**. The system takes two key visual outputs from the TPS as input: the dose distribution images (isodose curves overlaid on the patient's anatomy) and the dose-volume histograms (DVHs), which graphically represent the dose received by different structures [19]. GPT-4Vision analyzes these images and provides a "textual feedback" critique in natural language. This feedback explicitly states which clinical goals are being met and which are not. For example, it might say, "The PTV coverage is adequate, but the dose to the rectum exceeds the protocol limit of 31 Gy for 34% of the volume." This critique is then used to suggest specific adjustments to the planning parameters for the next iteration. The entire iterative process, with its sequence of visual inputs and textual critiques, serves as a transparent and auditable record of the plan's refinement.

Evaluation and Outcomes GPT-RadPlan was tested on 17 prostate cancer and 13 head & neck cancer cases. In all cases, the final plans generated by the automated system either matched or outperformed the clinical plans created by expert human planners [19]. Specifically, GPT-RadPlan demonstrated superior target coverage and achieved an average reduction of 5 Gy in OAR doses (a 15% reduction for prostate and 10–15% for head & neck). This was achieved without any domain-specific training of the model, relying only on in-context learning with three examples of approved clinical plans.

Strengths and Limitations The system's key strength is its ability to perform multi-modal reasoning, translating complex visual data into actionable and explainable textual feedback. It achieves expert-level performance without costly model training, making it easy to adapt to different clinical workflows. However, the system is not without limitations. Its performance is highly dependent on the quality of the in-context examples and the clarity of the clinical protocol provided in the prompt. The iterative process, while effective, can be computationally expensive. Most importantly, it has only been evaluated in a research setting and would require rigorous safety validation before clinical deployment.

4.6 DeepDR-LLM: Multi-Modal Diabetes Management

Managing chronic diseases like diabetes requires integrating diverse data streams, including clinical history, lab results, and imaging, to create a holistic, long-term treatment plan. A key complication of diabetes is diabetic retinopathy (DR), which is diagnosed through retinal fundus imaging. Jia et al. developed DeepDR-LLM, a pioneering hybrid system that combines a deep learning model for image analysis with an LLM for comprehensive diabetes management [20]. The goal was to create an AI assistant that could not only detect DR from images but also use that finding to generate a personalized, guideline-informed management plan covering glycemic control, referrals, and patient education.

Explainability Technique DeepDR-LLM's explainability is fundamentally multi-**modal and evidence-driven**. The system's reasoning process has two linked components. First, a vision transformer model (DeepDR-Transformer) analyzes a patient's fundus image and classifies the severity of DR. This classification serves as a piece of concrete, visual evidence. Second, this image-derived finding is passed to a fine-tuned LLM, which then generates a comprehensive management plan. The explanation is the direct linkage between the two: the LLM's recommendation for an ophthalmology referral is explicitly justified by the DR grade detected in the image. The textual part of the plan is further grounded in the clinical guidelines on which the LLM was fine-tuned, creating a transparent, two-part rationale that is both visually and textually verifiable.

Evaluation and Outcomes The DeepDR-LLM system was evaluated in both retrospective and prospective settings. In retrospective tests, its management recommendations were comparable to those of endocrinology residents and superior to primary care physicians (PCPs) [20]. When PCPs were assisted by the system, their accuracy in making appropriate DR referrals improved from 81.0 to 92.3%. More impressively, in a prospective trial involving 769 patients, those managed in the AI-assisted arm showed significantly higher treatment adherence and better self-management behaviors compared to the control group. This demonstrates a real-world impact on patient outcomes.

Strengths and Limitations The system's greatest strength is its proven clinical impact, demonstrating that a well-designed, explainable AI can improve not just physician decisions but also patient behavior. Its multi-modal grounding provides a very strong and intuitive form of explanation. However, the system has limitations. It requires large, high-quality, annotated image datasets for training, which can be expensive and difficult to obtain. The LLM component's performance depends on the quality of the clinical data it was fine-tuned on, which may contain biases. The complexity of integrating an imaging model with an LLM into a clinical workflow also presents significant technical and logistical challenges.

5 Comparative Analysis of Explainability Techniques for Treatment Planning LLMs

The six case studies presented reveal a diverse landscape of explainability techniques tailored to the specific demands of treatment planning. While no single method is universally superior, comparing the approaches highlights common principles and trade-offs. This analysis synthesizes the findings along three key dimensions: the core techniques for generating explanations, the domain-specific use cases, and the methods used for evaluation.

5.1 Techniques for Explainability

Retrieval-Augmented Generation (RAG): This was one of the most powerful techniques for ensuring verifiable and trustworthy explanations. Systems like **MEREDITH** [18] grounded their recommendations by retrieving and citing specific passages from curated knowledge bases (e.g., PubMed, clinical trials, FDA approvals). This approach directly addresses the "black-box" problem by providing a clear evidence trail, transforming the LLM from a knowledge oracle into a research assistant. The key advantage is high verifiability. The main limitation is dependence on the quality and currency of the retrieval corpus; an outdated or incomplete database leads to flawed explanations.

Reasoning Scaffolds and Chain-of-Thought (CoT): Several systems used structured prompting to force the LLM into a transparent, step-by-step reasoning process. ChatGPT-CARE [7] encoded an entire clinical guideline as a decision graph within its prompt, compelling the model to articulate its path through the guideline logic. Similarly, the RCT by Goh et al. [11] leveraged the inherent CoT capabilities of GPT-4, where the model's narrative output naturally follows a clinical reasoning structure. This technique excels at providing procedural transparency, showing how a conclusion was reached. Its weakness lies in its potential brittleness; the reasoning is only as good as the prompt's structure and the model's internal knowledge, without the external validation of RAG.

Multi-Modal Explainability: This advanced technique, showcased by GPT-RadPlan [19] and DeepDR-LLM [20], grounds textual explanations in non-textual data. GPT-RadPlan's critiques of radiation plans were based on its "vision" of dose-volume histograms and isodose curves, making its suggestions directly auditable against the visual evidence. DeepDR-LLM linked its diabetes management advice to the specific grade of retinopathy detected in a patient's fundus image. This provides an exceptionally strong form of justification by connecting abstract recommendations to concrete, observable data. The primary barrier is the high technical complexity and the need for large, annotated multi-modal datasets.

Comparative Explainability: The oncology tumor board study [17] used a more indirect form of explanation, where transparency came from analyzing concordance and discordance with human experts. While useful for evaluation, this method lacks the rich, narrative justification needed for complex, multi-faceted treatment decisions.

5.2 Domain-Specific Use Cases

Guideline-Adherent Standard Care: For domains with well-defined, protocolized care paths, such as outpatient COVID-19 treatment (ChatGPT-CARE), reasoning scaffolds that encode guideline logic are highly effective. They ensure consistency and safety for common clinical scenarios.

High-Complexity Specialty Care (Oncology): In oncology, where treatment depends on a vast and rapidly changing evidence base, RAG-based systems like **MEREDITH** shine. They can synthesize trial data, molecular markers, and off-label evidence to support decision-making in molecular tumor boards. The general oncology board study [17] showed that while standard LLMs can handle routine cases, they fail on the complex, individualized cases where systems like MEREDITH excel.

Technical and Optimization Tasks (Radiotherapy): For highly technical fields like radiotherapy planning, multi-modal explainability is most suitable. **GPT-RadPlan**'s ability to provide human-like critiques of visual data is a powerful paradigm, translating quantitative plan quality into an interpretable format.

Chronic Disease Management: The management of chronic diseases like diabetes benefits most from multi-modal systems like **DeepDR-LLM** that can integrate longitudinal data (e.g., imaging, labs) to provide holistic, long-term plans. Explainability here is crucial for patient engagement and adherence, as demonstrated by the improved outcomes in the DeepDR-LLM trial.

5.3 Evaluation Methods and Outcomes

Performance Gains in Randomized Controlled Trials (RCTs): The strongest evidence for clinical utility came from the RCT by Goh et al. [11], which demonstrated that LLM assistance led to statistically significant improvements in physician performance on management reasoning, validating the real-world value of explainable AI assistance.

Concordance with Expert Panels: A common evaluation method, used in the oncology board study [17] and by **MEREDITH** [18], was measuring agreement (e.g., using Cohen's Kappa or concordance rates) with multidisciplinary teams. These studies consistently found that specialized, evidence-grounded systems like MEREDITH achieved near-perfect concordance, while general LLMs performed well on standard cases but failed on nuanced ones.

Technical Performance Metrics: In radiotherapy, evaluation focused on technical dosimetric outcomes. GPT-RadPlan [19] produced plans that were technically superior to or on par with those of human experts, showing improved target conformity and/or OAR sparing. This demonstrates that explainable AI can lead to objectively better technical outcomes.

Beyond Accuracy—Adherence and Safety: Several studies moved beyond simple accuracy. **DeepDR-LLM** [20] measured patient adherence, a critical real-world outcome. The radiotherapy study highlighted safety, noting that multimodal critique reduces the likelihood of unsafe parameters, underscoring the need for rigorous validation. This shows a maturing evaluation landscape where explainability is linked to broader goals of equity, patient engagement, and safety.

6 Explainable Large Language Models in Patient Interaction

Patient interaction is one of the most impactful areas where explainable large language models (LLMs) can reshape healthcare delivery. Conversational agents and virtual assistants powered by LLMs are increasingly being deployed to support patients in symptom tracking, treatment adherence, chronic disease management, and health education. Their success, however, depends not only on accuracy but also on transparency. Patients are more likely to trust and follow recommendations when explanations are clear, tailored to their level of understanding, and verifiable by clinicians, as summarized in Fig. 2.

Unlike the preceding sections, which primarily examined academic case studies, this section also incorporates several **industry-driven implementations** and, importantly, **non-LLM comparators**. Systems such as AppliedVR and BioDigital Human are not LLM-based but are included here as critical benchmarks, providing contrastive insights into how other forms of explainability (e.g., mechanism-based or visual) operate in patient-facing technologies. Their inclusion highlights both the unique role of LLMs in patient interaction and the broader design space of explainability across digital health tools.

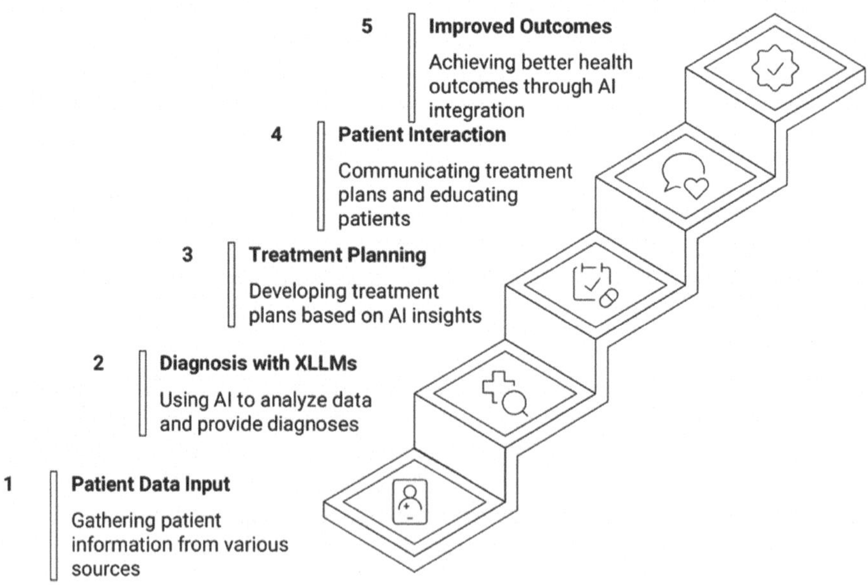

Fig. 2 Summary of explainable LLM mechanisms and their impacts across clinical domains

6.1 Case Studies

Woebot Health Woebot Health was developed by clinical psychologists at Stanford to address global mental healthcare disparities by providing scalable, stigma-free support through conversational interactions that draw on cognitive behavioural therapy (CBT), interpersonal psychotherapy (IPT) and dialectical behaviour therapy (DBT) ~ [21, 22]. It fosters trust via programmatic transparency: it discloses its artificial identity at the outset, offers real-time explanations for exercises (e.g., explaining how breathing activates the parasympathetic nervous system), and reminds users of its limitations while encouraging them to seek human care when needed [22]. Randomized controlled trials report significant reductions in PHQ-9 ($d = 0.44$) and GAD-7 ($d = 0.41$) scores, with improvements sustained at follow-up and approximately 70% retention at 30 days [21]. Strengths include scalability, accessibility, and adherence to validated therapy protocols; limitations involve an inability to manage crises, less nuanced emotional recognition than a human therapist, and dependence on user literacy.

Babylon Health Babylon Health integrates an AI-driven symptom checker with telehealth to alleviate pressure on primary care systems. Its proprietary knowledge graph synthesises clinical guidelines, medical literature and epidemiological data to analyse symptoms and identify red flags [23]. The platform explains its reasoning by clearly linking reported symptoms to clinical contexts, presenting likely differential diagnoses with confidence scores, and providing actionable recommendations, thereby enhancing user understanding and trust. Clinical validation efforts report high triage safety and improved adherence to appropriate care pathways [23]. Strengths include improved efficiency, access and patient education; limitations are the lack of a physical exam, reliance on self-reported data and the possibility of algorithmic bias.

Livongo Livongo combines FDA-cleared connected devices, AI-driven analytics and human coaching to deliver continuous, data-driven support for chronic disease management [24, 25]. It emphasises contextualised and actionable feedback by linking physiological measurements to behaviour (e.g., contrasting glucose spikes after different meals) and offering specific self-management suggestions, while highlighting trends and triggers [24]. Clinical trials demonstrate reductions in HbA1c and improved patient satisfaction, with real-world data supporting better blood-pressure outcomes [24, 25]. The programme's strengths lie in its personalised coaching and scalability; limitations include barriers due to digital literacy, subscription costs and variable electronic health record integration.

AppliedVR AppliedVR uses immersive virtual reality to deliver evidence-based pain management by competing for neural resources that would otherwise process nociceptive signals; it has received FDA clearance as a prescription digital therapeutic [26, 27]. Instead of an LLM, it relies on mechanism-based analogies that explain how the VR experience distracts the brain from pain (e.g., invoking gate-control theory) and often includes visual aids to reinforce the underlying neuroscience [26]. Clinical trials report 35–50% reductions in pain scores, reduced opioid consumption during wound-care procedures, and functional improvements in chronic pain [26, 27]. As an industry-driven comparator, its strengths include providing a non-pharmacologic option with high patient acceptance and opioid-sparing effects, whereas limitations involve equipment cost, insurance variability and unsuitability for motion-sensitive individuals.

BioDigital Human BioDigital Human offers cloud-based, interactive 3D anatomical models to enhance patient education, informed consent and comprehension for complex medical procedures [28]. Its explanatory power comes from visual storytelling: users can explore healthy versus diseased states and watch step-by-step animations of interventions (for example, comparing normal cardiac conduction with atrial fibrillation and observing ablation therapy) [28]. Studies and vendor reports suggest improved patient recall, reduced anxiety and greater efficiency in consent discussions [28]. As a non-LLM comparator, it demonstrates that visual explanation can be intuitive and effective, though it requires hardware and internet access, has limited reimbursement and cannot replace tactile modelsv.

For a comprehensive overview of the surveyed systems, see Table 3.

Table 3 Explainable large language models in patient interaction (case studies)

System	Primary use	Interaction modality	Explainability style	Evidence/ knowledge source
Woebot Health [21, 22]	Mental health coaching	Conversational app	Programmatic transparency; explains exercises and limits	CBT/IPT/DBT protocols
Babylon Health [23, 29]	Symptom triage & telehealth gateway	Chat + triage UI	Reasoning transparency; shows red flags & confidence	Proprietary KG; guidelines/ literature
Livongo [24, 25]	Chronic disease self-management (diabetes, BP)	Connected devices + coaching app	Contextualized, data-linked feedback; trend explanations	Device data; coaching protocols
AppliedVR [26, 27]	Pain management (digital therapeutic)	Immersive VR sessions	Mechanism-based analogies (e.g., gate-control); visual cues	Neuroscience rationale; clinical scripts
BioDigital Human [28, 30]	Patient education & consent	Interactive 3D anatomy models	Visual storytelling (healthy vs. diseased; step-by-step)	Anatomy library; procedure animations

6.2 Comparative Analysis

The case studies reviewed in this section encompass a spectrum of patient-facing technologies, ranging from academically evaluated LLM-driven conversational agents to industry-developed digital health platforms. Unlike the diagnostic and treatment planning sections, this discussion includes **non-LLM comparators** (e.g., AppliedVR and BioDigital Human) that employ different modalities, such as mechanism-based or visual explanations. These comparators serve as critical benchmarks for contrasting LLM-based approaches with alternative transparency paradigms. Moreover, several implementations (e.g., Livongo, Babylon Health) are primarily industry-driven rather than products of academic research. Their inclusion underscores the importance of real-world deployment and scalability, illustrating that the frontier of explainable patient interaction extends beyond controlled laboratory studies. Collectively, these cases demonstrate that explainability is not a one-size-fits-all construct; it must be tailored to the interaction modality, the clinical objective, and the cognitive and emotional needs of patients.

Key Findings and Insights Explainability is domain-specific. The LLM-based mental health agent Woebot builds therapeutic trust via programmatic transparency [21], while Babylon Health employs reasoning transparency to structure triage decisions [23]. AppliedVR, a non-LLM comparator, grounds its mechanism-based explanations in neuroscience [27]; BioDigital Human uses visual explanation to demystify complex anatomy [28]; and Livongo contextualizes physiological data into actionable lifestyle insights [24]. These varied techniques highlight that the choice of explainability method must match the interaction context and user expectations.

Complementary roles in ecosystems In a patient journey, these tools can function synergistically: Babylon may assist with initial assessment, BioDigital aids procedure education, AppliedVR offers adjunctive pain management during treatment, Livongo supports long-term disease management, and Woebot delivers ongoing mental health support. This complementary usage emphasizes the need for interoperability and consistent transparency across disparate systems.

Trust building through transparency Each system operationalizes transparency differently: logical reasoning in triage (Babylon), therapeutic self-disclosure (Woebot), mechanistic explanation grounded in science (AppliedVR), visual verification of medical information (BioDigital), and personalized data-driven feedback (Livongo). Effective trust-building hinges on aligning these explanations with user comprehension and expectations, demonstrating that both content and delivery matter.

Measurement of success Evaluation metrics vary according to clinical context: Woebot measures reductions in mental health scales [21], Babylon tracks triage safety and adherence to guidelines [23], AppliedVR monitors pain reduction and opioid use [26], BioDigital assesses patient comprehension and recall [28], and Livongo reports HbA1c improvements and engagement [24]. This diversity of metrics underscores the need for context-sensitive evaluation frameworks that link explainability not only to usability but also to clinically meaningful outcomes.

7 Conclusion

This chapter has demonstrated that explainable large language models (XLLMs) are fundamental to the responsible integration of AI in healthcare. Through case studies across diagnosis, treatment planning, and patient interaction, we have shown how techniques like chain-of-thought reasoning, retrieval-augmented generation, and tailored explanations serve dual purposes: they build essential clinical trust while simultaneously acting as cognitive scaffolding that enhances model accuracy and reduces errors. The interdependence of explainability, trust, and performance forms the foundation for effective XLLM deployment. While challenges in standardization and bias mitigation remain, the path forward requires continued focus on developing adaptive, clinically-aligned explanation systems. The evidence presented confirms that XLLMs are evolving from black-box tools into collaborative partners, ultimately enabling more transparent, accountable, and effective patient care.

References

1. D. McDuff *et al.*, "Towards accurate differential diagnosis with large language models," *Nature*, vol. 642, no. 8067, pp. 451–457, 2025, https://doi.org/10.1038/s41586-025-08869-4.
2. T. Savage, A. Agbarya, and U. Shalit, "Diagnostic reasoning prompts reveal the potential for large language model interpretability in medicine," *npj Digital Medicine*, vol. 7, no. 1, p. 20, 2024, https://doi.org/10.1038/s41746-024-01010-1.
3. T. Savage, A. Agbarya, A. Ben-Yosef, T. Ben-Yosef, A. Peled, and U. Shalit, "Large language model uncertainty proxies: discrimination and calibration for medical diagnosis and treatment," *Journal of the American Medical Informatics Association*, vol. 32, no. 1, pp. 139–149, 2025, https://doi.org/10.1093/jamia/ocae254.
4. Y. Gao and others, "Leveraging Medical Knowledge Graphs Into Large Language Models for Diagnosis Prediction: Design and Application Study," *JMIR AI*, vol. 4, no. 1, p. e58670, 2025, https://doi.org/10.2196/58670.
5. S.-A. Zhou, Z. Zhang, S. Zhang, C. Fivez, T. Demeester, and C. Develder, "Uncertainty-Aware Large Language Models for Explainable Disease Diagnosis." 2025. https://doi.org/10.48550/arXiv.2505.03467.
6. M. Saleh and A. Tabatabaei, "Building Trustworthy Multimodal AI: A Review of Fairness, Transparency, and Ethics in Vision-Language Tasks," *International Journal of Web Research*, vol. 8, no. 2, pp. 11–24, Apr. 2025, https://doi.org/10.22133/ijwr.2025.503147.1264.
7. Y. Wang, S. Visweswaran, S. Kapoor, S. Kooragayalu, and X. Wu, "ChatGPT-CARE: a Superior Decision Support Tool Enhancing ChatGPT with Clinical Practice Guidelines." 2024. https://doi.org/10.1101/2023.08.09.23293890.
8. K. Alami *et al.*, "Evaluation of ChatGPT-4's Performance in Therapeutic Decision-Making During Multidisciplinary Oncology Meetings for Head and Neck Squamous Cell Carcinoma," *Cureus*, vol. 16, no. 9, p. e68808, 2024, https://doi.org/10.7759/cureus.68808.

9. A. Tabatabaei and A. ZadZiabari, "Generative AI in Medical Imaging," in *Application of Generative AI in Healthcare Systems*, Springer, Cham, 2025, pp. 1–33. https://doi.org/10.1007/978-3-031-82963-5_1.
10. T. Tu and others, "Towards Conversational Diagnostic AI." 2024. [Online]. Available: https://arxiv.org/abs/2401.05654
11. E. Goh *et al.*, "GPT-4 assistance for improvement of physician performance on patient care tasks: a randomized controlled trial," *Nature Medicine*, vol. 31, pp. 1233–1238, 2025, https://doi.org/10.1038/s41591-024-03456-y.
12. F. Nazary, Y. Deldjoo, T. Di Noia, and E. Di Sciascio, "XAI4LLM: Let Machine Learning Models and LLMs Collaborate for Enhanced In-Context Learning in Healthcare." 2025. https://doi.org/10.48550/arXiv.2405.06270.
13. S. Zhou *et al.*, "Explainable differential diagnosis with dual-inference large language models," *npj Health Systems*, vol. 2, no. 1, p. 12, 2025, https://doi.org/10.1038/s44401-025-00015-6.
14. Z. Wang and others, "RetCare: Towards Interpretable Clinical Decision Making through LLM-Driven Medical Knowledge Retrieval." 2024. [Online]. Available: https://openreview.net/pdf/f0cd33ebf8ce9c69f36ade81fc4a6f0805671279.pdf
15. A. ZadZiabari and A. Tabatabaei, "Ethics and Regulations in Generative AI," in *Application of Generative AI in Healthcare Systems*, A. Zamanifar and M. Faezipour, Eds., Cham: Springer Nature Switzerland, 2025, pp. 197–211. https://doi.org/10.1007/978-3-031-82963-5_8.
16. B. Yang and others, "DrHouse: An LLM-empowered Diagnostic Reasoning System through Harnessing Outcomes from Sensor Data and Expert Knowledge." 2024. [Online]. Available: https://arxiv.org/abs/2405.12541
17. I. Dogan *et al.*, "Chat GPT Performance in Multi-Disciplinary Boards—Should AI Be a Member of Cancer Boards?," *Healthcare*, vol. 13, no. 18, p. 2254, 2025, https://doi.org/10.3390/healthcare13182254.
18. J. Lammert *et al.*, "Expert-Guided Large Language Models for Clinical Decision Support in Precision Oncology," *JCO Precision Oncology*, vol. 8, p. e2400478, 2024, https://doi.org/10.1200/PO.24.00478.
19. S. Liu *et al.*, "Automated radiotherapy treatment planning guided by GPT-4Vision." 2025. https://doi.org/10.48550/arXiv.2406.15609.
20. W. Jia and others, "Integrated image-based deep learning and language models for primary diabetes care," *Nature Medicine*, vol. 30, pp. 2886–2896, 2024, https://doi.org/10.1038/s41591-024-03139-8.
21. K. K. Fitzpatrick, A. Darcy, and M. Vierhile, "Delivering Cognitive Behavior Therapy to Young Adults With Symptoms of Depression and Anxiety Using a Fully Automated Conversational Agent (Woebot): A Randomized Controlled Trial," *JMIR Mental Health*, vol. 4, no. 2, p. e19, 2017, https://doi.org/10.2196/mental.7785.
22. "Woebot Health." Accessed: Sept. 25, 2025. [Online]. Available: https://woebothealth.com/
23. S. Razzaki and others, "A Comparative Study of Artificial Intelligence and Human Doctors for Triage and Diagnosis." 2018. [Online]. Available: https://arxiv.org/abs/1806.10698
24. D. J. Amante, T. S. Harlan, Y.-F. Ye, S. Olinsky, M. Sullivan, and others, "Evaluation of a Diabetes Remote Monitoring Program Based on Connected Glucose Meters: A Randomized Controlled Crossover Trial," *Journal of Medical Internet Research*, vol. 23, no. 4, p. e23402, 2021, https://doi.org/10.2196/23402.
25. J. B. Bollyky, S. T. Melton, T. Xu, S. L. Painter, J. Wellday, and others, "The Effect of a Cellular-Enabled Glucose Meter on HbA1c: A Prospective Study of the Livongo for Diabetes Program," *Journal of Diabetes Science and Technology*, vol. 13, no. 4, pp. 709–716, 2019, https://doi.org/10.1177/1932296819845670.
26. L. M. Garcia, B. J. Birckhead, P. Krishnamurthy, A. Collinsworth, B. D. Darnall, and others, "An 8-Week Self-Administered At-Home Behavioral Skills–Based Virtual Reality Program for Chronic Low Back Pain: Double-Blind, Randomized, Placebo-Controlled Trial," *Journal of Medical Internet Research*, vol. 23, no. 2, p. e26292, 2021, https://doi.org/10.2196/26292.
27. B. Mallari, E. K. Spaeth, H. Goh, and B. S. Boyd, "Virtual Reality as an Analgesic for Acute and Chronic Pain in Adults: A Systematic Review and Meta-Analysis," *Journal of Pain Research*, vol. 12, pp. 2053–2085, 2019, https://doi.org/10.2147/JPR.S200498.

28. BioDigital, Inc., "BioDigital Human: Interactive 3D Anatomy and Disease Platform." 2025. [Online]. Available: https://human.biodigital.com/
29. Z. Hussain and J. O'Donoghue, "The Role of AI-Based Chatbots in Triage and Management of Patients in Primary Care: A Review," *Journal of Medical Systems*, vol. 44, no. 11, p. 195, 2020, https://doi.org/10.1007/s10916-020-01657-7.
30. R. A. Kockro and P. Y. K. Hwang, "Virtualized Cadaver and 3D Models for Neurosurgical Training," *Neurosurgery Clinics of North America*, vol. 20, no. 2, pp. 209–219, 2009, https://doi.org/10.1016/j.nec.2008.11.007.

Reinforcement Learning in Healthcare: From Treatment Optimization to the Challenge of Explainability with Large Language Models

Amirfarhad Farhadi, Daniel Masghati Amoli, Amir Fallahzadeh, and Azadeh Zamnifar

Abbreviations

BCQ	Batch-Constrained Q-learning
BEAR	Bootstrapping Error Accumulation Reduction
CGM	Continuous Glucose Monitor
CQL	Conservative Q-Learning
DRL	Deep Reinforcement Learning
EHR	Electronic Health Records
FQE	Fitted Q Evaluation
GAN	Generative Adversarial Networks
GRU	Gated Recurrent Unit
ICU	Intensive Care Unit
LLM	Large Language Model
LRP	Layer-wise Relevance Propagation
LSTM	Long Short-Term Memory
MAP	Mean Arterial Pressure
MAPPO	Multi-Agent PPO
MDP	Markov Decision Process
OD	Overdose

A. Farhadi (✉)
School of Computer Engineering, Iran University of Science and Technology, Tehran, Iran
e-mail: am_farhadi@mail.iust.ac.ir

D. M. Amoli · A. Fallahzadeh · A. Zamnifar
Department of Computer Engineering, SR.C, Islamic Azad University, Tehran, Iran
e-mail: Daniel.masghati@iau.ir

A. Zamnifar
e-mail: azamanifar@iau.ac.ir

A. Zamanifar et al. (eds.), *Explainable Large Language Models in Healthcare Applications*, Bio-IT and AI, https://doi.org/10.1007/978-3-032-15088-2_9

OPE	Off-Policy Evaluation
OR	Odds Ratio
PER	Prioritized Experience Replay
PPO	Proximal Policy Optimization
Q-learning	Quality Learning
RAG	Retrieval-Augmented Generation
RL	Reinforcement Learning
RNN	Recurrent Neural Network
SHAP	SHapley Additive Explanations
SOFA	Sequential Organ Failure Assessment
T1D	Type 1 Diabetes
T2D	Type 2 Diabetes
TIR	Time in Range
VAE	Variational Autoencoder
XAI	Explainable Artificial Intelligence

1 Introduction

Contemporary healthcare increasingly faces complex, time-dependent decisions that benefit from adaptive, patient-specific strategies. Conditions at both ends of the spectrum—chronic diseases like type 2 diabetes and acute syndromes such as sepsis—show substantial inter-patient heterogeneity; consequently, no single fixed regimen is likely to be optimal across patients [1, 2]. In these settings, reinforcement learning (RL)—a subfield of AI in which an agent learns through trial and error in an environment to maximize long-term reward—is well-suited to optimizing dynamic treatment regimes (DTRs) and shows promising potential for supporting the shift toward adaptive, patient-specific care [3]. Recent advances—particularly Deep Reinforcement Learning (DRL), which uses deep neural networks to handle high-dimensional state spaces—have expanded the role of RL in medicine and advanced it toward applied, simulation-backed prototypes, including personalized controllers and decision-support frameworks [4, 5]. DRL algorithms can discern intricate patterns from vast clinical datasets, including electronic health records (EHRs), and propose treatment strategies that may exceed human intuition. The very complexity that makes deep learning approaches—including DRL—powerful also contributes to their opacity, often termed the "black-box" problem. When a model proposes clinically consequential actions (e.g., dose adjustments) without an interpretable rationale, clinician trust and adoption suffer. In high-stakes medicine, *why* matters alongside *what*; accordingly, research has increasingly shifted toward Explainable AI (XAI) to make complex models more transparent and clinically accountable [6, 7]. Alongside these efforts, Large Language Models (LLMs) offer a promising linguistic interface to mitigate opacity: by understanding and generating natural language, LLMs can render complex model outputs—and, when coupled

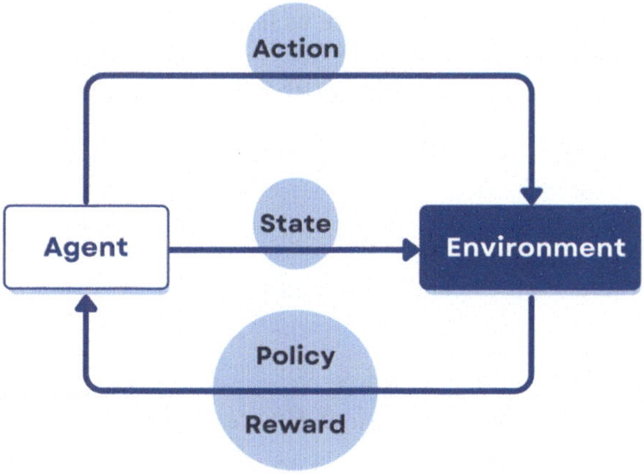

Fig. 1 This diagram illustrates how a Reinforcement Learning (RL) agent takes actions based on its current state, receives feedback in the form of rewards, and uses this to improve future decisions

with XAI/XRL tooling, articulate the state–action rationales of RL policies—into clinician-readable explanations. Such explanations, however, must be evaluated for faithfulness and safety prior to clinical use [8–15]. Recent clinical imaging work shows automated treatment planning from dental panoramic X-rays using YOLOv7, illustrating how vision models already propose multi-stage therapies—exactly the kind of recommendations our RL pipeline must keep transparent and explainable [16]. This chapter of the book primarily examines reinforcement learning, offering a thorough understanding of its applications and methodology within the healthcare sector. We will initially conduct a thorough analysis of the principles of reinforcement learning and its primary methodologies in medicine, including offline reinforcement learning. Subsequently, we examine the function of XAI and LLMs not as the ultimate objective, but as facilitating instruments for addressing the explainability challenge in RL. Subsequently, through comprehensive case studies on the treatment of sepsis and the management of diabetes, we illustrate the practical implementation of this synergy. Ultimately, by analyzing the technical, evaluative, and ethical problems, we will delineate a vision for the future of this domain. The primary objective of this chapter is to present a thorough and reinforcement learning-oriented perspective, whereby explainability is regarded as a crucial element for the application of this potent technology in contemporary medicine (Fig. 1).

2 Principles of Reinforcement Learning in Medical Applications

Reinforcement learning (RL) is a subset of machine learning in which an agent interacts with an environment to maximize the expected cumulative reward. Unlike supervised learning, which requires labeled data, RL relies on a scalar reward signal as feedback and is therefore well-suited to the sequential decision-making problems common in medicine [17].

Reinforcement learning (RL) objective is to maximize the discounted return:

$$G_t = \sum_{k=0}^{\infty} \gamma^k R_{t+k+1}$$

where $\gamma \in [0, 1]$ is the discount factor.

2.1 *Markov Decision Process (MDP)*

At the core of most RL problems is the Markov Decision Process (MDP)—a mathematical framework for decision-making in settings where outcomes are partly stochastic and partly governed by the agent's actions. An MDP is defined as the quintuple (S, A, P, R, γ) [17].

States ($\mathcal{S}$)

The set of possible environment configurations. In clinical settings, a state can be a patient feature vector (e.g., vital signs, laboratory results, demographics).

Examples

Diabetes—current glucose level and trend, recent carbohydrate intake, insulin on board [5]. ICU ventilation—PaO_2/FiO_2, CO_2, gas-exchange indices, airway pressures [18]. Sepsis—mean arterial pressure (MAP), lactate, urine output, SOFA components [1].

Actions ($\mathcal{A}$)

The set of decisions available to the agent in each state (e.g., dose adjustments, ordering diagnostics, selecting a treatment strategy) [17].

Examples

Diabetes—adjusting insulin dose/timing or recommending carbohydrate intake [5]. Ventilation—modifying V_T, PEEP, respiratory rate (RR), or FiO_2 [18]. Sepsis—administering fluids/vasopressors, adjusting antibiotics [1].

Transition kernel (P)

The environment dynamics $P(s' \mid s, a)$ specify the probability of transitioning to next state s' after taking action a in state s [17].

Examples

Diabetes—how blood glucose evolves after a bolus or a meal [5]. Ventilation—how oxygenation/ventilation responds to setting changes [18]. Sepsis—hemodynamic evolution following fluids or vasopressors [1].

Reward function (R)

A scalar feedback that encodes the clinical objective. A common formalization is

$$R(s, a, s') \triangleq \mathbb{E}\left[R_{t+1} \mid S_t = s, A_t = a, S_{t+1} = s'\right],$$

with task-specific shaping [17].

Examples

Diabetes—positive reward for time-in-range, penalties for hypo/hyperglycemia [5]. Ventilation—reward for adequate oxygenation, penalties for injurious pressures [18]. Sepsis—reward for organ-function improvement, penalties for hemodynamic instability [1].

Discount factor (γ)

A weight in [0, 1) that trades off immediate and future rewards; smaller γ emphasizes short-term outcomes [17].

The goal in RL is to learn an optimal policy π^* that maximizes the expected discounted return. A (stationary) policy maps states to actions; more generally, it maps states to distributions over actions, i.e., $\pi(a \mid s)$ with $\sum_a \pi(as)$= The (discounted) return from time t is

$$G_t = \sum_{k=0}^{\infty} \gamma^k R_{t+k+1}$$

With the value function $V^\pi(s) = \mathbb{E}[G_t \mid S_t = s]$, an optimal policy satisfies $\pi^* \in \operatorname{argmax}_\pi V^\pi(s)$ for all $s \in \mathcal{S}$, equivalently maximizing the objective $J(\pi) = \mathbb{E}[G_0]$ over policies [12, 17].

2.2 *Principal Reinforcement Learning Algorithms*

Reinforcement learning algorithms are commonly grouped into model-free and model-based approaches [17].

Model-Free Algorithms

These methods do not learn an explicit model of the environment's dynamics (and often no separate reward model); instead, they learn a policy or value functions directly from interaction (or offline logs). Q-learning is a classic value-based method that learns the action–value (state–action) function. Formally, $Q^{\pi}(s, a)$ is the expected discounted return after taking action a in state s and then following policy π; Q-learning seeks $Q^{*}(s, a)$ for the optimal policy [12, 17]. Policy-gradient methods directly optimize parameterized policies to maximize expected return; Proximal Policy Optimization (PPO) is a widely used, stable baseline in many applications [19].

Model-Based Algorithms
These methods first learn a model of the environment (transition dynamics—and, when needed, a reward model), then plan/simulate under that model to derive or improve the policy. This is typically more sample-efficient because synthetic rollouts augment limited real data—an advantage in medicine, where data collection can be costly or risky. In ICU mechanical ventilation, a model-based strategy (X-Vent) reports improved policy performance (via FQE) and leverages synthetic data to better understand system dynamics under data-constrained settings [18].

2.3 Challenges of Reinforcement Learning in Medicine and Corresponding Solutions

The application of standard reinforcement learning algorithms in medicine presents distinct challenges that necessitate specialized methodologies.

Offline Reinforcement Learning

Classic RL algorithms typically require online exploration to learn effectively; i.e., the agent must try actions to observe their consequences—an approach that can pose unacceptable patient-safety risks in clinical settings [20–22]. Offline (batch) RL addresses this by learning exclusively from previously collected, static datasets (e.g., EHR or ICU logs), with no online interaction during training [21]. Privacy-preserving synthetic clinical data with GenAI (e.g., GAN/VAE pipelines) can expand trajectories for offline RL while minimizing direct exposure of PHI, provided bias and validation checks are enforced [23]. A central challenge in offline RL is distributional shift: a learned policy may select actions rare or absent in the dataset, making value estimates unreliable. State-of-the-art methods mitigate this by penalizing or constraining out-of-distribution (OOD) actions. For example, Conservative Q-Learning (CQL) augments the Bellman objective with a conservatism penalty to avoid optimistic overestimation of OOD actions, yielding more reliable value estimates and policies in the offline setting [24]. Related approaches include BCQ [25] and BEAR [26], which explicitly constrain the learned policy toward the dataset's support. In intensive-care ventilation, a model-based strategy (X-Vent) illustrates how learning a dynamics model and using off-policy evaluation (e.g., Fitted Q Evaluation,

FQE) and synthetic rollouts can improve policy development under data-constrained conditions—an especially valuable property in medicine [18].

Simulated Environments (in Silico)

A practical way to avoid risky online interaction is to train and test policies in simulated environments. Physiology-based simulators let an RL agent learn in a safe, controlled setting. The UVA/Padova Type 1 Diabetes Simulator—FDA-accepted in 2008 as a substitute for certain animal trials—has been widely used to develop and assess glucose-control algorithms [27]. DTR-Bench was recently introduced with four healthcare environments—cancer chemotherapy, radiotherapy, glucose management in diabetes, and sepsis treatment—and supports evaluation under realistic challenges such as noise, missing data, and PK/PD variability in a standardized way [3].

Off-Policy Evaluation (OPE)

After learning a new policy—offline or in simulation—its performance must be assessed before any clinical deployment. Off-policy evaluation (OPE) methods, such as Fitted Q Evaluation (FQE), estimate a target policy's value using logged data without online testing. Technically, FQE iteratively regresses the Q-function toward Bellman targets under the evaluation policy (rather than learning an explicit environment model) [28, 29]. In ICU ventilation, X-Vent shows a model-based pipeline that uses FQE and synthetic rollouts to improve policy development under data-constrained conditions [18]. These ingredients—robust simulators, standardized benchmarks, and rigorous OPE—are key to safe, responsible RL in healthcare.

2.4 Important Factors in the Design of Reinforcement Learning Agents for Healthcare

Beyond the fundamentals, designing effective and safe RL systems for medical use requires attention to technical factors shaped by the domain's constraints—reward design, exploration management, use of simulators, and algorithm choice.

Design of the Reward Function

The reward function directly steers the agent's behavior. In medicine, mis-specified rewards can induce unsafe policies; for example, optimizing solely for lowering glucose may encourage hypoglycemia-prone behavior (illustrated in T1D controllers and ICU settings) [4, 18, 22]. Research therefore treats reward design as multi-objective, balancing patient outcomes, adverse events, and costs. A standard tool is reward shaping—adding guided intermediate rewards to accelerate learning—ideally in a potential-based form so the optimal policy remains invariant [30]. LLMs are a promising assistant for drafting reward templates from natural-language clinical guidelines, e.g., via LLM-guided credit assignment or language-based reward signals; however, these require clinical validation before deployment [31, 32]. Another complementary strategy is reward decomposition, splitting the total reward into interpretable components (e.g., "time-in-range" vs. "hypoglycemia avoidance"), which can improve learning diagnostics and explainability [12].

Exploration Management in Medical Settings

RL agents learn by exploring new actions, but in healthcare this can imply unsafe, unconventional treatments [20]. Hence algorithms must be conservative. Offline RL mitigates risk by training only on static EHR/ICU logs with no online interaction [21]. Dataset bias and support mismatch then become central issues; therefore, CQL penalizes over-optimistic values for out-of-distribution actions, while BCQ and BEAR constrain the learned policy toward the dataset's support [24–26]. In simulator-based training, exploration is safe but should be efficient; Prioritized Experience Replay (PER) improves sample-efficiency by replaying more informative transitions more often [33].

The Significance of Simulation and the Progression from Simulator to Real-World Application (Sim2Real)

Simulated environments (in silico) are essential for advancing reinforcement learning in healthcare. Platforms such as DTR-Bench provide standardized simulators across multiple diseases, enabling reproducible development and evaluation under realistic challenges such as noise, missing data, and patient variability [3]. A persistent hurdle is the simulation-to-reality gap: policies that excel in simulation may underperform clinically due to mismatches in dynamics and data distributions. Domain randomization helps narrow this gap by stochastically perturbing key environment/physiology parameters during training so the agent learns policies robust to uncertainty and variability [34–36]. In addition, hybrid, physics-informed approaches—for example, integrating physiological cardiovascular models with deep RL—can yield more robust and generalizable control policies [1]. The transition to real-world use should be staged and safety-first (e.g., extensive offline evaluation/OPE, shadow-mode deployment, prospective pilots) before any clinical rollout [21, 28, 29]. Energy-efficient city-scale communication schemes like EECASC highlight how to reduce end-to-end delay and device energy—constraints we mirror when streaming bedside telemetry to our RL/LLM services [37].

2.5 Practical Formalization Details and Robustness Considerations

Return and Control Horizon
In clinical RL, the objective is usually the discounted return

$$G_t = \sum_{k=0}^{\infty} \gamma^k R_{t+k+1}, \gamma \in [0, 1) \text{ (continuing)}; \gamma \in [0, 1] \text{ (episodic)},$$

Balancing near-term safety with long-term outcomes [17]. Choosing γ is not cosmetic: in sepsis, a myopic objective (small γ) may over-emphasize immediate MAP correction while neglecting downstream renal risk; an overly far-sighted one may tolerate unsafe short-term dips. Report results across a small grid of γ iand control horizons H (e.g., number/size of decision bins) as standard practice [17, 20, 21].

Partial Observability
Patient trajectories are rarely Markovian under observed variables; the problem is better seen as a POMDP. Recurrent function approximators (GRU/LSTM) help infer latent state but can also entangle confounding with memory. A pragmatic template is to (i) include short windows of summary statistics (trends/deltas) and (ii) run ablations to quantify how history length impacts policy/action distributions [17, 20].

Confounding and Action Support
In offline RL, policies are fragile when they select actions rarely supported by the logs. Support-constrained methods (BCQ/BEAR) and pessimistic value learning (CQL) explicitly discourage out-of-distribution (OOD) actions [24–26]. In clinical datasets with strong treatment conventions, always pair value estimates with coverage diagnostics (e.g., empirical propensity of the selected actions, overlap ratios) [21, 24].

Model-Based Versus Model-Free
Model-based pipelines (learned or hybrid physiological dynamics) can be more sample-efficient, enable counterfactual rollouts, and expose interpretable safety artifacts (e.g., dose–response curves) [1, 18]. Model-free baselines (e.g., PPO, DQN/DDQN) remain strong when dynamics modeling is unreliable, but should be paired with stringent OPE and sensitivity analyses [19, 20, 28, 29].

Domain Randomization and sim2real
To avoid overfitting to a single simulator regime, inject variability in key physiological parameters (PK/PD), measurement noise, and missingness during training. This improves robustness when moving from in silico to clinical logs or pilots [3, 34–36]. Hybridizing with mechanistic models—e.g., cardiovascular dynamics—further grounds the learned policy in biophysics [1].

3 The Role of LLMs and XAI in Reinforcement Learning Explainability

While reinforcement learning (RL) offers powerful algorithmic capabilities, limited transparency remains a major barrier to clinical adoption. This section outlines Explainable AI (XAI) as the goal and Large Language Models (LLMs) as a key tool to move RL-based medical systems toward clinician-trustworthy use.

3.1 Categorization of Explainability Approaches in Reinforcement Learning

The XRL literature can be organized by what is explained and how explanations are produced. A practical scheme groups methods into three levels—feature importance, learning process/MDP-level, and policy-level [12]. A complementary view classifies approaches by when explainability is applied: intrinsically interpretable vs post-hoc techniques [38]. In medical settings—where high-performing DRL models are often complex—post-hoc methods are frequently the most pragmatic option after training, provided their faithfulness and clinical relevance are evaluated [12–14, 38].

3.2 LLMs as a Tool for Enhancing Reinforcement Learning (LLM4RL)

The interaction between LLMs and RL is bidirectional [39]. Here we focus on LLM4RL—using LLMs to improve the RL learning process (as opposed to RL4LLM, which improves LLMs) [40]. Several roles are promising for medical applications:

LLM as a Reward Designer
Reward design in healthcare is challenging: narrowly optimizing a single metric (e.g., glucose reduction) can induce unsafe behaviors [4, 18, 22]. LLMs, grounded in clinical texts/guidelines, can help draft reward templates or shaping signals from natural-language objectives, improving alignment between agent behavior and clinical goals [31, 39, 40]. These designs still require validation and safety checks before deployment [13, 14].

LLM as a Policy Teacher (Teacher–Student Paradigm)
In complex environments, RL can be slow and data-hungry. An LLM acting as an expert "teacher" can provide high-level guidance, curricula, or demonstrations to accelerate a "student" RL agent, improving exploration and reducing costly interactions [40–42].

LLM-Assisted Safety and Reliability
LLMs can serve as supervisory modules that detect risks such as goal misgeneralization by analyzing trajectories and proposing corrective feedback (e.g., reward edits or constraints), thereby nudging the agent back toward intended clinical objectives [13, 14, 32].

3.3 LLM as a Generator of Explanations

A prominent use of LLMs in XRL is generating clinician-facing natural-language rationales. Many XAI methods (e.g., SHAP, LRP) output attribution scores that quantify each feature's contribution; these are informative for model developers but may be opaque to busy clinicians [43, 44]. LLMs can act as a "translator," turning attributions and clinical context into readable explanations—provided that faithfulness is monitored and hallucinations are controlled [10, 13, 14].

Pipeline (conceptual):

RL Decision
The RL agent proposes a treatment action (e.g., drug dose) based on the current patient state.

XAI Analysis
An attribution method (e.g., SHAP or LRP) identifies which inputs (labs, vitals, comorbidities) most influenced the policy's choice [43, 44].

LLM Explanation
The RL output, XAI attributions, and patient context are fed to an LLM, which converts them into a concise, clinician-readable rationale; frameworks such as Talk To Agent demonstrate LLM-mediated explanations for RL agents [10, 11, 13, 14]. (Figure 2 illustrates the RL–LLM–XAI workflow used throughout this chapter.)

This hybrid approach couples the optimization strength of RL with the communication ability of LLMs, allowing clinicians to consider AI recommendations with an accompanying rationale. When combined with measured faithfulness and user evaluation, such systems can improve trust and support clinical integration [13, 14, 45].

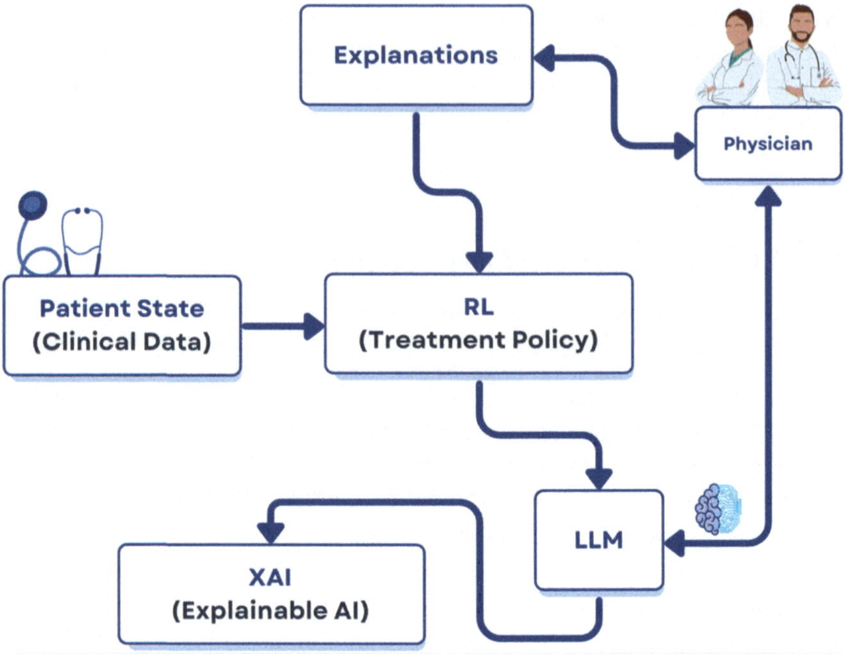

Fig. 2 This diagram depicts a clinical decision-support system where RL generates treatment policies, and LLM converts XAI attributions into explanations for the physician, who provides feedback for system improvement

3.4 *Faithfulness, Safety, and Clinical Usefulness of Explanations*

While LLMs can verbalize RL rationales, faithfulness—that an explanation truthfully reflects internal model evidence rather than a plausible story—remains the primary risk in high-stakes care [13, 14]. We outline a minimal, practical auditing suite:

Attribution-to-Text Consistency Tests
Generate post-hoc attributions (e.g., SHAP, LRP) for the exact state–action pair, then verify that tokens/phrases in the LLM rationale align with the top-k influential features. Track a mismatch rate (e.g., clinician-important features absent from the rationale) and set a deployment threshold [13, 43, 44].

Counterfactual Sensitivity
Perturb a clinically salient input (e.g., lactate + 10%) while holding others fixed (or using a physiology-consistent perturbation in a simulator), recompute the policy/

action, and regenerate the explanation. Safe systems should flip or qualify the rationale accordingly. Log monotonicity checks for variables with known clinical directionality (e.g., persistent MAP < 65 mmHg → vasopressor up titration per Surviving Sepsis) [13, 14, 22, 46].

Guideline Anchoring Via Retrieval
Use RAG to ground explanations in guideline snippets or textbooks so the LLM cites source fragments (internally) when making claims (e.g., "target MAP $\geq$ 65 mmHg in septic shock") [8, 45–47]. This reduces hallucination risk and aligns the narrative with established standards [13, 14].

Human Factors and Workload
User studies indicate perceived usefulness depends on brevity, signal-to-noise, and calibrated uncertainty. Provide a two-layer UI: a one-sentence rationale ("why now?") plus an expandable section with top features, confidence, and retrieved guidance [13, 45].

Safety Roles for LLMs beyond Narration
LLMs can act as watchdogs: scan trajectories for patterns of goal misgeneralization (e.g., proxy optimization of surrogate vitals) and trigger constraint edits or reward-shaping proposals for the RL agent. This teacher–critic setup shows training acceleration and safer exploration in non-clinical domains, and is promising for CDS when coupled with conservative offline RL [24, 31, 32, 41, 42].

Evidence Base in Healthcare
Scoping and empirical reviews report mixed reliability of LLMs as CDS and urge rigorous evaluation; oncology and specialty studies reveal inconsistency and occasional unsafe advice, reinforcing the need for faithfulness/auditing before clinical use [45, 48–51].

4 Case Studies: Applications in Sepsis Treatment and Diabetes Management

This section analyzes two representative case studies to illustrate the practical potential of integrating reinforcement learning (RL), large language models (LLMs), and explainable AI (XAI), with emphasis on the RL pipeline.

4.1 Management of Sepsis in the Intensive Care Unit (ICU)

Problem definition: Sepsis is a life-threatening dysregulated host response to infection that can cause organ dysfunction. In the ICU—especially with septic shock and hypotension—clinicians must make frequent decisions about intravenous fluids and vasopressors, where both under- and over-treatment can harm organs and increase mortality. The goal is to learn a dynamic treatment policy that keeps hemodynamics in a clinically desirable range [20, 22].

Formulated as an RL problem. The task maps naturally to an MDP:

State

A patient vector at fixed time steps (e.g., 4-h bins) containing vitals (MAP, HR, RR, SpO_2, temperature), neurological status (e.g., GCS), ventilation parameters, labs (lactate, creatinine, platelets), demographics, and recent medications [20, 22].

Action

A discrete action space over fluids and vasopressors; a common design is 5 levels $\times$ 5 levels (fluids $\times$ vasopressors) $=$ 25 actions [22].

Reward

Reward design is critical. A widely used sparse outcome is long-horizon mortality (e.g., $+$ 1 for 90-day survival, $-$ 1 for death), which aligns with the primary objective albeit with limited per-step guidance [22].

Learning Approaches

Because online exploration is unsafe, most work uses offline RL on large ICU datasets (e.g., MIMIC) [21, 22]. Data quality/coverage substantially affects learned policies. Lu et al. [20] showed that Dueling-DDQN policies are highly sensitive to feature sets, network architecture, reward design, and even random seeds—underscoring the need for thorough sensitivity analyses before clinical claims [20]. Model-based RL offers a complementary path: Nanayakkara et al. integrate a physiological cardiovascular model with DRL to learn uncertainty-aware controllers, aiming for more sample-efficient and robust policies aligned with domain knowledge [1].

Role of LLMs and XAI for explainability

Suppose the agent recommends increasing norepinephrine. A clinician asks *why*. XAI analysis. An attribution method (e.g., SHAP) identifies the most influential features—say, sustained MAP < 65 mmHg and declining urine output [43]. LLM explanation: The RL action, SHAP attributions, and current state feed an LLM, which produces a concise rationale:

"Increase in vasopressor dose is recommended because MAP has stayed below 65 mmHg with concurrent fall in urine output over the past two hours, suggesting organ hypoperfusion; projected hemodynamic response indicates higher norepinephrine is likely to restore MAP to target." Such explanations let clinicians rapidly vet recommendations; faithfulness should be audited before deployment [13, 14]. Implementation specifics: Sepsis RL commonly uses high-dimensional states aggregated every 4 h and the 25-action discretization above [22]. Partial observability is inherent; RNN-based function approximators (e.g., LSTM/GRU) can leverage history to infer latent state [17, 20]. Nanayakkara et al. [1] combine a physiology-driven recurrent autoencoder with DRL to map observations into an interpretable latent space with uncertainty, improving robustness and clinical plausibility [1]. Lu et al. [20] further highlight that reproducibility requires transparent reporting of features, rewards, seeds, and evaluation protocols [20].

4.2 Tailored Approaches to Diabetes Management

Problem definition. Effective management of type 1 and type 2 diabetes requires tight glucose control to prevent both acute (hypo/hyperglycemia) and chronic complications. Doses must be adapted continuously to meals, activity, and individual physiology. For type 1 diabetes, the long-term goal is an artificial pancreas that autonomously recommends optimal dosing [2, 4, 27].

Formulated as a reinforcement-learning problem.

State (S)

A patient vector comprising recent CGM readings (and trend), carbohydrate intake/announced meals, insulin-on-board, and (optionally) activity or stress signals [4, 27].

Action (A)

Adjusting insulin delivery (basal/bolus); in dual-hormone designs, glucagon delivery for hypoglycemia prevention is included [4].

Reward (R)

Multi-objective signals that increase time-in-range (70–180 mg/dL) and penalize hypo/hyperglycemia; sparse outcome terms (e.g., event avoidance) may be added [4].

Learning approaches

Because human online exploration is unsafe, most work trains/evaluates in silico using the UVA/Padova Type 1 Diabetes Simulator—FDA-accepted since 2008 as a substitute for certain preclinical trials [27]. Within this setting, DRL controllers (e.g., PPO/DQN families) have learned policies that raise TIR and reduce adverse events; Zhu et al. showed a dual-hormone DRL controller achieving strong in-silico performance with personalization to individual data [4, 19, 27]. After population-level training, models are typically fine-tuned for each patient.

LLMs and XAI for Explainability

If the system recommends a small insulin bolus at BG = 120 mg/dL, a patient may ask *why*.

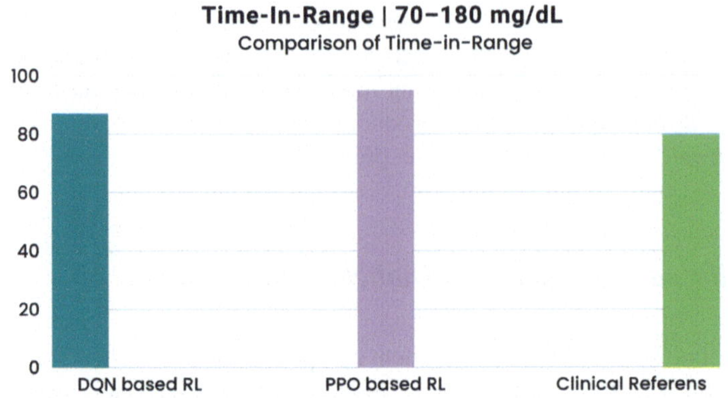

Fig. 3 Comparison of Time-in-Range (TIR; 70–180 mg/dL) across methods (DQN-based RL, PPO-based RL, and a clinical reference policy)

XAI Analysis
An attribution method (e.g., SHAP) shows the decision depends not only on the current BG but also on a rapid upward trend over the last 30 min and an imminent meal announcement [43].

LLM Explanation (Patient-Facing)
"Your glucose is currently in range, but it's rising quickly and you're about to eat; a small dose now helps keep levels within the target range and prevents a post-meal spike."

This transparent interaction supports acceptance, education, and engagement. Offline RL with clinical data. Beyond simulators, offline RL on CGM/EHR logs is a pragmatic path: CQL/BCQ/BEAR constrain policies toward the dataset's support to improve safety [24–26]. Nambiar et al. [52] report that deep offline RL (including CQL with re-balancing) can learn treatment policies aligned with clinical intent in real-world optimization scenarios [52]. These methods enable large-scale learning of personalized controllers from routine data (Fig. 3).

4.3 Comparative Table of Key Studies

To provide a comprehensive overview of the research landscape, the following (Table 1) summarizes the prominent studies reviewed in this chapter. This table compares the RL methods employed, the disease or application domain, key findings, and the evaluation metrics used, thereby facilitating a better understanding of the current state and trends in this field.

Table 1 Comparison of key studies on the application of RL in medicine and the challenge of explainability

Study (References)	Disease/ domain	Main method	Key finding/result	Evaluation metrics
Zhu et al. [4]	Type 1 diabetes	DRL controller with dilated RNN (dual-hormone insulin + glucagon)	In-silico dual-hormone policy achieved very high TIR ($\approx$ 93% adults; $\approx$83% adolescents) and reduced adverse events	Time-in-range, % hypo/hyper
Safaei et al. [18]	ICU mechanical ventilation	Model-based RL with learned dynamics; OPE via FQE; explainable	Model-based pipeline improved policy evaluation and transparency under data-limited settings vs baselines	FQE, action-distribution analyses
Nanayakkara et al. [1]	Sepsis treatment	Physiology-informed DRL with cardiovascular model; uncertainty-aware	Integrating a mechanistic CV model with DRL yielded more robust, interpretable, uncertainty-aware policies	Clinical plausibility/ qualitative alignment; uncertainty quantification
Lu et al. [20]	Sepsis management	Dueling DDQN (offline)	Learned policies are highly sensitive to features, architecture, reward, and random seed → rigorous sensitivity analysis needed	Policy action-distribution and sensitivity analyses
Sun et al. [2]	Type 2 diabetes	RL recommender (hybrid knowledge- and data-driven)	Recommendations associated with better glycemic/BP/ lipid control and lower long-term risk	ORs for control; complication-risk proxies
Nambiar et al. [52]	Diabetes and Sepsis	Offline RL (CQL) with re-balancing	Handling class imbalance improved outcome prediction and guideline concordance	MCR, WIS
Luo et al. [3]	Chemo, radiotherapy, diabetes, sepsis	DTR-bench simulators	RL performance varies under noise/ patient-variability; some algorithms fail to converge	Cumulative reward; degradation under noise/missing-data
Isakov et al. [53]	Healthcare scheduling	Explainable multi-agent system (BDI-based), not strictly RL	Reduced waiting time with interpretable evaluator agents	AWT, KL divergence

4.4 Case Study III: Oncology CDS with Guideline-Grounded LLMs

Problem Precision-oncology workflows combine pathology, biomarkers, staging, and evolving guidelines. Decisions span multi-step regimens, dose adjustments, and toxicity management—an appealing but risky target for RL given sparse outcomes and high confounding.

LLM-Anchored CDS Landscape

Recent oncology-focused evaluations show that general-purpose LLMs can produce reasonable summaries but vary in guideline adherence and may omit contraindications [45, 48, 50]. Expert-guided fine-tuning and guideline-grounded RAG have improved alignment in pilot studies [47, 49, 54].

RL Opportunity and Constraints

Rather than issuing direct treatment policies, a conservative starting point is sequential recommendation support: (i) encode tumor-board decisions as logged sequences; (ii) learn an offline value function estimating downstream outcomes (e.g., a time-to-progression proxy); (iii) restrict the action set to guideline-compliant options with strong dataset support [21, 28]. CQL/BEAR help maintain conservatism under support mismatch [24, 26].

LLM4RL Integration

Reward templates from guidance

Use LLMs to draft multi-objective reward components (tumor response, toxicity avoidance, dose-intensity adherence), grounded by retrieved guideline fragments; clinicians review/approve before training [39, 40, 47].

Teacher prompts for credit assignment

In multi-agent tumor-board simulations (pathology /medical oncology/radiation oncology "agents"), LLM prompts can structure stage-wise curricula and shape credit assignment—ideas explored in LLM-guided MARL and credit-assignment work [31, 42].

Explanations to clinicians

Post-hoc SHAP/LRP + LLM yield concise rationales—e.g., "Dose reduction suggested due to cumulative neuropathy grade and marginal projected PFS gain this cycle"—with expandable extracted guideline lines and uncertainty disclaimers [13, 14, 43, 44].

1. Evaluation plan: Before any prospective test:
2. OPE using FQE with confidence sets; report value distributions and worst-case bounds [28, 29].

3. Action-support heatmaps across strata (age, stage, comorbidity) to detect coverage gaps [21, 24].
4. Guideline concordance vs audited baselines, plus error analysis on discordant cases [45, 47–49].
5. Human-factors studies: time-to-decision, workload, trust, and override rates in vignette experiments [45].

5 Challenges, Limitations, and Future Directions

Although integrating RL with LLMs for explainable AI in medicine is promising, several technical, evaluative, and ethical hurdles remain before broad real-world deployment. Sector-wide reviews show GenAI permeating imaging, admin workflows, and patient engagement, while underscoring the need for robust governance and safety scaffolding around decision support [55].

5.1 Unique Challenges in Medical RL

LLM accuracy and hallucination

Hallucinations and variable accuracy in LLMs pose safety risks; even with retrieval-augmented generation (RAG), guaranteeing clinical correctness is unresolved. Reviews and empirical studies report inconsistent or incorrect outputs in oncology and CDS evaluations [9, 45, 48, 50, 51, 56].

Data quality and bias for offline RL

Offline RL depends on historical EHR/ICU logs that are noisy, incomplete, and biased; coverage gaps and confounding can mislead value estimation and policy learning [21]. Practical case studies show that dataset imbalance and quality strongly affect results and require mitigation [52].

Learning from Imperfect Demonstrations

If models simply mimic historical suboptimal care, they may learn unsafe policies. Conservative and support-constrained offline methods (e.g., CQL/BCQ/BEAR) explicitly address this risk by penalizing or constraining out-of-distribution actions [24–26].

Non-stationarity and partial observability

Many RL algorithms assume stationary Markovian dynamics, while patient physiology can shift over time due to unmeasured factors; this violates assumptions and complicates policy learning [17, 21].

Long-Term Credit Assignment

Clinical effects often unfold over long horizons, making it difficult to attribute outcomes to past actions—an inherent RL challenge amplified in healthcare [17].

Sensitivity and reproducibility

Sepsis studies show high sensitivity of learned policies to features, architecture, rewards, and even random seeds, underscoring the need for rigorous sensitivity analysis and transparent reporting before clinical claims [20].

5.2 *Future Directions*

Robust evaluation before deployment: Before any clinical use, estimate the learned policy's value with standardized off-policy evaluation (OPE). Fitted Q-Evaluation (FQE), paired with uncertainty analysis (e.g., bootstrapping for confidence intervals), offers a transparent baseline: it quantifies how much the learned policy is expected to outperform the historical/behavior policy—and with what margin of confidence. Report FQE alongside complementary importance-sampling estimators (e.g., WIS) so conclusions are not overly sensitive to overlap assumptions [28, 29].

Physiology-informed RL

Marrying mechanistic physiological models with RL—e.g., as hybrid controllers—brings three concrete benefits: better sample efficiency when labeled clinical data are scarce, stronger robustness to distribution shift, and more interpretable decisions because intermediate physiological variables remain explicit. In critical-care settings such as sepsis, these hybrids help constrain learned behaviors and make them easier to audit [1, 18].

Sim-to-Real and robustness

Before any prospective pilot, stress-test policies in benchmarked in-silico environments. Domain randomization over key parameters—measurement noise, missingness, and PK/PD variability—reduces overfitting to a single simulator regime and prepares the policy for clinical heterogeneity. Coupling this practice with dynamic-treatment benchmarks lowers the risk of unpleasant surprises in the real world [3, 34–36].

Human-in-the-loop and trust

Explanations should be measured, concise, and auditable. A practical pattern is a one-line rationale answering "why this, now?" plus an expandable panel with top-influential features, uncertainty, and retrieved guideline snippets. To ensure explanations reflect the model's actual evidence, include faithfulness audits—e.g., alignment between attribution maps and the textual rationale, and counterfactual sensitivity tests. User studies and scoping reviews in CDS repeatedly emphasize these needs [13, 14, 45].

Data governance and bias mitigation

Because most medical RL projects are offline, rigorous dataset documentation (datasheets), quality checks, and handling of class/cohort imbalance are part of the safety case. In offline RL specifically, monitor action support and the behavior policy to avoid overly optimistic OPE when the learned policy proposes rarely observed actions [21, 52].

Domain-specialized medical LLMs

To reduce hallucinations and improve clinical accuracy, prefer LLMs trained or adapted on high-quality medical corpora. Evidence from surveys and field evaluations shows that general-purpose models are less stable on CDS tasks; targeted corpora and task-specific evaluations improve both knowledge coverage and safety-critical reliability [8, 9, 51, 56, 57].

Trustworthy AI beyond explainability

Good explanations are necessary but not sufficient. Trustworthy clinical AI also requires robustness to distribution shift, safety mechanisms, and algorithmic fairness—from data design through post-deployment monitoring [51, 58].

Standardized benchmarks

Sustainable progress depends on shared datasets and evaluation protocols. The community should converge on benchmarks that capture task-specific clinical outcomes and explicit faithfulness tests for explanations, enabling apples-to-apples comparisons on both performance and explanation quality [3, 13, 14, 45]. GenAI is accelerating de-novo molecule design; coupling generators with RL (policy-guided search) and XAI keeps candidate selection traceable across efficacy/toxicity objectives [23].

Governance and oversight

Even excellent models should not enter clinical workflows without clear pathways for approval, monitoring, and accountability. Define pre-deployment evaluation gates, authorization processes, post-market surveillance, incident reporting, and independent review mechanisms to maintain professional and public trust [51]. A BERT-enhanced blockchain pipeline for health-insurance fraud shows how transformer evidence can be chained on-ledger—an approach we can mirror to record RL rationales, clinician overrides, and outcomes [59].

Executive takeaway

A mature path for RL + LLM in healthcare runs through uncertainty-aware offline evaluation, physiology-aware modeling, Sim-to-Real stress-testing, faithful, clinician-centered explanations with humans in the loop, and strong data/model governance—all within standardized benchmarks and oversight frameworks [1, 3, 13, 14, 21, 28, 29, 34–36, 45, 51, 58].

5.3 Ethical and Practical Challenges

Data privacy is paramount: using sensitive patient data calls for privacy-preserving strategies (e.g., de-identification, secure governance; often discussed within trustworthy-AI frameworks) [51, 58]. Current guidance emphasizes privacy, security, fairness, and explainability checkpoints across the GenAI lifecycle—principles we integrate into our RL pipeline's pre-deployment validation and continuous monitoring [60]. Algorithmic bias is a major risk—models can inherit and even amplify existing health inequities if training data are biased [51, 58]. Accountability is also complex: when AI-supported decisions cause harm, responsibility among developers, institutions, and clinicians must be clearly specified [51]. Finally, clinical trust and acceptance remain practical barriers; even with explanations, clinicians may be reluctant to rely on algorithms without rigorous clinical validation and human-centered design [45, 51].

5.4 Challenges Unique to Explainable RL (XRL)

Temporal dependencies

Unlike single-shot classifiers, RL decisions depend on sequences of past observations and actions. Explanations should answer "Why this action now?," capturing long-term effects and causal relations along a trajectory. Emerging methods such as behavior discovery/attribution begin to surface consistent behavioral patterns but remain early-stage for clinical use [61].

Policy versus value explanations

Much XRL work explains value functions (e.g., features that increase QQQ). For clinicians, policy-level understanding ("under what conditions does the policy escalate a dose?") can be more actionable. Policy distillation (e.g., training a simpler surrogate such as a decision tree) can improve interpretability, but inevitably trades off fidelity vs simplicity [12, 38].

Measuring explanation quality

What makes a "good" explanation in RL? Beyond quantitative fidelity, explanations must be clear, useful, and reliable for end users. This requires rigorous human studies measuring effects on decision-making, trust, and workload. The absence of widely accepted human-evaluation benchmarks/metrics hampers comparison across XRL methods [12–14, 38, 45].

5.5 A Deployment Playbook for RL + LLM + XAI CDS

Phase 0—Problem scoping and governance

Define a narrow decision locus (e.g., vasopressor titration bands or insulin bolus magnitude) with clear exclusions. Establish data governance, de-identification, and model risk management artifacts (datasheets, model cards) [51, 58, 62, 63]. Lightweight blockchain-based authentication and key-agreement patterns—proven in high-mobility settings—can be adapted to IoMT so every RL action and LLM rationale is tamper-evident and auditable [64].

Phase 1—Data curation for offline RL

Cohort building with explicit inclusion/exclusion, timestamp alignment, and leakage checks. Support diagnostics: estimate behavior-policy propensities; prune states with insufficient action coverage to avoid spurious OPE [21, 24, 28, 29]. Fairness and imbalance handling: reweighting or stratified sampling to stabilize rare but critical subpopulations (e.g., renal failure), as evidenced in offline RL applications [52].

Phase 2—In-silico and hybrid modeling

Train and stress-test in trustworthy simulators (e.g., UVA/Padova for T1D, DTR-Bench for DTR problems) with domain randomization to probe robustness [3, 27, 34–36]. Where feasible, incorporate mechanistic models (e.g., cardiovascular dynamics) to constrain learning and expose interpretable intermediates [1, 18].

Phase 3—Conservative learning and evaluation

Prefer pessimistic offline RL (CQL) or support-constrained methods (BCQ/BEAR) to reduce OOD risk [24–26].

OPE stack: FQE with bootstrap CIs; complement with importance-sampling estimators (e.g., WIS) for sanity checks under overlap [28, 29]. Sensitivity analysis: replicate Lu et al.—vary features, seeds, architectures, reward weights; report action-distribution stability, not only scalar returns [20].

Phase 4—Explanation and safety auditing

Faithfulness suite per §3.3**:** attribution alignment, counterfactual flips, and RAG anchoring to guidelines [13, 14, 43–45, 47].

Watchdog LLM to monitor goal misgeneralization patterns and propose constraint edits; changes are reviewed offline by clinicians before retraining [32, 39–41]. Human-in-the-loop UI: collapsed one-liner rationale with expandable evidence (top features, confidence, retrieved guideline lines) [45].

Phase 5—Shadow mode → prospective pilot

Shadow deployment: run the policy in parallel without affecting care; capture disagreement with clinicians, near-misses, and explanation-override reasons. Prospective pilots under IRB with pre-specified stopping rules and clinician override

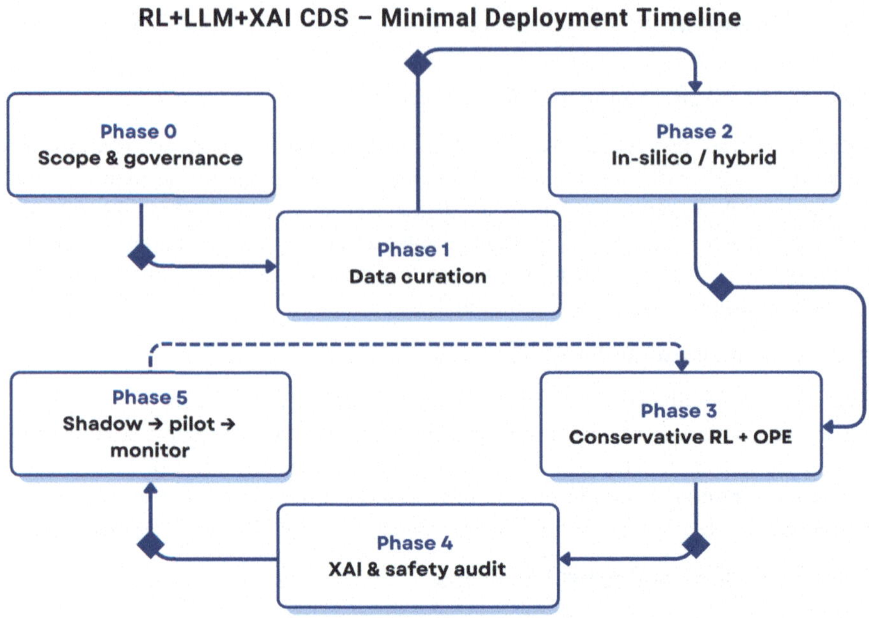

Fig. 4 Minimal deployment timeline for an RL + LLM + XAI clinical decision support system. Solid arrows indicate the forward staging from Phase 0 (scoping and governance) through Phase 5 (shadow → pilot → monitoring). Diamonds (◆) mark pass/fail gates before advancing. The dashed arrow denotes the feedback loop from Phase 5 back to Phase 3 to trigger conservative retraining and offline policy evaluation under governance

as first-class control. Post-market monitoring for drift, inequities, and rare adverse patterns; retrain with governance checkpoints [51, 58] (Fig. 4).

6 Conclusion

This chapter explored the intersection of reinforcement learning (RL), large language models (LLMs), and explainable AI (XAI) in clinical decision support. RL shows promise for optimizing treatment decisions, but its "black-box" nature hinders clinical adoption [1–4, 18, 20, 22, 27]. LLMs can enhance transparency by translating model outputs into understandable rationales, especially when integrated with XAI [8, 9, 11, 13, 14, 39, 40]. Through various frameworks, LLMs not only generate explanations for RL policies but can also assist in designing rewards or guiding learning processes to improve data efficiency and safety [11, 13, 14, 31, 40–42]. Case studies in sepsis and diabetes have demonstrated the potential of this synergy [1, 2, 4, 20, 27, 52]. Challenges remain, including LLM accuracy, data biases in RL, and ethical concerns around privacy and accountability [3, 9, 13, 14, 21, 45, 48, 51,

52, 56, 58]. Moving forward, clinical validation, robust AI frameworks, and collaboration between clinicians and AI developers will be crucial [1, 18, 28, 29, 45, 51, 58]. Ultimately, RL, LLM, and XAI could provide transparent and safe decision support in complex clinical settings, but will require careful design and ongoing validation to become reliable tools in practice.

References

1. Nanayakkara T, Clermont G, Langmead CJ, Swigon D. Unifying Cardiovascular Modelling with Deep Reinforcement Learning for Uncertainty Aware Control of Sepsis Treatment. PLOS Digit Health. 2022;1:e0000012. doi:https://doi.org/10.1371/journal.pdig.0000012
2. Sun X, Bee YM, Lam SW, Liu Z, Zhao W, Chia SY, et al. Effective Treatment Recommendations for Type 2 Diabetes Management Using Reinforcement Learning: Treatment Recommendation Model Development and Validation. J Med Internet Res. 2021;23:e27858. doi:https://doi.org/10.2196/27858
3. Luo Z, Zhu M, Liu F, Li J, Pan Y, Zhou J, et al. DTR-Bench: An in silico Environment and Benchmark Platform for Reinforcement Learning Based Dynamic Treatment Regime [Internet]. arXiv; 2024 [cited 2025 Aug 16]. https://doi.org/10.48550/arXiv.2405.18610
4. Zhu T, Li K, Georgiou P. Personalized Dual-Hormone Control for Type 1 Diabetes Using Deep Reinforcement Learning. In: Shaban-Nejad A, Michalowski M, Buckeridge DL, editors. Explainable AI in Healthcare and Medicine [Internet]. Cham: Springer International Publishing; 2021 [cited 2025 Aug 16]. p. 45–53. doi:https://doi.org/10.1007/978-3-030-53352-6_5
5. Singh Rathore SP, Prajapat D, Sharma S, Pooja S, Seth R, Sharma G. Personalized Diabetes Management and Treatment Planning using Reinforcement Learning. 2024 4th International Conference on Advancement in Electronics & amp; Communication Engineering (AECE) [Internet]. GHAZIABAD, India: IEEE; 2024 [cited 2025 Aug 16]. p. 357–63. doi:https://doi.org/10.1109/AECE62803.2024.10911581
6. Thakur R. Explainable AI: Developing Interpretable Deep Learning Models for Medical Diagnosis.
7. Prentzas N, Kakas A, Pattichis CS. Explainable AI applications in the Medical Domain: a systematic review.
8. Wang W, Ma Z, Wang Z, Wu C, Ji J, Chen W, et al. A Survey of LLM-based Agents in Medicine: How far are we from Baymax? [Internet]. arXiv; 2025 [cited 2025 Aug 16]. https://doi.org/10.48550/arXiv.2502.11211
9. A Survey on Using Large Language Models in Healthcare. International Journal of Science and Engineering Applications [Internet]. 2025 [cited 2025 Aug 16]; doi:https://doi.org/10.7753/IJSEA1406.1001
10. Bilal A, Ebert D, Lin B. LLMs for Explainable AI: A Comprehensive Survey [Internet]. arXiv; 2025 [cited 2025 Oct 18]. https://doi.org/10.48550/arXiv.2504.00125
11. Kim H, Chen H, Li C, Lee JM. TalkToAgent: A Human-centric Explanation of Reinforcement Learning Agents with Large Language Models [Internet]. arXiv; 2025 [cited 2025 Oct 18]. https://doi.org/10.48550/arXiv.2509.04809
12. Milani S, Topin N, Veloso M, Fang F. Explainable Reinforcement Learning: A Survey and Comparative Review. ACM Comput Surv. 2024;56:1–36. doi:https://doi.org/10.1145/3616864
13. Lyu Q, Apidianaki M, Callison-Burch C. Towards Faithful Model Explanation in NLP: A Survey. Computational Linguistics. 2024;50:657–723. doi:https://doi.org/10.1162/coli_a_00511
14. Zhao H, Chen H, Yang F, Liu N, Deng H, Cai H, et al. Explainability for Large Language Models: A Survey. ACM Trans Intell Syst Technol. 2024;15:1–38. doi:https://doi.org/10.1145/3639372

15. Cheng Z, Yu J, Xing X. A Survey on Explainable Deep Reinforcement Learning [Internet]. arXiv; 2025 [cited 2025 Oct 18]. https://doi.org/10.48550/arXiv.2502.06869
16. Rashidi Ranjbar F, Zamanifar A. Autonomous dental treatment planning on panoramic x-ray using deep learning based object detection algorithm. Multimed Tools Appl. 2023;83:42999–3033. doi:https://doi.org/10.1007/s11042-023-17048-4
17. Sutton RS, Barto A. Reinforcement learning: an introduction. Nachdruck. Cambridge, Massachusetts: The MIT Press; 2014.
18. Safaei F, Nenadović M, Liessner R, Theilen R, Wittenstein J, Lehmann J, et al. X-Vent: ICU Ventilation with Explainable Model-Based Reinforcement Learning. In: Endriss U, Melo FS, Bach K, Bugarín-Diz A, Alonso-Moral JM, Barro S, et al., editors. Frontiers in Artificial Intelligence and Applications [Internet]. IOS Press; 2024 [cited 2025 Aug 16]. doi:https://doi.org/10.3233/FAIA241069
19. Schulman J, Wolski F, Dhariwal P, Radford A, Klimov O. Proximal Policy Optimization Algorithms [Internet]. arXiv; 2017 [cited 2025 Oct 18]. https://doi.org/10.48550/arXiv.1707.06347
20. Lu M, Shahn Z, Sow D, Doshi-Velez F, Lehman LH. Is Deep Reinforcement Learning Ready for Practical Applications in Healthcare? A Sensitivity Analysis of Duel-DDQN for Hemodynamic Management in Sepsis Patients.
21. Levine S, Kumar A, Tucker G, Fu J. Offline Reinforcement Learning: Tutorial, Review, and Perspectives on Open Problems [Internet]. arXiv; 2020 [cited 2025 Oct 18]. https://doi.org/10.48550/arXiv.2005.01643
22. Komorowski M, Celi LA, Badawi O, Gordon AC, Faisal AA. The Artificial Intelligence Clinician learns optimal treatment strategies for sepsis in intensive care. Nat Med. 2018;24:1716–20. doi:https://doi.org/10.1038/s41591-018-0213-5
23. Zamanifar A, Faezipour M, editors. Application of Generative AI in Healthcare Systems [Internet]. Cham: Springer Nature Switzerland; 2025 [cited 2025 Oct 18]. doi:https://doi.org/10.1007/978-3-031-82963-5
24. Kumar A, Zhou A, Tucker G, Levine S. Conservative Q-Learning for Offline Reinforcement Learning [Internet]. arXiv; 2020 [cited 2025 Oct 18]. https://doi.org/10.48550/arXiv.2006.04779
25. Fujimoto S, Meger D, Precup D. Off-Policy Deep Reinforcement Learning without Exploration [Internet]. arXiv; 2019 [cited 2025 Oct 18]. https://doi.org/10.48550/arXiv.1812.02900
26. Kumar A, Fu J, Tucker G, Levine S. Stabilizing Off-Policy Q-Learning via Bootstrapping Error Reduction [Internet]. arXiv; 2019 [cited 2025 Oct 18]. https://doi.org/10.48550/arXiv.1906.00949
27. Cobelli C, Kovatchev B. Developing the UVA/Padova Type 1 Diabetes Simulator: Modeling, Validation, Refinements, and Utility. J Diabetes Sci Technol. 2023;17:1493–505. doi:https://doi.org/10.1177/19322968231195081
28. Le HM, Voloshin C, Yue Y. Batch Policy Learning under Constraints.
29. Zhang R, Zhang X, Ni C, Wang M. Off-Policy Fitted Q-Evaluation with Differentiable Function Approximators: Z-Estimation and Inference Theory.
30. Ng AY, Harada D, Russell S. Policy invariance under reward transformations: Theory and application to reward shaping.
31. Lin M, Shi S, Guo Y, Tadiparthi V, Chalaki B, Pari EM, et al. Speaking the Language of Teamwork: LLM-Guided Credit Assignment in Multi-Agent Reinforcement Learning [Internet]. arXiv; 2025 [cited 2025 Aug 16]. https://doi.org/10.48550/arXiv.2502.03723
32. Barj HNE, Sautory T. Reinforcement Learning from LLM Feedback to Counteract Goal Misgeneralization [Internet]. arXiv; 2024 [cited 2025 Aug 16]. https://doi.org/10.48550/arXiv.2401.07181
33. Schaul T, Quan J, Antonoglou I, Silver D. Prioritized Experience Replay [Internet]. arXiv; 2016 [cited 2025 Oct 18]. https://doi.org/10.48550/arXiv.1511.05952
34. Tobin J, Fong R, Ray A, Schneider J, Zaremba W, Abbeel P. Domain randomization for transferring deep neural networks from simulation to the real world. 2017 IEEE/RSJ International Conference on Intelligent Robots and Systems (IROS) [Internet]. Vancouver, BC: IEEE; 2017 [cited 2025 Oct 18]. p. 23–30. doi:https://doi.org/10.1109/IROS.2017.8202133

35. Peng XB, Andrychowicz M, Zaremba W, Abbeel P. Sim-to-Real Transfer of Robotic Control with Dynamics Randomization. 2018 IEEE International Conference on Robotics and Automation (ICRA) [Internet]. Brisbane, QLD: IEEE; 2018 [cited 2025 Oct 18]. p. 3803–10. doi:https://doi.org/10.1109/ICRA.2018.8460528
36. Packer C, Gao K, Kos J, Krähenbühl P, Koltun V, Song D. Assessing Generalization in Deep Reinforcement Learning [Internet]. arXiv; 2019 [cited 2025 Oct 18]. https://doi.org/10.48550/arXiv.1810.12282
37. Zamanifar A, Nazemi E. EECASC: an energy efficient communication approach in smart cities. Wireless Netw. 2020;26:925–40. doi:https://doi.org/10.1007/s11276-018-1838-5
38. Bekkemoen Y. Explainable reinforcement learning (XRL): a systematic literature review and taxonomy. Mach Learn. 2024;113:355–441. doi:https://doi.org/10.1007/s10994-023-06479-7
39. Pternea M, Singh P, Chakraborty A, Oruganti Y, Milletari M, Bapat S, et al. The RL/LLM Taxonomy Tree: Reviewing Synergies Between Reinforcement Learning and Large Language Models. jair. 2024;80:1525–73. doi:https://doi.org/10.1613/jair.1.15960
40. Cao Y, Zhao H, Cheng Y, Shu T, Chen Y, Liu G, et al. Survey on Large Language Model-Enhanced Reinforcement Learning: Concept, Taxonomy, and Methods. IEEE Trans Neural Netw Learning Syst. 2025;36:9737–57. doi:https://doi.org/10.1109/TNNLS.2024.3497992
41. Zhou Z, Hu B, Zhao C, Zhang P, Liu B. Large Language Model as a Policy Teacher for Training Reinforcement Learning Agents [Internet]. arXiv; 2024 [cited 2025 Aug 16]. https://doi.org/10.48550/arXiv.2311.13373
42. Sun C, Huang S, Pompili D. LLM-based Multi-Agent Reinforcement Learning: Current and Future Directions [Internet]. arXiv; 2024 [cited 2025 Aug 16]. https://doi.org/10.48550/arXiv.2405.11106
43. Lundberg S, Lee S-I. A Unified Approach to Interpreting Model Predictions [Internet]. arXiv; 2017 [cited 2025 Oct 18]. https://doi.org/10.48550/arXiv.1705.07874
44. Bach S, Binder A, Montavon G, Klauschen F, Müller K-R, Samek W. On Pixel-Wise Explanations for Non-Linear Classifier Decisions by Layer-Wise Relevance Propagation. Suarez OD, editor. PLoS ONE. 2015;10:e0130140. doi:10.1371/journal.pone.0130140
45. Delourme S, Redjdal A, Bouaud J, Seroussi B. Measured Performance and Healthcare Professional Perception of Large Language Models Used as Clinical Decision Support Systems: A Scoping Review. In: Mantas J, Hasman A, Demiris G, Saranto K, Marschollek M, Arvanitis TN, et al., editors. Studies in Health Technology and Informatics [Internet]. IOS Press; 2024 [cited 2025 Aug 16]. doi:https://doi.org/10.3233/SHTI240543
46. Evans L, Rhodes A, Alhazzani W, Antonelli M, Coopersmith CM, French C, et al. Surviving sepsis campaign: international guidelines for management of sepsis and septic shock 2021. Intensive Care Med. 2021;47:1181–247. doi:https://doi.org/10.1007/s00134-021-06506-y
47. Oniani D, Wu X, Visweswaran S, Kapoor S, Kooragayalu S, Polanska K, et al. Enhancing Large Language Models for Clinical Decision Support by Incorporating Clinical Practice Guidelines [Internet]. arXiv; 2024 [cited 2025 Aug 16]. https://doi.org/10.48550/arXiv.2401.11120
48. Kaiser KN, Hughes AJ, Yang AD, Turk AA, Mohanty S, Gonzalez AA, et al. Accuracy and consistency of publicly available Large Language Models as clinical decision support tools for the management of colon cancer. Journal of Surgical Oncology. 2024;130:1104–10. doi:https://doi.org/10.1002/jso.27821
49. Lammert J, Dreyer T, Mathes S, Kuligin L, Borm KJ, Schatz UA, et al. Expert-Guided Large Language Models for Clinical Decision Support in Precision Oncology. JCO Precis Oncol. 2024;e2400478. doi:https://doi.org/10.1200/PO-24-00478
50. Reicher L, Lutsker G, Michaan N, Grisaru D, Laskov I. Exploring the role of artificial intelligence, large language models: Comparing patient-focused information and clinical decision support capabilities to the gynecologic oncology guidelines. Intl J Gynecology & Obste. 2025;168:419–27. doi:https://doi.org/10.1002/ijgo.15869
51. Jung K-H. Large Language Models in Medicine: Clinical Applications, Technical Challenges, and Ethical Considerations. Healthc Inform Res. 2025;31:114–24. doi:https://doi.org/10.4258/hir.2025.31.2.114

52. Nambiar M, Ghosh S, Ong P, Chan YE, Bee YM, Krishnaswamy P. Deep Offline Reinforcement Learning for Real-world Treatment Optimization Applications [Internet]. arXiv; 2023 [cited 2025 Aug 16]. https://doi.org/10.48550/arXiv.2302.07549
53. Isakov A, Peregorodiev D, Tomilov I, Ye C, Gusarova N, Vatian A, et al. Real-Time Scheduling with Independent Evaluators: Explainable Multi-Agent Approach. Technologies. 2024;12:259. doi:https://doi.org/10.3390/technologies12120259
54. Lammert J, Dreyer TF, Lörsch AM, Jung J, Lange S, Pfarr N, et al. Large language models for precision oncology: Clinical decision support through expert-guided learning. JCO. 2024;42:e13609–e13609. doi:https://doi.org/10.1200/JCO.2024.42.16_suppl.e13609
55. Rashidieranjbar F, Farhadi A, Zamanifar A. Revolutionizing Healthcare with Generative Artificial Intelligence Technologies. In: Vajjhala NR, Roy SS, Taşcı B, Hoque Chowdhury ME, editors. Generative Artificial Intelligence (AI) Approaches for Industrial Applications [Internet]. Cham: Springer Nature Switzerland; 2025 [cited 2025 Oct 18]. p. 189–221. doi:https://doi.org/10.1007/978-3-031-76710-4_10
56. Chen X, Xiang J, Lu S, Liu Y, He M, Shi D, et al. Evaluating large language models in medical applications: a survey.
57. Vrdoljak J, Boban Z, Vilović M, Kumrić M, Božić J. A Review of Large Language Models in Medical Education, Clinical Decision Support, and Healthcare Administration. Healthcare. 2025;13:603. doi:https://doi.org/10.3390/healthcare13060603
58. Teng Z, Li L, Xin Z, Xiang D, Huang J, Zhou H, et al. A literature review of artificial intelligence (AI) for medical image segmentation: from AI and explainable AI to trustworthy AI. Quant Imaging Med Surg. 2024;14:9620–52. https://doi.org/10.21037/qims-24-723
59. Developing a BERT-Enhanced Blockchain Model for HealthInsurance Fraud Detection | Journal of Information Systems Research and Practice [Internet]. [cited 2025 Oct 19]. https://ijps.um.edu.my/index.php/JISRP/article/view/54256. Accessed 19 Oct 2025
60. ZadZiabari A, Tabatabaei A. Ethics and Regulations in Generative AI. In: Zamanifar A, Faezipour M, editors. Application of Generative AI in Healthcare Systems [Internet]. Cham: Springer Nature Switzerland; 2025 [cited 2025 Oct 19]. p. 197–211. doi:https://doi.org/10.1007/978-3-031-82963-5_8
61. Rishav R, Nath S, Michalski V, Kahou SE. Behaviour Discovery and Attribution for Explainable Reinforcement Learning [Internet]. arXiv; 2025 [cited 2025 Oct 18]. https://doi.org/10.48550/arXiv.2503.14973
62. Mitchell M, Wu S, Zaldivar A, Barnes P, Vasserman L, Hutchinson B, et al. Model Cards for Model Reporting. Proceedings of the Conference on Fairness, Accountability, and Transparency [Internet]. 2019 [cited 2025 Oct 18]. p. 220–9. doi:10.1145/3287560.3287596
63. Gebru T, Morgenstern J, Vecchione B, Vaughan JW, Wallach H, Iii HD, et al. Datasheets for datasets. Commun ACM. 2021;64:86–92. doi:https://doi.org/10.1145/3458723
64. Shahparian J, Erfani SH, Zamanifar A. A secure and efficient authentication and key agreement protocol in blockchain-enabled VANETs. Computers and Electrical Engineering. 2025;122:109947. doi:https://doi.org/10.1016/j.compeleceng.2024.109947

Evaluating Explainability: Metrics, Benchmarks, and Human-Centered Evaluation Methods

Neda Yousefi and Azadeh Zamanifar

1 Introduction

The rapid integration of machine learning (ML) has marked a profound paradigm shift in computational science and engineering. Among the domains poised for transformation, healthcare stands as one of the most significant. Advanced ML models, particularly deep neural networks, demonstrate unprecedented performance across a spectrum of clinical applications. Their capacity to learn intricate patterns from high-dimensional data is revolutionizing medical imaging diagnostics [1], enabling the prediction of disease progression from complex electronic health records [2], optimizing treatment regimens [3] and accelerating drug discovery [4]. Collectively, these advancements promise to augment and revolutionize clinical practice [5]. The advent of Large Language Models (LLMs) represents a paradigm shift in artificial intelligence, with healthcare poised to be one of its most profoundly impacted domains [6–9]. The capacity of models like GPT-4 [10–13], Med-PaLM 2 [14, 15], and their contemporaries to understand, process, and generate human-like text at scale offers transformative potential. Applications range from ambient clinical documentation that alleviates physician burnout [16, 17], to the summarization of complex electronic health records (EHRs) for rapid clinical review [18, 19], to the augmentation of differential diagnosis through interactive dialogue, and the drafting of empathetic, personalized patient communications.

However, the very architectural and computational scale that underpins these capabilities also gives rise to an unprecedented level of opacity [20–22]. The internal mechanisms through which an LLM generates a specific clinical summary or diagnostic hypothesis are vastly more complex and less scrutable than those of traditional

N. Yousefi · A. Zamanifar (✉)
Department of Computer Engineering, SR.C, Islamic Azad University, Tehran, Iran
e-mail: azamanifar@iau.ac.ir

N. Yousefi
e-mail: neda.yousefi@srbiau.ac.ir

A. Zamanifar et al. (eds.), *Explainable Large Language Models in Healthcare Applications*, Bio-IT and AI, https://doi.org/10.1007/978-3-032-15088-2_10

machine learning classifiers [23]. This operational opacity presents a formidable barrier to their translation and adoption into clinical practice—a domain governed by inviolable principles of accountability, patient safety, and ethical oversight [24]. This "black box" problem is exacerbated in LLMs by their generative nature and their propensity for "hallucination"—the confident generation of factually incorrect or nonsensical information [25, 26]. In a domain governed by the Hippocratic Oath, where decisions directly impact patient well-being, this opacity is not merely a technical inconvenience; it is a fundamental barrier to safe and ethical adoption [27]. A clinician cannot responsibly act upon a recommendation from a system whose reasoning is unknown and potentially flawed [28].

This imperative has catalyzed the emergence of Explainable LLMs (XLLMs), a specialized subfield of Explainable AI (XAI) focused on elucidating the behavior of these models [29–31]. Yet, the rapid development of XLLM techniques has critically outpaced the creation of rigorous, standardized, and context-aware methodologies for their evaluation. An explanation—whether it be a highlighted passage of text, a natural language justification, or a visualized chain of thought—is not inherently valuable. A misleading, unfaithful, or clinically implausible explanation can be more dangerous than no explanation at all, as it may induce over-reliance and mask critical model failures [32].

Therefore, the central challenge for the field is no longer simply generating explanations, but rather, assessing their quality. This chapter directly addresses the challenges by presenting a comprehensive and structured framework for the evaluation of explainability in LLMs, with particular emphasis on the unique requirements of the healthcare ecosystem. The discussion proceeds to systematically examine the two pillars of robust evaluation: technical (computer-centered) metrics and rigorous human-centered methods. The abstract of this chapter presented in Fig. 1.

2 The Unique Challenges of Evaluating Explainability in LLMs for Healthcare

The evaluation of explainability for LLMs is not a simple extension of traditional Explainable AI (XAI) paradigms. The unique architectural and functional properties of LLMs, especially when deployed in the high-stakes clinical domain, introduce a distinct and more complex set of challenges that demand a bespoke evaluation framework.

2.1 Semantic and Compositional Complexity of Language

Traditional XAI methods often operate on structured or semi-structured data where "features" are well-defined, and their interactions can be approximated (e.g., pixel

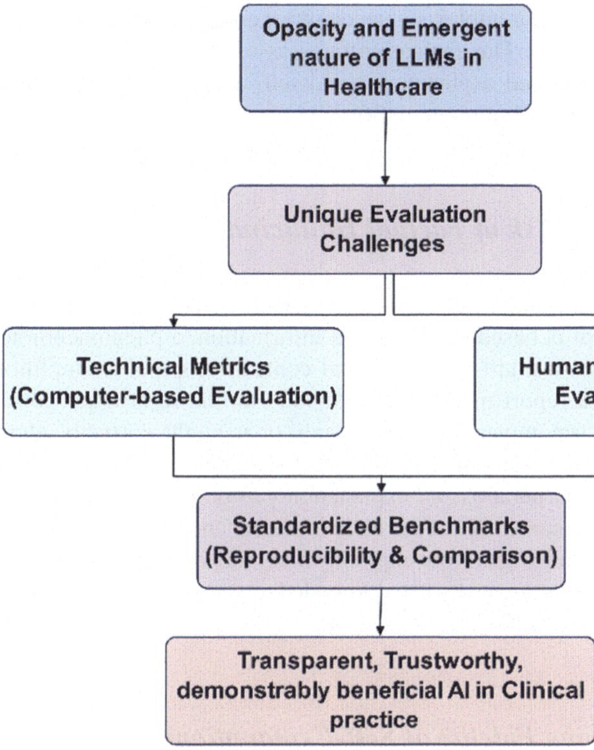

Fig. 1 Conceptual framework

intensity, blood pressure value). In contrast, multi-modal LLMs operate on a joint space of natural language and medical imagery, domains characterized by high dimensionality, sequential dependency, and complex semantic compositionality. The concept of a "feature" becomes ambiguous; its importance can reside at the level of a single text token, an image pixel/patch, or an abstract, cross-modal concept. Perturbing a single word can invert the overall meaning of a clinical note, while subtle alterations in image texture can affect diagnostic conclusions. As a result, the evaluation of feature importance in clinical settings becomes highly complex and warrants careful methodological consideration [33–35].

2.2 *The Shift from Discriminative to Generative Tasks*

The vast majority of traditional XAI research focuses on explaining discriminative models that map an input to a discrete class or a continuous value. LLMs, however, are fundamentally generative. They produce novel, variable-length sequences of text, often conditioned on multi-modal inputs. Therefore, an explanation must justify not just a single decision, but a complex sequence of token-level decisions made from a near-infinite action space. The evaluation must assess the logic, coherence, and factual basis of the generated explanatory text itself, a fundamentally harder task than validating a feature attribution map [36, 37].

2.3 The Pervasive Risk of Factual Hallucination

An LLM can generate a syntactically flawless, stylistically coherent, and seemingly logical explanation that is based on fabricated information, a phenomenon termed "hallucination" [25]. In a multi-modal clinical context, this risk is amplified. An LLM might generate a report mentioning a "nodule in the right upper lobe" that does not exist in the accompanying chest X-ray, or it might correctly identify a finding in an image but attribute it to an incorrect statement in the clinical history. This creates a dangerous paradox: an explanation can be perfectly faithful to the model's flawed internal reasoning but factually incorrect and clinically catastrophic. Evaluation frameworks must therefore explicitly and independently measure factual correctness and grounding against trusted knowledge sources, a step not typically required for traditional XAI [24].

2.4 The Ambiguous Fidelity of Self-Explanations (Chain-of-Thought)

Modern LLMs can be prompted to generate their own "chain-of-thought" (CoT) reasoning before providing a final answer [38]. In a multi-modal setting, this can involve interleaving textual and visual analysis. While these CoTs appear to be a form of built-in explanation, recent research demonstrates they are not guaranteed to be a faithful representation of the model's true computational path. The CoT can be a post-hoc rationalization that the model generates to justify an answer it arrived at through other, inscrutable means [39, 40].

2.5 The Problem of Scale and Emergent Abilities

The sheer scale of state-of-the-art LLMs makes direct inspection of their internal states computationally prohibitive. Furthermore, these models exhibit "emergent abilities"—capabilities that are not present in smaller models and were not explicitly designed or predicted [38, 41]. Explaining a behavior that emerged unpredictably from scale is a profound challenge, as traditional methods often assume a more direct link between model architecture and function.

2.6 *Implicit Knowledge and Unverifiable Provenance*

LLMs are pre-trained on vast, uncurated swathes of the internet, encoding a massive amount of implicit knowledge. An LLM's explanation may be based on a statistical correlation learned from this opaque data corpus, with no verifiable provenance or causal basis. This is particularly problematic in medicine, where evidence-based practice is paramount. The explanation may be correct, but if its origin cannot be traced to established medical literature or the provided patient data, it cannot be fully trusted [42].

2.7 *The Exacerbated Demands of the Healthcare Context*

The generic challenges above are amplified by the specific requirements of clinical practice. An explanation must not only be technically sound but also:

- **Clinically Plausible**: It must align with established pathophysiological and clinical reasoning. An explanation highlighting a statistically valid but clinically nonsensical correlation is not useful and erodes trust [43].
- **Actionable**: It must provide insights that a clinician can safely incorporate into their workflow to make a decision, such as ordering a specific test or modifying a treatment plan [44].
- **Contextually Aware**: It must be relevant to the specific patient's unique context and the clinical question being asked, not a generic recitation of textbook knowledge [45, 46].

3 Technical (Computer-Centered) Evaluation Methods for Explainable Multi-Modal LLMs

Technical metrics offer objective, scalable, and reproducible means of assessing specific properties of multi-modal LLM explanations without direct human involvement. They are indispensable for algorithmic development, rigorous comparison of XLLM methods, and verifying that an explanation adheres to fundamental logical and functional principles. We organize these metrics into eight primary categories: Fidelity, Truthfulness, Factual correctness, Robustness, Complexity and Interpretability Proxies, Axiomatic Properties, computational efficiency, and Semantic Plausibility. To correctly apply evaluation metrics, we first establish a formal classification of explanation methods, recognizing the hybrid nature of XLLMs.

3.1 A Formal Taxonomy of Explainability Methods for LLMs

The evaluation of explainability for any LLM must begin with a precise taxonomy of the explanation methods themselves, as the form and origin of an explanation dictate the applicable evaluation metrics. The landscape of explainability for LLMs in healthcare is a hybrid domain, comprising both the adaptation of classical paradigms from traditional Explainable AI (XAI) and the emergence of novel, LLM-native explanation forms [47]. These methods apply the foundational principles of traditional XAI, originally developed for unimodal structured or image data, to the complex, multi-modal domain of clinical text and images.

Attribution-Based Explanations

Attribution-based methods assign a scalar attribution score $\phi_i \in R$ to each input modality component, quantifying its contribution to a specific model output. These methods answer the question: "Which parts of the multi-modal input were most influential on the model's output?" [48]. In a healthcare context, this is invaluable for grounding an LLM's output in the source clinical data [49]. Examples of attribution techniques include Attention-based Mechanisms [50, 51], Gradient-based Methods [52, 53], and Perturbation-based Methods [54–56].

Example-Based Explanations

These methods retrieve specific multi-modal instances from a dataset to explain or contextualize the model's behavior for a given prediction. This paradigm closely mimics a common practice in clinical reasoning and medical education: the use of case-based reasoning. An explanation can find similar patient cases to contextualize the current LLM prediction. Example-based techniques include influential instances [57], and retrieval augmentation as explanation [58].

Counterfactual Explanations

A multi-modal counterfactual explanation identifies a minimal, semantically coherent edit across one or more modalities of an input x that results in a desired alternative output y'. It answers the clinical question: "What is the key piece of information that, if different, would change the outcome?" [59–63]

This is highly valuable in healthcare for "what-if" analysis and understanding model sensitivity. For example, a multi-modal counterfactual could state: "To change the diagnosis from 'Pneumonia Likely' to 'Heart Failure Likely,' the patient's report would need to mention 'orthopnea' instead of 'fever,' AND the cardiothoracic ratio on the chest X-ray would need to be greater than 0.5" [44, 64–67].

Concept-Based Explanations

Concept-based methods explain a model's decision not in terms of low-level input features (e.g., pixels, tokens), but in terms of high-level, human-understandable concepts that are meaningful to domain experts [68]. They answer the critical question: "Did the model base its decision on clinically relevant concepts that I, as a physician, understand and trust?" [69, 70].

This is arguably one of the most important frontiers for clinical XAI. Clinicians reason using a vocabulary of established medical concepts and an explanation that operates at this conceptual level is far more intuitive, useful, and verifiable than a raw saliency map or a list of important tokens. It allows for a direct assessment of whether the model has learned medically sound correlations.

- **Adapting Discriminative Methods (e.g., TCAV)**: Adapting discriminative interpretability techniques, such as Testing with Concept Activation Vectors (TCAV), to multimodal large language models to quantify the global importance of a concept for a classification or decision made by the model. TCAV originally provided a framework for assessing how much a human-defined concept influences a model's predictions in vision architectures [71]. Extending this paradigm to multimodal LLMs involves defining concept activation in the joint textual-visual (and potentially other modalities) representation space and evaluating how concept directions affect model outputs across tasks and inputs.
- **Probing for Latent Concepts**: A broad class of techniques from the NLP literature, known as "probing," involves training simple linear models on the internal embeddings of an LLM to test if they encode specific concepts [72]. For example, one could train a probe to detect the concept of "clinical uncertainty" from the LLM's internal representation of a clinical note [73]. A successful probe serves as an explanation that the model has indeed learned to represent this concept internally.
- **LLM-Generated Concepts (Self-Explanation)**: A powerful, LLM-native approach is to directly prompt the model to identify the key concepts it used. This reframes concept explanation as a generative task [74].

LLM-Native Explanation Paradigms

These are novel forms of explanation that are uniquely enabled by the generative and meta-reasoning capabilities of modern LLMs, and they naturally extend to multimodal inputs.

- **Natural Language Explanations (NLEs)**: An NLE is a free-text justification, generated by the LLM itself, to explain its own reasoning, prediction, or behavior, which can be grounded in multi-modal evidence [75, 76]. This is arguably the most intuitive and directly usable form of explanation for clinicians, as it mimics the natural language of human consultation and medical discourse [77]. The LLM can be prompted to synthesize its reasoning by referring to both textual and visual evidence. For example: "Based on the patient's reported symptom of 'persistent cough' from the clinical note and the observation of opacities in the upper right lobe of the accompanying chest X-ray, the primary differential diagnosis is pneumonia" [78, 79].

- **Chain-of-Thought (CoT) Explanations**: A CoT explanation is a sequence of intermediate, explicit reasoning steps that the LLM generates before arriving at a final answer. In a multi-modal context, these steps can involve interleaving observations from different modalities. A multi-modal CoT can expose the LLM's logical flow in a way that closely mimics a clinician's diagnostic process. This allows for the identification of logical fallacies or misinterpretations of data across modalities [38]. The fidelity of these traces remains an open research question, as the generated CoT may be a post-hoc rationalization rather than a true reflection of the computational path [39].

Global Surrogate Models

Training a single, globally interpretable model (e.g., a decision tree) to accurately mimic a state-of-the-art multi-modal LLM is generally considered intractable. The function mapping the vast space of possible text-image pairs to output texts is far too complex to be faithfully approximated by a simple model [56].

Direct Rule Extraction

Similarly, directly extracting a comprehensive set of global if–then rules from a multi-billion parameter multi-modal LLM is computationally infeasible. However, the concept of rule-based explanation is re-framed: an LLM can be prompted to generate a set of multi-modal rules as its NLE (e.g., "Based on this case, what is a general clinical rule for identifying this condition?") [80–82]. This transforms the problem from a technical challenge of extraction to a new evaluation challenge: assessing the fidelity and clinical correctness of the LLM-generated rules [83].

3.2 Detailed Evaluation Metrics by Property

Let $f : X_{text} \times X_{image} \rightarrow Y_{text}$ be the multi-modal LLM, $\left(X_{text}, X_{image}\right)$ be the input, and $\phi\left(f, X_{text}, X_{image}\right)$ be the explanation. This section summarizes the technical evaluation metrics illustrated in Fig. 2.

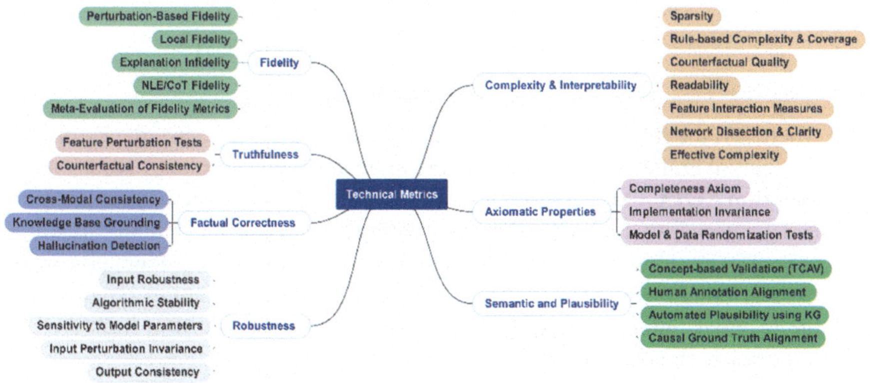

Fig. 2 Technical (computer-based) metrics

Fidelity Metrics: Faithfulness to the Model's Process

Fidelity, or faithfulness, is arguably the most critical technical property. It quantifies how accurately an explanation reflects the true reasoning process of the underlying model [84]. A low-fidelity explanation, no matter how intuitive, is a misleading rationalization and is fundamentally untrustworthy. The primary challenge in measuring fidelity is the absence of a "ground-truth" explanation for a complex, non-linear model. To circumvent this, the predominant approach relies on a core assumption: a faithful explanation should correctly identify the input features that are most influential on the model's output. Consequently, most fidelity metrics are implemented via input perturbation: features are systematically altered based on their attributed importance, and the resulting impact on the model's output is measured [85–87].

- **Perturbation-Based Fidelity (for Attributions)**: These metrics assess the contribution of the features highlighted by an explanation to the model's prediction. They are typically evaluated by perturbing the input based on the explanation [88]. Fidelity is assessed by systematically perturbing features based on their assigned importance and measuring the impact on the model's output [89]. The Area Over the Perturbation Curve (AOPC) is a common aggregate metric [90].
  - **Insertion AOPC (Sufficiency/Inclusion/Preservation)**: This measures if the features deemed most important are sufficient to reconstruct the prediction. Starting from a baseline input x_{base}, features are added in order of importance. Let S_k denote the set of the top-k features, which can include both text tokens and image patches. x_{S_k} denote the input where only features in S_k are revealed, other features remain at their baseline values. $f(.)$ denote the predictive function whose output is being analyzed. The Insertion AOPC (Sufficiency) is defined as follows.

$$AOPC_{\text{Insertion}} = \frac{1}{\text{K}} \sum_{k=1}^{\text{k}} \left(f\left(x_{S_k}\right) - f\left(x_{\text{base}}\right) \right) \tag{1}$$

where k is the total number of features (or channels, pixels, etc., depending on the feature granularity). A high AOPC indicates high sufficiency [87].

– **Deletion AOPC (Necessity)**: This measures if the features deemed most important are necessary for the prediction. Starting from the original input x, features are removed (masked) one by one in order of importance. $x_{\neg s_k}$ denote the input after deleting (removing) features in S_k, while all other features remain as in the original input. A large drop in model output, leading to a high AOPC, indicates high necessity [87].

$$AOPC_{\text{Deletion}} = \frac{1}{\text{K}} \sum_{k=1}^{\text{K}} \left(f(x) - f\left(x_{\neg s_k}\right) \right) \tag{2}$$

– **Remove and Retrain (ROAR)**: A computationally expensive but more robust measure of necessity. Features deemed most important are removed from the entire training dataset, and the model is retrained. A large drop in the retrained model's performance compared to the original indicates high fidelity of the identified features [91].

- **Local Fidelity (Local Accuracy)**: Local Fidelity is defined as a measure of how well a surrogate or local model g matches the original LLM f within a neighborhood around a given instance x. For a neighborhood $\mathcal{N}(x)$ around instance x, fidelity is the weighted accuracy of g with respect to f.

$$Fidelity_{Local}(f, g, x) = E_{z \sim \mathcal{N}(x)} \left[\mathcal{L}(f(z), g(z)) \right] \tag{3}$$

where $z \sim \mathcal{N}(x)$ indicates that z is drawn from the neighborhood $\mathcal{N}(x)$, $\ell(\cdot, \cdot)$ is an appropriate loss or indicator function that measures discrepancy between the outputs of f and g.

- **Explanation Infidelity**: Explanation infidelity measures the expected discrepancy between the response of the explanation to a perturbation and the actual change in the model's output. Specifically, it quantifies how far the dot product between a perturbation p and the explanation $\phi(x)$ is from the true change in the model's prediction when the input is perturbed by p [92].

$$Infidelity(x) = \mathbb{E}_{P \sim D} \left[\left(p^T \phi(x) - (f(x) - f(x - p)) \right)^2 \right] \tag{4}$$

where, $x \in R^d$ is the input, $\varphi(x) \in R^d$ is the explanation (e.g., feature attribution vector), $f(\cdot)$ is the predictive function, p is a perturbation drawn from distribution D. Lower infidelity indicates that the explanation better accounts for the observed change in the model's output under perturbations.

- **Natural Language Explanation (NLE) or the Chain-Of-Thought (CoT) Fidelity (Explanation-Driven Simulation)**: NLE/CoT Fidelity refers to an evaluation metric that assesses whether the NLE or the CoT produced by a LLM, when fed back into the model, can accurately predict its output on similar multi-modal instances. This approach treats explanations as a proximal surrogate task: if the elicited reasoning consistently yields outputs aligned with those produced for analogous inputs, then the fidelity of the explanation to the model's underlying decision process is supported. Practically, this involves reusing the generated NLE/CoT as part of the prompt for related multi-modal inputs and measuring concordance between the predicted outputs and the original outputs, thereby quantifying the extent to which the explanation channels preserve or reveal the model's reasoning trajectory across related cases. This methodology provides a formal mechanism for evaluating whether explanations faithfully reflect the model's internal reasoning when applied to comparable data distributions [93–96].
- **Meta-Evaluation of Fidelity Metrics**: Recent research has shown that fidelity metrics themselves can be unreliable and often disagree [97], motivating the rise of meta-evaluation to assess the quality, reliability, validity and stability of evaluation metrics[98].

  - **Reliability Checks**: A reliable metric should be consistent, evaluated through [99, 100]:

    Inter-rater Reliability: Consistent scores for explanations of similar inputs, typically measured by Krippendorff's α [101].
    Inter-method Reliability: Consistent ranking of different high-quality XAI methods, typically measured by Spearman's ρ [102].

Truthfulness (Accuracy of Explanation)

Truthfulness is the empirical verification of the explanation's claims [55, 56, 68]. Truthfulness, in the context of XAI, complements fidelity by focusing on specific empirical tests designed to verify that the explanation accurately represents the model's decision process, rather than offering a misleading post-hoc rationalization [55, 56]. It is about ensuring that the explanation's claims about feature importance or model behavior are empirically verifiable by observing the model's response to input perturbations.

- **Feature Perturbation Tests**: These tests directly manipulate input features across modalities to ascertain if the model's output changes in a manner consistent with the explanation's claims. This provides empirical evidence for the truthfulness of the explanation.

  - **Feature Removal/Masking Consistency**: If an explanation identifies a feature x_j as highly important, removing or masking or modifying x_j from the input should significantly alter the model's prediction. This is closely related to

"Necessity" (Sect. 3.2.1) but emphasizes the direct verification of the explanation's claim. Let $\varphi(x)_j$ be the attribution for feature j and $x_{\neg j}$ be the input with feature j masked.

$$\Delta P_j = |F(x) - F(x_{\neg j})| \tag{5}$$

A larger ΔP_j for a feature ranked high by the explanation supports its truthfulness. For classification, one might observe if the predicted class changes or if the probability of the original class significantly drops [91].

- **Feature Addition/Modification Consistency**: If an explanation states that a feature contributes positively to a prediction (e.g., a higher value of a feature increases the risk), then increasing that feature's value should indeed increase the prediction (and vice-versa for negative contributions). Let $\varphi(x)_j$ be a feature with a positive importance score from the explanation. where $\delta > 0$ is a small increment and u_j is a unit vector for feature j: $x'' = x + \delta \cdot u_j$.

$$\Delta Pj'' = f(x'') - f(x) \tag{6}$$

If $\varphi(x)_j$ is positive, we expect $\Delta P_i'' > 0$. If $\varphi(x)_j$ is negative, we expect $\Delta P_i'' < 0$. The magnitude of $\Delta P_i''$ should also correlate with the $\varphi(x)_j$.

- **Counterfactual Consistency**: Verifies that a generated multi-modal counterfactual (X'_{text}, X'_{image}) truly yields the desired target output y' when fed into the model. For counterfactual explanations, which propose minimal changes to an input x to achieve a desired alternative outcome Y_{target} [44], truthfulness means verifying that these proposed changes actually lead to Y_{target} when fed into the original AI model. Let x_{cf} be the counterfactual input generated by the explanation, claiming to result in Y_{target}. $TruthfulnessCF = I\left(f\left(X'_{text}, X'_{image}\right) = y'\right)$, where I[·] is the indicator function.

  The explanation is truthful if $f(x_{cf}) = Y_{target}$. This is often a binary measure. Additionally, the counterfactual explanation often claims to be "minimal," which can be further verified by ensuring no smaller perturbation achieves the same outcome.

Factual Correctness and Grounding

This is a critical category for generative multi-modal LLMs, assessing whether the explanation is based on verifiable facts and accurately reflects cross-modal evidence [103–108].

Cross-Modal Consistency: Verifies that claims made in the generated text are supported by the provided image, and vice-versa.

- Object-level Grounding [109–111]: At the object level, grounding evaluates the alignment between textual phrases and localized visual entities, such as

determining that a statement like "nodule in the right upper lobe" corresponds to a detected nodule within the specified region of the image, thereby establishing a direct mapping between linguistic description and spatial localization.
- Attribute-level Grounding [112–114]: At the attribute level, grounding extends to verifiable image-derived attributes, where a claim such as "cardiomegaly" is grounded in quantitative or qualitative image analysis features (for example, measurements of the cardiothoracic ratio) that substantiate the textual assertion.

Knowledge Base Grounding: Extracts factual claims from the NLE/CoT and cross-references them against a trusted medical knowledge base (e.g., UMLS [115], SNOMED-CT [116]). The metric is the proportion of verifiable claims [117–120].
Hallucination Detection: Refers to the use of specialized models or human annotation to predict the factual consistency of a generated statement against all source modalities. The goal is to determine whether the output is supported by the accompanying sources and to quantify the degree of factual alignment [25, 121, 122].

Robustness and Consistency

These metrics evaluate the stability, reliability and consistency of an explanation method. An explanation should not change erratically in response to minor, irrelevant, and insignificant changes in the input or the algorithm's internal randomness [85, 123].

- **Input Robustness (Local Lipschitz Estimate)**: Measures the maximum change in an explanation for a small input perturbation of size ϵ. Let x' be a perturbed version of x. A stable explanation ϕ should satisfy $\phi(f,x) \approx \phi\left(f,x'\right)$ and a lower value indicates higher stability [85].

$$\mathcal{R}(\phi, f, x, \epsilon) = \max_{x' : x - x'_p < \epsilon} \frac{\phi(f,x) - \phi\left(f,x'\right)_q}{x - x'_p} \tag{7}$$

- **Algorithmic Stability (Repeatability)**: For stochastic methods (e.g., LIME), this measures consistency across multiple runs on the exact same input. Let $\phi_i(f,x)$ be the explanation from run i.

$$Repeatability(\phi, f, x) = \frac{2}{N(N-1)} \sum_{i=1}^{N-1} \sum_{j=i+1}^{N} d\left(\phi_i(f,x), \phi_j(f,x)\right) \tag{8}$$

where, N is the number of independent runs, $d(\cdot,\cdot)$ is a similarity metric (e.g., Spearman correlation, Jaccard index, cosine similarity) between explanations. A higher average similarity indicates higher stability of the explanations across runs.

- **Sensitivity to Model Parameters**: This metric assesses how an explanation changes if the underlying model's internal parameters are slightly perturbed.
- **Multi-Modal Input Perturbation Invariance**:
  - **Paraphrastic Invariance (Text)**: The explanation for two semantically equivalent paraphrased representations for the same input text should be semantically equivalent.
  - **Visual Invariance (Image)**: The explanation must be robust to semantically irrelevant changes in the image (e.g., small rotations, brightness changes, or minor noise) while the text remains constant.
  - **Cross-Modal Alignment Robustness**: Measures if the identified cross-modal correspondences remain stable under minor perturbations to either modality.

- **Output Consistency (Self-Consistency)**: For a single input, generate multiple explanations using stochastic sampling. A robust method should yield semantically similar explanations across runs, indicating stable reasoning and reducing random variability in outputs.

Complexity and Interpretability Proxies

These metrics use computational proxies to estimate properties related to human comprehension. True interpretability can only be measured with humans, but these metrics provide scalable estimates.

- **Sparsity**: Counts the number of non-zero attribution scores across both text tokens and image patches.

$$Sparsity(\phi) = ||\phi_{text}||_0 + ||\phi_{image}||_0 \tag{9}$$

- **Rule-based Complexity and Coverage**: For rule-based explanations, complexity is measured by the structural burden of the explanation [124].

- **Rule Complexity**: Total rules ($|R|$) or average clauses per rule in an LLM-generated rule set.
- **Rule Coverage**: The fraction of data points in a dataset D that are covered by at least one rule in the set R.
- **Rule Overlap**: Fraction of multi-modal data points covered by multiple rules. High overlap can indicate ambiguity or redundancy in the rule set, potentially reducing clarity[125].

- **Counterfactual Quality**: For counterfactual explanations, several metrics assess their utility beyond simple validity.
  - **Proximity**: The distance between the original and counterfactual inputs, measured across modalities [44].

$$proximity(x, x_{cf}) = w_{\text{text}} \cdot \|x_{\text{text}} - x'_{\text{text}}\|_p + w_{\text{image}} \cdot \|x_{\text{image}} - x'_{\text{image}}\|_q \quad (10)$$

where w_{text} and w_{image} are weights for each modality.

- **Plausibility (Data Manifold Closeness)**: Measures if x_{cf} is a realistic, in-distribution instance. This is often approximated using the reconstruction error from an autoencoder trained on the data, or by checking its density in the original data distribution [126].

- **Readability (Textual)**: Standard linguistic scores (e.g., Flesch-Kincaid [127]) for the generated NLE/CoT [128–130].
- **Feature Interaction Measures (H-statistic)**: Quantifies the strength of interactions between textual and visual features in contributing to a prediction [131].
- **Network Dissection and Clarity (Visual)**: Quantifies the alignment between individual neural network units in the visual encoder of the LLM and human-labeled semantic concepts in medical images [132].
- **Effective Complexity**: The minimum number of attribution-ordered multi-modal features required to achieve a certain level of predictive performance [133].

Axiomatic Properties and Sanity Checks

These are principled tests for the logical consistency of the XAI method itself, ensuring it adheres to fundamental theoretical desiderata [134].

- **Completeness Axiom**: For attribution methods, the sum of feature importances should equal the model's output difference from a baseline [52]. Formally, for input x and baseline x_{base}, with attribution vector $a \in R^d$ satisfying

$$\sum_{i \in \text{text}} \varphi_i(f, x) + \sum_{j \in \text{image}} \varphi_j(f, x) = f(x) - f(x_{base}) \quad (11)$$

- **Implementation Invariance**: Functionally equivalent LLMs should produce identical explanations [52].
- **Sensitivity**: A feature to which the model is functionally dependent must receive non-zero attribution. If altering feature x_j changes the output $f(x)$, then the corresponding attribution $\phi_j(f, x)$ should reflect a non-negligible magnitude [92, 135].
- **Model and Data Randomization Tests**: These are crucial sanity checks to ensure the explanation method is actually sensitive to the learned patterns in the model and data, rather than just input structure [134, 136].
  - **Model Randomization**: A valid explanation method should produce significantly different results for a trained model versus a randomly initialized model [134]. Let f_{train} be a trained model and f_{rand} a randomly initialized model, then

explanations $E(f_{train}, x)$ and $E(f_{rand}, x)$ should differ beyond chance for typical inputs x.
  - **Data Randomization**: Explanations from a model trained on true labels f_{true} should differ from explanations from a model trained on randomized labels $f_{rand-labels}$.

Computational Efficiency

Computational efficiency is a crucial practical consideration for XAI, particularly in healthcare settings where real-time or near real-time decision support is often required [137].

Runtime Complexity and Actual Execution Time: Time taken to generate an explanation.
Memory Footprint: Peak memory usage during multi-modal explanation generation.
Scalability: Performance with increasing model/data complexity.

Semantic and Plausibility Metrics

While the aforementioned metrics capably assess an explanation's functional characteristics, they fundamentally operate within a semantic vacuum. What does this mean? It means they are inherently incapable of discerning whether a model's purported justification—or indeed, the explanation itself—possesses genuine semantic meaningfulness, clinical plausibility, or critical consistency with established real-world knowledge. This particular lacuna becomes acutely problematic, even dangerous, in high-stakes domains such as medicine. Here, an explanation's trustworthiness and practical utility hinge entirely upon its demonstrable clinical coherence [24]. Consequently, semantic and plausibility metrics emerge as indispensable tools. They precisely address this gap by deliberately incorporating external, domain-specific knowledge into the evaluation process, thereby rigorously assessing an explanation's true alignment with both human intuition and verified scientific facts [138].

- **Concept-based Validation (TCAV)**: Quantifies global importance of a human-understandable concept (e.g., "spiculation" for a mammogram classifier) [71]. This can apply to visual concepts in the image modality or textual concepts in the text modality.
- **Human Annotation Alignment (IoU/AUC)**: Measures overlap/alignment of saliency maps (visual attributions) with expert-defined Regions of Interest (ROIs) in medical images.
- **Automated Plausibility using Knowledge Graphs**: This advanced approach attempts to quantify plausibility by leveraging structured domain knowledge

encoded in ontologies and knowledge graphs (KGs) without direct human intervention, offering scalability and consistency [139–141]. An overall plausibility score can be derived by weighting each feature's plausibility (based on graph-based semantic similarity to the target concept) by its attribution score. Let the biomedical knowledge graph be represented as $G = (V, E)$, where V is the set of medical concepts and E is the set of relations between them. Let an explanation for a prediction Y highlight a set of important features $F_{imp} = f_1, \ldots, f_k$.

- **Concept Mapping**: The first step is to map the features and the outcome to standardized concepts in the KG. Let $\mu : F \cup \{Y\} \rightarrow V$ be the mapping function. This yields a set of feature concepts $\{v_1, ..., v_k\}$ and a target concept v_Y.
- **Knowledge Graph Validation**: Once mapped, the relationships can be validated against the KG's structure.
- **Path-based Metrics**: These metrics assume that stronger semantic relationships are represented by shorter paths in the graph. Let $d(v_i, v_Y)$ be the shortest path length between a feature concept v_i and the target concept v_Y. A basic plausibility score for a single feature can be defined as an inverse function of this distance:

$$Plausibility_{path}(v_i, v_Y) = \frac{1}{1 + d(v_i, v_Y)} \tag{12}$$

  This score approaches 1 for directly connected concepts and decays as the path length increases.
- **Embedding-based Metrics**: These metrics leverage knowledge graph embeddings, which represent concepts as vectors in a continuous space such that semantic relationships are preserved.

  Let $e_{v_i} \in \mathbb{R}^d$ and $e_{v_Y} \in \mathbb{R}^d$ be the embedding vectors for the feature and target concepts, respectively, learned using a model like TransE or Node2Vec [139]. Semantic similarity is then computed, typically using cosine similarity.

$$Plausibility_{embed}(v_i, v_Y) = \frac{e_{v_i} \cdot e_{v_Y}}{||e_{v_i}|| \cdot ||e_{v_Y}||} \tag{13}$$

  A similarity scores closer to 1 indicates higher semantic relatedness and thus higher plausibility.
- **Aggregated Plausibility Score**: The individual plausibility scores for each salient feature can, in fact, be aggregated. This yields an overall automated plausibility score for the explanation in its entirety. Typically, this aggregation is achieved by judiciously weighting each feature's plausibility with its corresponding attribution score a_i, as provided by the XLLM:

$$Plausibility_{overal} = \frac{\sum_{i=1}^{k} a_i.PlausibilityScore(v_i, v_Y)}{\sum_{i=1}^{k} a_i} \tag{14}$$

This aggregated score then offers a singular, quantitative metric for gauging the explanation's intrinsic consistency with established domain knowledge.

- **Causal Ground Truth Alignment**: Consider settings where a definitive causal model is either known a priori or can be reliably approximated—think synthetic data environments or robust physics-based simulations. In such invaluable scenarios, the explanation's efficacy can be rigorously evaluated based on how accurately it identifies the true causal features or how precisely it aligns with critical causal counterfactuals. This is a powerful, albeit specialized, form of validation [89].

4 Human-Centered Evaluation Methods for LLM Explanations in Healthcare

While technical metrics—and they are essential—provide an objective, scalable, and entirely reproducible foundation for assessing explainability, they remain, at their core, fundamentally incomplete. Why? Because they primarily illuminate the algorithmic properties of an explanation: its fidelity to the model, its mathematical robustness. What they utterly fail to capture is the ultimate purpose of an explanation, especially within a high-stakes domain like healthcare. Its true mission, after all, is to function as a vital cognitive and collaborative bridge, connecting an intricate computational system with its human stakeholders. The final arbiter of an explanation's value is not a formula, but the diverse group of individuals—clinicians, patients, ethicists, and regulators—who must use it to make decisions, understand outcomes, and ensure safety [142, 143]. Human-centered evaluation is therefore not an ancillary step, but a co-equal and indispensable pillar of the validation process, essential for determining if an explanation is understandable, trustworthy, and clinically useful [143–148].

This section provides a comprehensive, multi-level framework for conducting rigorous human-centered evaluations of multi-modal LLM explanations in healthcare. We move beyond simple user satisfaction surveys to a structured methodology that assesses the full lifecycle of an explanation's impact, from its initial comprehensibility to its ultimate effect on clinical outcomes and ethical standards, all while considering the critical variable of the evaluator's own expertise and role.

4.1 The Spectrum of Human Evaluators in Healthcare XAI

The primary objective of XAI in healthcare is to create a synergistic partnership between human intellect and artificial intelligence, empowering users to make more informed, confident, and safer decisions [2, 149]. Technical metrics alone cannot capture this synergy. An explanation might be perfectly faithful to a model's internal logic but be clinically irrelevant, cognitively overwhelming to a physician, or misleadingly persuasive, fostering misplaced trust [150, 151]. Human-centered evaluation bridges this gap by directly measuring an explanation's impact on its intended audience. A critical flaw in many XAI studies is treating "the user" as a monolith; in healthcare, the evaluator's expertise and goals profoundly shape an explanation's requirements [152]. A robust evaluation framework must therefore be tailored to a spectrum of stakeholders [143, 148, 153].

Domain Experts (Clinicians)

This group forms the primary audience for most clinical decision support systems. Yet, even within this seemingly singular category, crucial, distinct sub-groups demand our attention:

- **Sub-Specialists (e.g., Cardiothoracic Surgeons)**: These experts require highly detailed, technically precise explanations that align with their deep but narrow domain knowledge. They are essential for evaluating the clinical plausibility, factual correctness, and subtle relevance of highly technical multi-modal explanations [152].
- **Generalists (e.g., General Practitioners, Internists)**: These clinicians require explanations that synthesize information effectively and are presented with high clarity and conciseness. They are key evaluators of utility, actionability, and workflow integration [154].
- **Trainees (e.g., Residents, Fellows)**: This group may use explanations for educational purposes. For them, the explanation's ability to teach a diagnostic process is a key quality [32].

Technical Experts (AI/ML Practitioners)

This group consists of the developers and researchers building the XLLM systems. Their overarching mission is defined by three critical objectives: meticulous debugging, profound comprehension of insidious failure modes, and the relentless pursuit of enhanced algorithmic performance. Consequently, their expertise positions them uniquely to undertake the rigorous evaluation of an explanation's fidelity and its inherent robustness. This often necessitates the deployment of sophisticated human-in-the-loop methodologies, through which technical metrics are verified against the perceived, real-world behavioral intricacies of the underlying model [155].

End-Users (Patients and Caregivers)

As multi-modal LLMs power patient-facing applications, this group's perspective is paramount. They evaluate explanations based on lay-term understandability, emotional impact (e.g., reassurance vs. anxiety), and the ability to facilitate shared decision-making with their clinical team [156, 157].

Oversight and Governance Experts (Ethicists, Regulators, Hospital Administrators)

This essential group consists of ethicists, regulators, and clinical administrators, each bringing their own, unique, important views on safety, equity, and compliance. They have the responsibility to ensure that any AI system operating under their oversight functions safety, equitably, free of bias, and with total, diligent, consistent adherence to every possible legal and ethical standard (e.g., HIPAA, FDA guidelines). They are the chief third-party assessors of fairness and inequity, and will take an intense look to see if an explanation exposes or conceals inequitable rationale. They also will be looking for an audit and accountability trail as well, provided in part by the explanation.

4.2 A Multi-Dimensional Framework for Human-Centered Evaluation

We structure the evaluation of multi-modal LLM explanations across a comprehensive set of dimensions, each containing specific, measurable metrics. This framework allows for a holistic assessment of an explanation's quality and impact, as illustrated in Fig. 3.

Dimension: Understandability and Cognitive Load

This dimension assesses the ease with which a user can process and comprehend the explanation's logic and the cognitive effort required to do so. An effective explanation must form an accurate mental model of the AI's reasoning without imposing an undue cognitive burden [158–160]. Evaluation must assess not just the readability of text and clarity of image highlights, but the user's ability to understand the synergistic link between them. For example, can the user comprehend why a specific phrase in the patient history ("chronic smoker") is linked to a specific visual feature (a spiculated lung nodule)?[161]

Associated metrics with understandability and cognitive load are presented in Table 2.

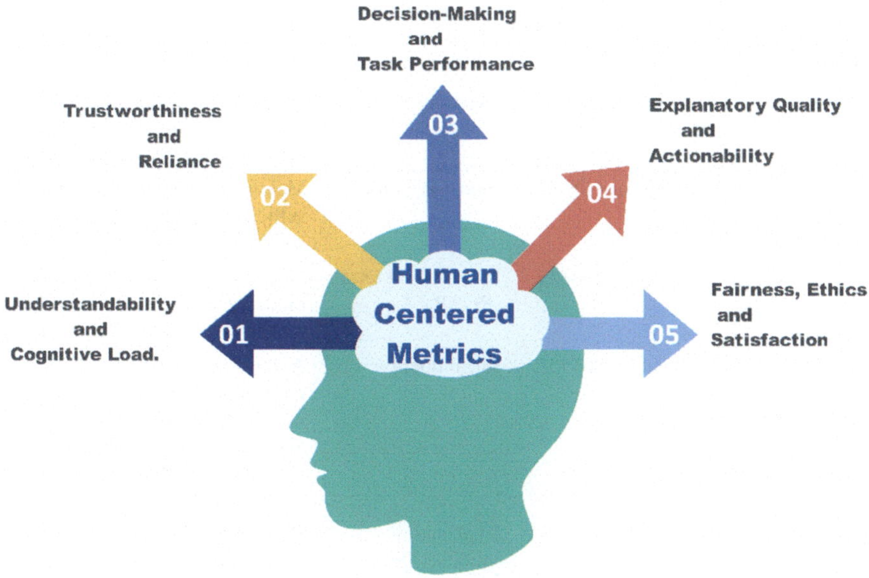

Fig. 3 Human-cantered metrics

Dimension: Trustworthiness and Reliance

This dimension quantifies the user's belief in the AI's competence and reliability, and how that belief translates into reliance on its outputs. The goal is not to maximize trust, but to achieve calibration—fostering appropriate reliance when the AI is correct and appropriate skepticism when it is wrong [87, 88]. Over-reliance can lead clinicians to accept incorrect AI suggestions, while under-reliance can lead them to ignore valuable insights. Both scenarios compromise patient safety [9, 36]. A compelling visual explanation (e.g., a bright heat map) can disproportionately influence trust, potentially causing a user to overlook contradictory evidence in the text. Evaluation must assess the holistic impact. Associated metrics with trustworthiness and reliance are presented in Table 3.

Dimension: Decision-Making and Task Performance

This dimension provides the ultimate test of an explanation's value: does it lead to better, safer, or more efficient human performance on clinically relevant tasks? [175]. The primary purpose of a clinical AI tool is to improve outcomes. Explanations must be evaluated on their ability to contribute to this goal, whether by improving diagnostic accuracy, speeding up report generation, or helping to catch human errors [143, 176]. Associated metrics with decision making and task performance are presented in Table 4.

Table 1 Summary of technical evaluation metrics

Category	Sub-category	Metric name	Definition
Fidelity	Perturbation-based [86]	Deletion AOPC	Measures impact on output by removing/adding highest-ranked text tokens and/or image patches
	Perturbation-based [86]	Insertion AOPC	Measures if top features are sufficient for the prediction by adding them
	Perturbation-based [91]	Remove and retrain (ROAR)	Measures necessity by retraining model on dataset with important multi-modal features removed
	Surrogate-based [55, 56]	Local fidelity	Agreement between local multi-modal surrogate and LLM in neighborhood. Measures agreement between a surrogate g and black-box f in a neighborhood
		Explanation infidelity [92]	Discrepancy between an explanation's predicted impact and the true impact of a perturbation
Truthfulness	Feature perturbation [91]	Feature removal/ modification consistency	Verifies if model output changes empirically consistent with explanation's claims
	Counterfactual [44]	Counterfactual consistency	Verifies if generated counterfactual x_{cf} truly yields Y_{target} in the model
Factual correctness	Cross-modal grounding [109–114]	Object/ attribute-level grounding score	Checks if textual claims about objects/attributes are localized/ supported in the image
	Knowledge-based [115–120]	Knowledge base grounding score	Proportion of factual claims in explanation verifiable against medical KG
	Hallucination [25, 121, 122]	Automated hallucination detection	Specialized models predict factual consistency of generated explanation against all sources
Robustness	Input stability [85]	Local Lipschitz estimate	Measures the maximum change in an explanation for a small input perturbation
	Algorithmic stability [92]	Repeatability	Measures consistency of a stochastic explanation method across multiple runs
	Model stability [85]	Sensitivity to model parameters	Measures how explanation changes with small perturbations to model weights

(continued)

Table 1 (continued)

Category	Sub-category	Metric name	Definition
Complexity		Sparsity	Counts non-zero attributions across text tokens and image patches
		Rule Complexity	Number of rules or average number of clauses in a rule-based explanation
	Counterfactual quality [44]	Proximity	The magnitude of change required to generate the counterfactual
	Counterfactual quality [126]	Plausibility (data manifold closeness)	Measures if a counterfactual is a realistic, in-distribution instance
		Network dissection clarity (visual) [132]	Quantifies alignment between CNN units and human-labeled semantic concepts
	Readability (textual) [127]	Scores	Linguistic complexity of textual explanations (e.g., Flesch-Kincaid)
	Feature interactions [131]	H-statistic	Quantifies the strength of interactions between features
Axiomatic properties		Completeness axiom [52]	Checks if attributions sum to the model's output difference from a baseline
		Implementation invariance [52]	Functionally equivalent models produce identical explanations
		Sensitivity [92, 135]	A functionally dependent feature must receive non-zero attribution
		Model and data randomization test [134, 136]	These are crucial sanity checks to ensure the explanation method is actually sensitive to the learned patterns in the model and data, rather than just input structure
Computational efficiency	Runtime memory scalability [137]	Actual execution time	Time taken to generate an explanation for a single instance
Semantic plausibility	Concept-based [71]	TCAV score	Global importance of a human-understandable concept for a class
	Annotation-based	Human annotation IoU/AUC	Measures overlap/alignment of saliency map with expert ROI

(continued)

Table 1 (continued)

Category	Sub-category	Metric name	Definition
	Knowledge-based [139–141]	Knowledge graph alignment	Correlation between feature importance and semantic relatedness in a KG
	Causal alignment [89]	Causal feature inclusion	Measures if known causal features receive high attribution

Table 2 Understandability and cognitive load

Metric name	Definition	Clinical significance
Perceived understanding [162, 163]	Subjective rating of how well a user believes they comprehend the AI's reasoning after seeing an explanation	A direct measure of user confidence in their own comprehension, which is a prerequisite for trust and adoption
Actual understanding [148, 164]	Objective measure of a user's ability to correctly answer comprehension questions about the model's logic or the case	Moves beyond perception to verify that knowledge was effectively transferred, crucial for ensuring safe and correct application of AI insights
Predictability of AI behavior [71, 164]	The user's accuracy in predicting the AI's output on new, unseen multi-modal instances after being trained by explanations	A high score indicates the user has formed a robust and accurate mental model of the AI's function, enabling effective collaboration
User prediction of model failure [165, 166]	The user's ability to correctly identify cases where the AI is likely to be incorrect, demonstrating an understanding of its limitations	Essential for trust calibration. Enables clinicians to know when to override the AI, a critical safety function
Subjective cognitive load [162, 163, 167]	Self-reported mental effort, frustration, or difficulty required to process an explanation, often measured using validated scales like NASA-TLX	Explanations must be cognitively efficient. High load can lead to errors and user abandonment in busy clinical workflows
Objective cognitive load [160, 168, 169]	Physiological or behavioral measures of mental effort, such as gaze duration on specific explanation components, pupil dilation, or response time	Provides objective evidence of which parts of a multi-modal explanation are confusing or demanding, guiding interface design

Dimension: Explanatory Quality and Actionability

This dimension assesses the inherent properties of the explanation itself, such as its coherence and usefulness for guiding subsequent actions [44, 170, 181]. An explanation is not merely for passive understanding; it should inform the next step in

Table 3 Trustworthiness and reliance

Metric name	Definition	Clinical significance
Perceived trustworthiness [170, 171]	Subjective belief in the reliability, integrity, and competence of the AI system, measured using validated multi-item scales	A fundamental prerequisite for adoption. Without trust, even the most accurate AI system will not be used in clinical practice
Correct reliance rate [172, 173]	The frequency with which a user correctly agrees with a correct AI suggestion. An effective explanation should increase this rate	Measures the ability of the explanation to build justified confidence in the AI's strengths. It predicts actual usage; indicates if explanations foster sufficient confidence for clinicians to accept recommendations
Evolution of user trust [174]	The change in trust scores over repeated interactions, capturing the dynamic process of trust calibration over time $\Delta T = \overline{T}_{tN} - \overline{T}_{t0}$ (Change in average trust from initial t_0 to final t_N)	Assesses if users are learning to calibrate their trust appropriately through experience, essential for long-term system integration. It captures dynamic trust in longitudinal studies, crucial for real-world clinical integration and continuous monitoring

Table 4 Decision making and task performance

Metric name	Definition	Clinical significance
Decision accuracy improvement [176–178]	The change in a user's task accuracy (e.g., diagnostic correctness) when aided by an explanation versus a baseline (no AI or AI without explanation)	The gold-standard measure of utility. Demonstrates a tangible improvement in clinical performance
Task completion efficiency [179]	The change in time required for a user to complete a task	In a busy clinical setting, time is a critical resource. Explanations should ideally improve accuracy without increasing time cost
Error identification rate [180]	The percentage of AI model errors that a user successfully identifies and corrects, enabled by the explanation	Measures the explanation's utility as a safety and debugging tool, empowering clinicians to act as the final check on the AI
Clinical relevance rating [2, 43, 84]	An expert's subjective rating of how pertinent, insightful, and aligned the explanation's highlighted features are with established clinical knowledge	Ensures that the explanation is not just technically faithful but also medically meaningful and useful to a practicing clinician

Table 5 Explanatory quality and actionability

Metric name	Definition	Clinical significance
Plausibility/coherence [153, 184–186]	A user's subjective rating of how logical, consistent, and believable the explanation's narrative is	An implausible explanation, even if technically faithful, will be rejected by clinicians and will erode trust in the entire system
Actionability rating [187, 188]	A user's rating of the extent to which the explanation provides clear, concrete guidance for their next steps	Directly measures if the explanation bridges the gap between insight and clinical action, a key determinant of real-world utility
Informativeness/ sufficiency [189]	A user's rating of whether the explanation provides enough detail to justify the AI's output without being overwhelming	Captures the critical balance between completeness and conciseness, ensuring the explanation is both convincing and efficient
Alignment with clinical guidelines [190–192]	An expert's assessment of whether the reasoning presented in the explanation is consistent with established, evidence-based clinical guidelines	A crucial check to ensure that the AI's reasoning, even if learned from data, does not contradict best practices in medicine

the patient care pathway, such as ordering a specific follow-up test or considering a differential diagnosis [91]. Actionability in a multi-modal context could mean the explanation prompts a radiologist to re-examine a specific region of a CT scan or cross-reference a finding with the patient's lab results [182, 183]. Associated metrics with explanatory quality and actionability are presented in Table 5.

Dimension: Fairness, Ethics, and Satisfaction

This dimension addresses the broader user experience and the critical ethical implications of using explainable AI in healthcare. AI models can inherit and amplify societal biases present in medical data. Explanations are a key tool for auditing these biases and ensuring that decisions are made equitably [193, 194]. Bias can manifest in subtle, cross-modal ways, such as a model performing differently on dermatological images of different skin tones, or interpreting clinical notes differently based on gendered language. Explanations must be able to surface these issues [195, 196]. Associated metrics with fairness, ethics, and satisfaction are presented in Table 6.

Table 6 Fairness, ethics, and satisfaction

Metric name	Definition	Clinical significance
Perceived fairness [193, 194]	A user's subjective assessment of whether an AI's decision and explanation appear unbiased, equitable, and just across different patient groups. Mean Likert scale rating, often compared across demographic subgroups	Perceived bias, even if not present algorithmically, can destroy trust. Explanations must promote a sense of procedural justice
User satisfaction [197, 198]	The user's overall subjective experience with the XAI system, often measured with standardized usability scales like SUS. Standardized score from a multi-item questionnaire (e.g., System Usability Scale score from 0 to 100)	A holistic measure of the user experience. Low satisfaction is a strong predictor of system abandonment
Perceived usefulness [199, 200]	The degree to which a user believes that using the system would enhance their job performance, a key component of technology acceptance models	A strong predictor of a clinician's intention to integrate the AI tool into their regular workflow

5 Benchmarks for Evaluating Explainable Multi-Modal LLMs in Healthcare

The development of standardized benchmarks is crucial for reproducible research, fostering innovation, and enabling fair comparisons between different XLLM methods. For multi-modal LLMs in healthcare, this area is particularly nascent but represents a critical frontier for the field.

5.1 The Critical Role of Benchmarks

Benchmarks serve several vital functions [84, 143]:

- **Reproducibility**: It offers shared datasets, models, and evaluation protocols and allows the results of a research process to be validated by the independent examination of the research procedure.
- **Progress Tracking**: It allows the systematic comparison of new XLLM methods against the state of the art, so it can accelerate research and development.
- **Fair Comparison**: It alleviates biases that can come from ad-hoc evaluation setups, creating a more impartial consideration of the strengths and weaknesses of a method.
- **Identifying Gaps**: It points to areas where there could be a deficiency in the XLLM methods or evaluation metrics available.

5.2 *The Current Landscape: A Critical Gap for Multi-Modal XLLMs*

Currently, no single, comprehensive, and widely accepted benchmark specifically designed for evaluating explainability in clinical multi-modal LLMs exists. Researchers typically rely on adaptations of existing resources, which often fall short:

Clinical NLP Datasets (Unimodal Text)

Datasets like MIMIC-III/IV [201] and i2b2 challenges [202] provide vast corpora of de-identified Electronic Health Records (EHR) notes. They are very useful to train and test LLM performance on text-driven tasks (e.g., summarization, information extraction); however, they lack ground truth explanation or multi-modal pairings. Therefore, these settings are not suitable to evaluate the quality of an XLLM's reasoning or attributions.

Clinical Vision Datasets (Unimodal Image)

Collections of medical images with expert annotations, for example, CheXpert [203] for chest X-rays, ISIC [204] for dermatoscopic images, and ADNI [205] for brain MRI, can be used for training and evaluating the visual components of VLMs; however, they do not typically provide explanations associated with the annotations, nor are paired/linked rationales available for each diagnostic label.

Multi-Modal Clinical Datasets (Limited Explainability Focus)

Emerging datasets such as MIMIC-CXR [206] and ROCO [207] offer paired medical images with a corresponding radiology report. These are ideal for evaluating multi-modal performance; however, they do not typically contain explicit human-annotated explanation for a particular diagnosis made based on both modalities.

General LLM Performance Benchmarks

Benchmarks like HELM [208] and clinical variants like MedQA [209] evaluate the LLM's output accuracy on various tasks (e.g., medical question answering), but they do not directly assess the quality of the explanations generated by these models.

General XAI Benchmarks (Unimodal)

The ERASER benchmark [210] provides a model for evaluating rationale extraction in text but is not designed for the generative, multi-modal, or conversational nature of modern LLMs.

The critical gap is the absence of multi-modal clinical datasets augmented with expert-annotated ground-truth explanations that explicitly link textual and visual evidence to a specific clinical outcome or reasoning step.

5.3 Architecting a Future Benchmark for Explainable Multi-Modal LLMs in Healthcare

A future-proof benchmark for this domain must be a multi-component suite for hybrid evaluation taking in both technical aspects and human-centered perspectives.

Component 1: Curated, Multi-Modal Datasets with Expert-Annotated Explanations

- **Data Sources**: It is comprehensive collections of de-identified Electronic Health Records (EHRs), clinical conversations, multi-modal medical imaging, and potentially genomic data.
- **Ground-Truth Annotations**: This is the most important area of the benchmarks and the most resource-intensive aspect. Datasets must contain explicit, expert-annotated "gold standard" rationales or explanations. For example:
- To walk through a specific diagnostic task, a panel of physicians would provide the diagnosis, but they would also provide the ideal explanation, as to why the diagnosis is correct. As part of this, they would identify specific textual evidence (e.g., phrases included in clinical notes) and precise evidence from the visual evidence (e.g., specific regions in an X-ray or CT scans) in support of their reasoning.
- For a prognosis task, experts would annotate the key multi-modal factors and their causal relationships.
- For an EHR summarization task, experts would annotate the most important sentences/sections in the input text and the corresponding visual findings that must be included in a summary.
- **Multi-Granularity Annotations**: Explanations should be annotated at different granularity levels (e.g., for text, token-level, sentence-level, and paragraph-level; for images, pixel-level, patch-level, and region-level; and cross-modal links between them).

Component 2: A Suite of Reference Multi-Modal LLMs

A collection of pre-trained and fine-tuned multi-modal LLMs (e.g., Med-Flamingo [47], CheXzero [211, 212], open-source models like Llama fine-tuned on clinical multi-modal data) to serve as a common set of models to be explained. This allows for fair comparisons of XLLM methods across different underlying model architectures.

Component 3: Standardized Multi-Modal Clinical Tasks

- **Multi-Modal Differential Diagnosis**: Evaluate if the explanation helps clinicians correctly re-rank a differential diagnosis list or identify the most likely diagnosis based on both textual and visual evidence.
- **Radiology Report Generation with Grounding**: Evaluate if the generated report is factually consistent with the image, and if the explanation correctly grounds each statement in the corresponding image region.
- **Multi-Modal Clinical Question Answering (CQA)**: Evaluate if the explanation for an LLM's answer to a complex clinical question is factually correct, clinically plausible, and at an appropriate level of detail.
- **EHR Summarization with Critical Information Extraction**: Assess whether the explanation helps to identify and amend missing or incorrect elements to the AI-generated summary based on the summarized critical multi-modal evidence.

Component 4: A Hybrid Evaluation Toolkit

- An open-source software library implementing the key technical metrics (as detailed in Sect. 3) adapted for multi-modal inputs. This would include:
  - Automated factual consistency checkers against UMLS/SNOMED-CT.
  - Cross-modal grounding metrics (e.g., IoU between textual claim and image region).
  - Robustness-to-multi-modal-paraphrase scorers.
- A set of pre-validated human-centered evaluation protocols (as detailed in Sect. 4), including:
  - Standardized task interfaces for multi-modal clinician studies.
  - Validated survey instruments for trust calibration, cognitive load, and clinical plausibility.
  - Templates for qualitative data collection and analysis.

6 Conclusion: The Evaluation Framework for Clinical Practice

The analysis throughout this chapter reveals that no single metric or evaluation paradigm is sufficient for the responsible deployment of multi-modal LLMs in healthcare. The inherent complexities of reasoning, generative outputs, and high-stakes clinical decision-making necessitate a robust, holistic approach. This chapter therefore proposes this evaluation framework for the holistic assessment of XLLM systems. This framework mandates a multi-modal approach, integrating evidence from all pillars:

1. Pillar 1: Technical Validation. The XLLM system must undergo a thorough suite of automated tests to quantify its fidelity, robustness, and, most critically, its factual correctness against not only the input multi-modal data but also trusted medical knowledge bases. This provides a measure of technical reliability, cross-modal consistency, and safety.
2. Pillar 2: Clinical Utility Validation. The system must be evaluated in controlled studies with target clinical users (e.g., physicians, nurses). This evaluation must measure the explanation's tangible impact on task performance (e.g., diagnostic accuracy, report generation quality), trust calibration, cognitive load, and workflow efficiency. Crucially, this pillar assesses the explanation's clinical plausibility and actionability as perceived by experts.
3. Pillar 3: Comparative Benchmarking. The performance of the system on technical and human-centered metrics should be evaluated against state-of-the-art methods on existing public benchmarks. This enables claims of improvement to be based on empirical evidence that is reproducible and also places evaluation and overall performance in the context of discourse in the scientific community. This pillar also supports a continual improvement plan for an emergent and standardized approach to evaluation.

Together, these pillars can help the field to shift towards a more mature and principled science of explainability for LLMs. They also promote a transition from developing new explainability algorithms to a more systematic engineering and validation approach for actual clinical contexts. Ultimately, what we want to see is that explainability methods are not just technically sound but are plausible in a clinical sense, factually valid across modalities, and ultimately improve the practice of medicine and the safety of patients.

References

1. Esteva, A., et al., *Dermatologist-level classification of skin cancer with deep neural networks.* nature, 2017. **542**(7639): p. 115–118.
2. Topol, E.J., *High-performance medicine: the convergence of human and artificial intelligence.* Nature medicine, 2019. **25**(1): p. 44–56.

3. Komorowski, M., et al., *The artificial intelligence clinician learns optimal treatment strategies for sepsis in intensive care.* Nature medicine, 2018. **24**(11): p. 1716–1720.
4. Gupta, R., et al., *Artificial intelligence to deep learning: machine intelligence approach for drug discovery.* Molecular diversity, 2021. **25**(3): p. 1315–1360.
5. Alowais, S.A., et al., *Revolutionizing healthcare: the role of artificial intelligence in clinical practice.* BMC medical education, 2023. **23**(1): p. 689.
6. Thirunavukarasu, A.J., et al., *Large language models in medicine.* Nature medicine, 2023. **29**(8): p. 1930–1940.
7. Naveed, H., et al., *A comprehensive overview of large language models.* ACM Transactions on Intelligent Systems and Technology, 2025. **16**(5): p. 1–72.
8. Zhao, W.X., et al., *A survey of large language models.* arXiv preprint arXiv:2303.18223, 2023. **1**(2).
9. Yang, X., et al., *A large language model for electronic health records.* NPJ digital medicine, 2022. **5**(1): p. 194.
10. Achiam, J., et al., *Gpt-4 technical report.* arXiv preprint arXiv:2303.08774, 2023.
11. Peng, B., et al., *Instruction tuning with gpt-4.* arXiv preprint arXiv:2304.03277, 2023.
12. Nori, H., et al., *Capabilities of gpt-4 on medical challenge problems.* arXiv preprint arXiv: 2303.13375, 2023.
13. Zhu, D., et al., *Minigpt-4: Enhancing vision-language understanding with advanced large language models.* arXiv preprint arXiv:2304.10592, 2023.
14. Tu, T., et al., *Towards generalist biomedical AI.* Nejm Ai, 2024. **1**(3): p. AIoa2300138.
15. Singhal, K., et al., *Toward expert-level medical question answering with large language models.* Nature Medicine, 2025. **31**(3): p. 943–950.
16. Sallam, M. *ChatGPT utility in healthcare education, research, and practice: systematic review on the promising perspectives and valid concerns.* in *Healthcare*. 2023. MDPI.
17. Sallam, M., *ChatGPT utility in healthcare education, research, and practice: systematic review on the promising perspectives and valid concerns. Healthcare (Basel). 2023; 11 (6): 887.* Nature, 2023. **616**(7956): p. 259–265.
18. Singhal, K., et al., *Large language models encode clinical knowledge.* Nature, 2023. **620**(7972): p. 172–180.
19. Singhal, K., et al., *Large language models encode clinical knowledge.* arXiv preprint arXiv: 2212.13138, 2022.
20. Price, W.N., *Big data and black-box medical algorithms.* Science translational medicine, 2018. **10**(471): p. eaao5333.
21. Von Eschenbach, W.J., *Transparency and the black box problem: Why we do not trust AI.* Philosophy & technology, 2021. **34**(4): p. 1607–1622.
22. Castelvecchi, D., *Can we open the black box of AI?* Nature News, 2016. **538**(7623): p. 20.
23. Schwartz, I.S., K.E. Link, R. Daneshjou, and N. Cortés-Penfield, *Black box warning: large language models and the future of infectious diseases consultation.* Clinical infectious diseases, 2024. **78**(4): p. 860–866.
24. Ghassemi, M., L. Oakden-Rayner, and A.L. Beam, *The false hope of current approaches to explainable artificial intelligence in health care.* The lancet digital health, 2021. **3**(11): p. e745–e750.
25. Ji, Z., et al., *Survey of hallucination in natural language generation.* ACM computing surveys, 2023. **55**(12): p. 1–38.
26. Bai, Z., et al., *Hallucination of multimodal large language models: A survey.* arXiv preprint arXiv:2404.18930, 2024.
27. Ong, J.C.L., et al., *Ethical and regulatory challenges of large language models in medicine.* The Lancet Digital Health, 2024. **6**(6): p. e428–e432.
28. Ullah, E., A. Parwani, M.M. Baig, and R. Singh, *Challenges and barriers of using large language models (LLM) such as ChatGPT for diagnostic medicine with a focus on digital pathology–a recent scoping review.* Diagnostic pathology, 2024. **19**(1): p. 43.
29. Gao, Y., et al., *Chat-rec: Towards interactive and explainable llms-augmented recommender system.* arXiv preprint arXiv:2303.14524, 2023.

30. Hsieh, W., et al., *A comprehensive guide to explainable ai: From classical models to llms.* arXiv preprint arXiv:2412.00800, 2024.
31. Ali, T. and P. Kostakos, *Huntgpt: Integrating machine learning-based anomaly detection and explainable ai with large language models (llms).* arXiv preprint arXiv:2309.16021, 2023.
32. Bansal, G., et al. *Does the whole exceed its parts? the effect of ai explanations on complementary team performance.* in *Proceedings of the 2021 CHI conference on human factors in computing systems.* 2021.
33. Jin, X., et al., *Epidemiological, clinical and virological characteristics of 74 cases of coronavirus-infected disease 2019 (COVID-19) with gastrointestinal symptoms.* Gut, 2020. **69**(6): p. 1002–1009.
34. Huang, C., et al., *Clinical features of patients infected with 2019 novel coronavirus in Wuhan, China.* The lancet, 2020. **395**(10223): p. 497–506.
35. Danilevsky, M., et al., *A survey of the state of explainable AI for natural language processing.* arXiv preprint arXiv:2010.00711, 2020.
36. Ye, X., Q. Chen, I. Dillig, and G. Durrett, *Satlm: Satisfiability-aided language models using declarative prompting.* Advances in Neural Information Processing Systems, 2023. **36**: p. 45548–45580.
37. Sprague, Z., et al., *Musr: Testing the limits of chain-of-thought with multistep soft reasoning.* arXiv preprint arXiv:2310.16049, 2023.
38. Wei, J., et al., *Chain-of-thought prompting elicits reasoning in large language models.* Advances in neural information processing systems, 2022. **35**: p. 24824–24837.
39. Turpin, M., J. Michael, E. Perez, and S. Bowman, *Language models don't always say what they think: Unfaithful explanations in chain-of-thought prompting.* Advances in Neural Information Processing Systems, 2023. **36**: p. 74952–74965.
40. Yang, J., et al. *Thinking in space: How multimodal large language models see, remember, and recall spaces.* in *Proceedings of the Computer Vision and Pattern Recognition Conference.* 2025.
41. Wei, J., et al., *Emergent abilities of large language models.* arXiv preprint arXiv:2206.07682, 2022.
42. Bender, E.M., T. Gebru, A. McMillan-Major, and S. Shmitchell. *On the dangers of stochastic parrots: Can language models be too big??* in *Proceedings of the 2021 ACM conference on fairness, accountability, and transparency.* 2021.
43. Caruana, R., et al. *Intelligible models for healthcare: Predicting pneumonia risk and hospital 30-day readmission.* in *Proceedings of the 21th ACM SIGKDD international conference on knowledge discovery and data mining.* 2015.
44. Wachter, S., B. Mittelstadt, and C. Russell, *Counterfactual explanations without opening the black box: Automated decisions and the GDPR.* Harv. JL & Tech., 2017. **31**: p. 841.
45. Baldauf, M., S. Dustdar, and F. Rosenberg, *A survey on context-aware systems.* International Journal of ad Hoc and ubiquitous Computing, 2007. **2**(4): p. 263–277.
46. Bricon-Souf, N. and C.R. Newman, *Context awareness in health care: A review.* international journal of medical informatics, 2007. **76**(1): p. 2–12.
47. Moor, M., et al., *Foundation models for generalist medical artificial intelligence.* Nature, 2023. **616**(7956): p. 259–265.
48. Abhishek, K. and D. Kamath, *Attribution-based XAI methods in computer vision: A review.* arXiv preprint arXiv:2211.14736, 2022.
49. Alam, K.N., P.B. Zadeh, and A. Sheikh-Akbari, *Attribution-Based Explainability in Medical Imaging: A Critical Review on Explainable Computer Vision (X-CV) Techniques and Their Applications in Medical AI.* Electronics, 2025. **14**(15): p. 3024.
50. Jain, S. and B.C. Wallace, *Attention is not explanation.* arXiv preprint arXiv:1902.10186, 2019.
51. Wiegreffe, S. and Y. Pinter, *Attention is not not explanation.* arXiv preprint arXiv:1908.04626, 2019.
52. Sundararajan, M., A. Taly, and Q. Yan. *Axiomatic attribution for deep networks.* in *International conference on machine learning.* 2017. PMLR.

53. Shrikumar, A., P. Greenside, and A. Kundaje. *Learning important features through propagating activation differences.* in *International conference on machine learning.* 2017. PMIR.
54. Lundberg, S.M. and S.-I. Lee, *A unified approach to interpreting model predictions.* Advances in neural information processing systems, 2017. **30**.
55. Ribeiro, M.T., S. Singh, and C. Guestrin. *"Why should i trust you?" Explaining the predictions of any classifier.* in *Proceedings of the 22nd ACM SIGKDD international conference on knowledge discovery and data mining.* 2016.
56. Ribeiro, M.T., S. Singh, and C. Guestrin, *Model-agnostic interpretability of machine learning.* arXiv preprint arXiv:1606.05386, 2016.
57. Koh, P.W. and P. Liang. *Understanding black-box predictions via influence functions.* in *International conference on machine learning.* 2017. PMLR.
58. Guu, K., et al. *Retrieval augmented language model pre-training.* in *International conference on machine learning.* 2020. PMLR.
59. Guidotti, R., *Counterfactual explanations and how to find them: literature review and benchmarking.* Data Mining and Knowledge Discovery, 2024. **38**(5): p. 2770–2824.
60. Roese, N.J., *Counterfactual thinking.* Psychological bulletin, 1997. **121**(1): p. 133.
61. Mothilal, R.K., A. Sharma, and C. Tan. *Explaining machine learning classifiers through diverse counterfactual explanations.* in *Proceedings of the 2020 conference on fairness, accountability, and transparency.* 2020.
62. Goyal, Y., et al. *Counterfactual visual explanations.* in *International Conference on Machine Learning.* 2019. PMLR.
63. Verma, S., et al., *Counterfactual explanations and algorithmic recourses for machine learning: A review.* ACM Computing Surveys, 2024. **56**(12): p. 1–42.
64. Wang, Z., I. Samsten, and P. Papapetrou. *Counterfactual explanations for survival prediction of cardiovascular ICU patients.* in *International conference on artificial intelligence in medicine.* 2021. Springer.
65. Prosperi, M., et al., *Causal inference and counterfactual prediction in machine learning for actionable healthcare.* Nature Machine Intelligence, 2020. **2**(7): p. 369–375.
66. Wu, T., M.T. Ribeiro, J. Heer, and D.S. Weld, *Polyjuice: Generating counterfactuals for explaining, evaluating, and improving models.* arXiv preprint arXiv:2101.00288, 2021.
67. Wu, Z., et al. *Reasoning or reciting? exploring the capabilities and limitations of language models through counterfactual tasks.* 2024. Association for Computational Linguistics.
68. Ghorbani, A., J. Wexler, J.Y. Zou, and B. Kim, *Towards automatic concept-based explanations.* Advances in neural information processing systems, 2019. **32**.
69. Mincu, D., et al. *Concept-based model explanations for electronic health records.* in *Proceedings of the Conference on Health, Inference, and Learning.* 2021.
70. Patrício, C., J.C. Neves, and L.F. Teixeira. *Coherent concept-based explanations in medical image and its application to skin lesion diagnosis.* in *Proceedings of the IEEE/CVF conference on computer vision and pattern recognition.* 2023.
71. Kim, B., et al. *Interpretability beyond feature attribution: Quantitative testing with concept activation vectors (tcav).* in *International conference on machine learning.* 2018. PMLR.
72. Belinkov, Y., *Probing classifiers: Promises, shortcomings, and advances.* Computational Linguistics, 2022. **48**(1): p. 207–219.
73. Manigrasso, F., S. Schouten, L. Morra, and P. Bloem. *Probing llms for logical reasoning.* in *International Conference on Neural-Symbolic Learning and Reasoning.* 2024. Springer.
74. Huang, S., et al., *Can large language models explain themselves? a study of llm-generated self-explanations.* arXiv preprint arXiv:2310.11207, 2023.
75. Camburu, O.-M., T. Rocktäschel, T. Lukasiewicz, and P. Blunsom, *e-snli: Natural language inference with natural language explanations.* Advances in Neural Information Processing Systems, 2018. **31**.
76. Kumar, S. and P. Talukdar, *NILE: Natural language inference with faithful natural language explanations.* arXiv preprint arXiv:2005.12116, 2020.

77. Velupillai, S., et al., *Using clinical natural language processing for health outcomes research: overview and actionable suggestions for future advances.* Journal of biomedical informatics, 2018. **88**: p. 11–19.
78. Lampinen, A.K., et al., *Can language models learn from explanations in context?* arXiv preprint arXiv:2204.02329, 2022.
79. Meldo, A., L. Utkin, M. Kovalev, and E. Kasimov, *The natural language explanation algorithms for the lung cancer computer-aided diagnosis system.* Artificial intelligence in medicine, 2020. **108**: p. 101952.
80. Chen, K., et al. *LLM-DR: A Novel LLM-Aided Diffusion Model for Rule Generation on Temporal Knowledge Graphs.* in *Proceedings of the AAAI Conference on Artificial Intelligence.* 2025.
81. Liu, P., et al., *A survey on open information extraction from rule-based model to large language model.* arXiv preprint arXiv:2208.08690, 2022.
82. Sheikh, S.J. and S. Haider. *On Comparing LLM-Generated Causal Networks with a Rule-Based Approach.* in *Asian Conference on Intelligent Information and Database Systems.* 2025. Springer.
83. Zin, M.M., K. Satoh, and G. Borges, *Leveraging LLM for Identification and Extraction of Normative Statements*, in *Legal Knowledge and Information Systems.* 2024, IOS Press. p. 215–225.
84. Mohseni, S., N. Zarei, and E.D. Ragan, *A multidisciplinary survey and framework for design and evaluation of explainable AI systems.* ACM Transactions on Interactive Intelligent Systems (TiiS), 2021. **11**(3–4): p. 1–45.
85. Alvarez-Melis, D. and T.S. Jaakkola, *On the robustness of interpretability methods.* arXiv preprint arXiv:1806.08049, 2018.
86. Samek, W., T. Wiegand, and K.-R. Müller, *Explainable artificial intelligence: Understanding, visualizing and interpreting deep learning models.* arXiv preprint arXiv:1708.08296, 2017.
87. Samek, W., et al., *Evaluating the visualization of what a deep neural network has learned.* IEEE transactions on neural networks and learning systems, 2016. **28**(11): p. 2660–2673.
88. Nayfeh, A.H., *Perturbation methods.* 2024: John Wiley & Sons.
89. Pearl, J., *Causality.* 2009: Cambridge university press.
90. Wang, Y., T. Zhang, X. Guo, and Z. Shen, *Gradient based feature attribution in explainable ai: A technical review.* arXiv preprint arXiv:2403.10415, 2024.
91. Hooker, S., D. Erhan, P.-J. Kindermans, and B. Kim, *A benchmark for interpretability methods in deep neural networks.* Advances in neural information processing systems, 2019. **32**.
92. Yeh, C.-K., et al., *On the (in) fidelity and sensitivity of explanations.* Advances in neural information processing systems, 2019. **32**.
93. Lyu, Q., et al. *Faithful chain-of-thought reasoning.* in *The 13th International Joint Conference on Natural Language Processing and the 3rd Conference of the Asia-Pacific Chapter of the Association for Computational Linguistics (IJCNLP-AACL 2023).* 2023.
94. Lanham, T., et al., *Measuring faithfulness in chain-of-thought reasoning.* arXiv preprint arXiv: 2307.13702, 2023.
95. Paul, D., R. West, A. Bosselut, and B. Faltings, *Making reasoning matter: Measuring and improving faithfulness of chain-of-thought reasoning.* arXiv preprint arXiv:2402.13950, 2024.
96. Jacovi, A. and Y. Goldberg, *Towards faithfully interpretable NLP systems: How should we define and evaluate faithfulness?* arXiv preprint arXiv:2004.03685, 2020.
97. Krishna, S., et al., *The disagreement problem in explainable machine learning: A practitioner's perspective.* arXiv preprint arXiv:2202.01602, 2022.
98. Hedström, A., et al., *The meta-evaluation problem in explainable AI: identifying reliable estimators with MetaQuantus.* arXiv preprint arXiv:2302.07265, 2023.
99. Tomsett, R., et al. *Sanity checks for saliency metrics.* in *Proceedings of the AAAI conference on artificial intelligence.* 2020.
100. Tomsett, R., et al., *Rapid trust calibration through interpretable and uncertainty-aware AI.* Patterns, 2020. **1**(4).
101. Krippendorff, K., *Computing Krippendorff's alpha-reliability.* 2011.

102. Hauke, J. and T. Kossowski, *Comparison of values of Pearson's and Spearman's correlation coefficients on the same sets of data.* Quaestiones geographicae, 2011. **30**(2): p. 87–93.
103. Tang, L., P. Laban, and G. Durrett, *Minicheck: Efficient fact-checking of llms on grounding documents.* arXiv preprint arXiv:2404.10774, 2024.
104. Li, H., et al., *Cmmlu: Measuring massive multitask language understanding in chinese.* arXiv preprint arXiv:2306.09212, 2023.
105. Lin, S., J. Hilton, and O. Evans, *Truthfulqa: Measuring how models mimic human falsehoods.* arXiv preprint arXiv:2109.07958, 2021.
106. Huang, Y., et al., *C-eval: A multi-level multi-discipline chinese evaluation suite for foundation models.* Advances in Neural Information Processing Systems, 2023. **36**: p. 62991–63010.
107. Lee, N., et al., *Factuality enhanced language models for open-ended text generation.* Advances in Neural Information Processing Systems, 2022. **35**: p. 34586–34599.
108. Moiseev, F., Z. Dong, E. Alfonseca, and M. Jaggi, *SKILL: Structured knowledge infusion for large language models.* arXiv preprint arXiv:2205.08184, 2022.
109. Chen, Y., et al., *Grounded 3d-llm with referent tokens.* arXiv preprint arXiv:2405.10370, 2024.
110. Wang, T., et al., *From Objects to Anywhere: A Holistic Benchmark for Multi-level Visual Grounding in 3D Scenes.* arXiv preprint arXiv:2506.04897, 2025.
111. Wang, Y., Y.-L. Li, W.E. ZY, and S. Wang. *Liba: Language instructed multi-granularity bridge assistant for 3d visual grounding.* in *Proceedings of the AAAI Conference on Artificial Intelligence.* 2025.
112. Han, Y., et al. *GroundingFace: Fine-grained Face Understanding via Pixel Grounding Multi-modal Large Language Model.* in *Proceedings of the Computer Vision and Pattern Recognition Conference.* 2025.
113. Jiang, Y., et al., *CLEAR: A Clinically-Grounded Tabular Framework for Radiology Report Evaluation.* arXiv preprint arXiv:2505.16325, 2025.
114. Xu, C., et al. *Multi-attribute interactions matter for 3d visual grounding.* in *Proceedings of the IEEE/CVF Conference on Computer Vision and Pattern Recognition.* 2024.
115. Bodenreider, O., *The unified medical language system (UMLS): integrating biomedical terminology.* Nucleic acids research, 2004. **32**(suppl_1): p. D267–D270.
116. Donnelly, K., *SNOMED-CT: The advanced terminology and coding system for eHealth.* Studies in health technology and informatics, 2006. **121**: p. 279.
117. Feng, B., et al., *Caption matters: a new perspective for knowledge-based visual question answering.* Knowledge and Information Systems, 2024. **66**(11): p. 6975–7003.
118. Amayuelas, A., J. Sain, S. Kaur, and C. Smiley, *Grounding LLM Reasoning with Knowledge Graphs.* arXiv preprint arXiv:2502.13247, 2025.
119. Jokinen, K. *The need for grounding in llm-based dialogue systems.* in *Proceedings of the Workshop: Bridging Neurons and Symbols for Natural Language Processing and Knowledge Graphs Reasoning (NeusymBridge)@ LREC-COLING-2024.* 2024.
120. Li, Z., et al. *Flexkbqa: A flexible llm-powered framework for few-shot knowledge base question answering.* in *Proceedings of the AAAI conference on artificial intelligence.* 2024.
121. Manakul, P., A. Liusie, and M.J. Gales, *Selfcheckgpt: Zero-resource black-box hallucination detection for generative large language models.* arXiv preprint arXiv:2303.08896, 2023.
122. Tonmoy, S., et al., *A comprehensive survey of hallucination mitigation techniques in large language models.* arXiv preprint arXiv:2401.01313, 2024. **6**.
123. Slack, D., et al. *Fooling lime and shap: Adversarial attacks on post hoc explanation methods.* in *Proceedings of the AAAI/ACM Conference on AI, Ethics, and Society.* 2020.
124. Lakkaraju, H., E. Kamar, R. Caruana, and J. Leskovec, *Interpretable & explorable approximations of black box models.* arXiv preprint arXiv:1707.01154, 2017.
125. Lakkaraju, H., S.H. Bach, and J. Leskovec. *Interpretable decision sets: A joint framework for description and prediction.* in *Proceedings of the 22nd ACM SIGKDD international conference on knowledge discovery and data mining.* 2016.
126. Dandl, S., C. Molnar, M. Binder, and B. Bischl. *Multi-objective counterfactual explanations.* in *International conference on parallel problem solving from nature.* 2020. Springer.
127. Flesch, R., *Flesch-Kincaid readability test.* Retrieved October, 2007. **26**(3): p. 2007.

128. Marulli, F., et al., *Understanding readability of large language models output: an empirical analysis.* Procedia Computer Science, 2024. **246**: p. 5273–5282.
129. Liu, F., T. Jin, and J.S. Lee, *Automatic readability assessment for sentences: neural, hybrid and large language models.* Language Resources and Evaluation, 2025: p. 1–32.
130. Swisher, A.R., et al., *Enhancing health literacy: Evaluating the readability of patient handouts revised by ChatGPT's large language model.* Otolaryngology–Head and Neck Surgery, 2024. **171**(6): p. 1751–1757.
131. Friedman, J.H. and B.E. Popescu, *Predictive learning via rule ensembles.* 2008.
132. Bau, D., et al. *Network dissection: Quantifying interpretability of deep visual representations.* in *Proceedings of the IEEE conference on computer vision and pattern recognition.* 2017.
133. Gell-Mann, M. and S. Lloyd, *Effective complexity.* Nonextensive entropy, 2004: p. 387–398.
134. Adebayo, J., et al., *Sanity checks for saliency maps.* Advances in neural information processing systems, 2018. **31**.
135. Van Stein, B., et al., *A comparison of global sensitivity analysis methods for explainable AI with an application in genomic prediction.* IEEE Access, 2022. **10**: p. 103364–103381.
136. Adebayo, J., M. Muelly, I. Liccardi, and B. Kim, *Debugging tests for model explanations.* arXiv preprint arXiv:2011.05429, 2020.
137. Sendak, M., et al. *"The human body is a black box" supporting clinical decision-making with deep learning.* in *Proceedings of the 2020 conference on fairness, accountability, and transparency.* 2020.
138. Zhou, J., A.H. Gandomi, F. Chen, and A. Holzinger, *Evaluating the quality of machine learning explanations: A survey on methods and metrics.* Electronics, 2021. **10**(5): p. 593.
139. Wang, Q., Z. Mao, B. Wang, and L. Guo, *Knowledge graph embedding: A survey of approaches and applications.* IEEE transactions on knowledge and data engineering, 2017. **29**(12): p. 2724–2743.
140. Lecue, F., *On the role of knowledge graphs in explainable AI.* Semantic Web, 2020. **11**(1): p. 41–51.
141. Rajabi, E. and K. Etminani, *Knowledge-graph-based explainable AI: A systematic review.* Journal of information science, 2024. **50**(4): p. 1019–1029.
142. Gilpin, L.H., et al. *Explaining explanations: An overview of interpretability of machine learning.* in *2018 IEEE 5th International Conference on data science and advanced analytics (DSAA).* 2018. IEEE.
143. Doshi-Velez, F. and B. Kim, *Towards a rigorous science of interpretable machine learning.* arXiv preprint arXiv:1702.08608, 2017.
144. Gunning, D. and D. Aha, *DARPA's explainable artificial intelligence (XAI) program.* AI magazine, 2019. **40**(2): p. 44–58.
145. Gunning, D., et al., *XAI—Explainable artificial intelligence.* Science robotics, 2019. **4**(37): p. eaay7120.
146. Hoffman, R.R., S.T. Mueller, G. Klein, and J. Litman, *Metrics for explainable AI: Challenges and prospects.* arXiv preprint arXiv:1812.04608, 2018.
147. Miller, T., *Explanation in artificial intelligence: Insights from the social sciences.* Artificial intelligence, 2019. **267**: p. 1–38.
148. Liao, Q.V. and K.R. Varshney, *Human-centered explainable ai (xai): From algorithms to user experiences.* arXiv preprint arXiv:2110.10790, 2021.
149. Bussone, A., S. Stumpf, and D. O'Sullivan. *The role of explanations on trust and reliance in clinical decision support systems.* in *2015 international conference on healthcare informatics.* 2015. IEEE.
150. Kulesza, T., M. Burnett, W.-K. Wong, and S. Stumpf. *Principles of explanatory debugging to personalize interactive machine learning.* in *Proceedings of the 20th international conference on intelligent user interfaces.* 2015.
151. Amann, J., et al., *Explainability for artificial intelligence in healthcare: a multidisciplinary perspective.* BMC medical informatics and decision making, 2020. **20**(1): p. 310.
152. Likert, R., *A technique for the measurement of attitudes.* Archives of psychology, 1932.

153. Harte, R., et al., *A human-centered design methodology to enhance the usability, human factors, and user experience of connected health systems: a three-phase methodology.* JMIR human factors, 2017. **4**(1): p. e5443.
154. Ross, K.G., J.L. Shafer, and G. Klein, *Professional judgments and "naturalistic decision making".* The Cambridge handbook of expertise and expert performance, 2006: p. 403–419.
155. Casey, B., A. Farhangi, and R. Vogl, *Rethinking explainable machines.* Berkeley Technology Law Journal, 2019. **34**(1): p. 143–188.
156. Neo, J.R.E., J.S. Ser, and S.S. Tay, *Use of large language model-based chatbots in managing the rehabilitation concerns and education needs of outpatient stroke survivors and caregivers.* Frontiers in Digital Health, 2024. **6**: p. 1395501.
157. Mavrepis, P., et al., *XAI for all: Can large language models simplify explainable AI?* arXiv preprint arXiv:2401.13110, 2024.
158. Jin, W., et al., *EUCA: The end-user-centered explainable AI framework.* arXiv preprint arXiv: 2102.02437, 2021.
159. Hsiao, J.H.-w., et al., *Roadmap of designing cognitive metrics for explainable artificial intelligence (XAI).* arXiv preprint arXiv:2108.01737, 2021.
160. Chen, S., J. Epps, and F. Chen. *A comparison of four methods for cognitive load measurement.* in *Proceedings of the 23rd Australian computer-human interaction conference.* 2011.
161. Jacob, R.J. and K.S. Karn, *Eye tracking in human-computer interaction and usability research: Ready to deliver the promises,* in *The mind's eye.* 2003, Elsevier. p. 573–605.
162. Joshi, A., S. Kale, S. Chandel, and D.K. Pal, *Likert scale: Explored and explained.* British journal of applied science & technology, 2015. **7**(4): p. 396.
163. Nourani, M., S. Kabir, S. Mohseni, and E.D. Ragan. *The effects of meaningful and meaningless explanations on trust and perceived system accuracy in intelligent systems.* in *Proceedings of the AAAI Conference on Human Computation and Crowdsourcing.* 2019.
164. Ribeiro, M.T., S. Singh, and C. Guestrin. *Anchors: High-precision model-agnostic explanations.* in *Proceedings of the AAAI conference on artificial intelligence.* 2018.
165. Shen, H. and T.-H. Huang. *How useful are the machine-generated interpretations to general users? a human evaluation on guessing the incorrectly predicted labels.* in *Proceedings of the AAAI Conference on Human Computation and Crowdsourcing.* 2020.
166. Nushi, B., E. Kamar, and E. Horvitz. *Towards accountable ai: Hybrid human-machine analyses for characterizing system failure.* in *Proceedings of the AAAI Conference on Human Computation and Crowdsourcing.* 2018.
167. Hart, S.G. *NASA-task load index (NASA-TLX); 20 years later.* in *Proceedings of the human factors and ergonomics society annual meeting.* 2006. Sage publications Sage CA: Los Angeles, CA.
168. Kosch, T., et al., *A survey on measuring cognitive workload in human-computer interaction.* ACM Computing Surveys, 2023. **55**(13s): p. 1–39.
169. Wang, J.T.-y., *Pupil dilation and eye tracking.* A handbook of process tracing methods for decision research: A critical review and user's guide, 2011: p. 185–204.
170. Lee, J.D. and K.A. See, *Trust in automation: Designing for appropriate reliance.* Human factors, 2004. **46**(1): p. 50–80.
171. Jian, J.-Y., A.M. Bisantz, and C.G. Drury, *Foundations for an empirically determined scale of trust in automated systems.* International journal of cognitive ergonomics, 2000. **4**(1): p. 53–71.
172. Dzindolet, M.T., et al., *The role of trust in automation reliance.* International journal of human-computer studies, 2003. **58**(6): p. 697–718.
173. Parasuraman, R. and V. Riley, *Humans and automation: Use, misuse, disuse, abuse.* Human factors, 1997. **39**(2): p. 230–253.
174. Holliday, D., S. Wilson, and S. Stumpf. *User trust in intelligent systems: A journey over time.* in *Proceedings of the 21st international conference on intelligent user interfaces.* 2016.
175. Schemmer, M., et al. *A meta-analysis of the utility of explainable artificial intelligence in human-AI decision-making.* in *Proceedings of the 2022 AAAI/ACM Conference on AI, Ethics, and Society.* 2022.

176. Coussement, K., et al., *Explainable AI for enhanced decision-making*. 2024, Elsevier. p. 114276.
177. Lai, V., et al. *Towards a science of human-AI decision making: An overview of design space in empirical human-subject studies*. in *Proceedings of the 2023 ACM conference on fairness, accountability, and transparency*. 2023.
178. Alufaisan, Y., et al. *Does explainable artificial intelligence improve human decision-making?* in *Proceedings of the AAAI Conference on Artificial Intelligence*. 2021.
179. Senoner, J., et al., *Explainable AI improves task performance in human–AI collaboration*. Scientific reports, 2024. **14**(1): p. 31150.
180. Brito, L.C., G.A. Susto, J.N. Brito, and M.A. Duarte, *An explainable artificial intelligence approach for unsupervised fault detection and diagnosis in rotating machinery*. Mechanical Systems and Signal Processing, 2022. **163**: p. 108105.
181. Nauta, M., et al., *From anecdotal evidence to quantitative evaluation methods: A systematic review on evaluating explainable ai*. ACM Computing Surveys, 2023. **55**(13s): p. 1–42.
182. Singh, R., et al., *An Actionability Assessment Tool for Explainable AI*. arXiv preprint arXiv: 2407.09516, 2024.
183. Bergomi, L., et al., *Which explanations do clinicians prefer? A comparative evaluation of XAI understandability and actionability in predicting the need for hospitalization*. BMC Medical Informatics and Decision Making, 2025. **25**(1): p. 1–13.
184. Agarwal, C., S.H. Tanneru, and H. Lakkaraju, *Faithfulness vs. plausibility: On the (un) reliability of explanations from large language models*. arXiv preprint arXiv:2402.04614, 2024.
185. Gentile, D., G. Jamieson, and B. Donmez. *Evaluating human understanding in XAI systems*. in *ACM CHI XCXAI Workshop*. 2021.
186. Hoffman, R.R., S.T. Mueller, G. Klein, and J. Litman, *Measures for explainable AI: Explanation goodness, user satisfaction, mental models, curiosity, trust, and human-AI performance*. Frontiers in Computer Science, 2023. **5**: p. 1096257.
187. Watkins, E.A., et al., *ACE, Action and Control via Explanations: A Proposal for LLMs to Provide Human-Centered Explainability for Multimodal AI Assistants*. arXiv preprint arXiv: 2503.16466, 2025.
188. Ehsan, U., et al. *Human-centered explainable AI (HCXAI): Reloading explainability in the era of large language models (LLMs)*. in *Extended Abstracts of the CHI Conference on Human Factors in Computing Systems*. 2024.
189. Kunz, J. and M. Kuhlmann, *Properties and challenges of llm-generated explanations*. arXiv preprint arXiv:2402.10532, 2024.
190. Liu, Y., et al., *Trustworthy llms: a survey and guideline for evaluating large language models' alignment*. arXiv preprint arXiv:2308.05374, 2023.
191. Huang, Y., et al., *Trustllm: Trustworthiness in large language models*. arXiv preprint arXiv: 2401.05561, 2024.
192. Nazi, Z.A. and W. Peng. *Large language models in healthcare and medical domain: A review*. in *Informatics*. 2024. MDPI.
193. Shulner-Tal, A., T. Kuflik, and D. Kliger, *Enhancing fairness perception–Towards human-centred AI and personalized explanations understanding the factors influencing laypeople's fairness perceptions of algorithmic decisions*. International Journal of Human–Computer Interaction, 2023. **39**(7): p. 1455–1482.
194. Mehrabi, N., et al., *A survey on bias and fairness in machine learning*. ACM computing surveys (CSUR), 2021. **54**(6): p. 1–35.
195. Jobin, A., M. Ienca, and E. Vayena, *The global landscape of AI ethics guidelines*. Nature machine intelligence, 2019. **1**(9): p. 389–399.
196. Mittelstadt, B.D., et al., *The ethics of algorithms: Mapping the debate*. Big Data & Society, 2016. **3**(2): p. 2053951716679679.
197. Lewis, J.R., *The system usability scale: past, present, and future*. International Journal of Human–Computer Interaction, 2018. **34**(7): p. 577–590.

198. Davis, F.D., *Perceived usefulness, perceived ease of use, and user acceptance of information technology.* MIS quarterly, 1989: p. 319–340.
199. Venkatesh, V. and F.D. Davis, *A theoretical extension of the technology acceptance model: Four longitudinal field studies.* Management science, 2000. **46**(2): p. 186–204.
200. Arrieta, A.B., et al., *Explainable Artificial Intelligence (XAI): Concepts, taxonomies, opportunities and challenges toward responsible AI.* Information fusion, 2020. **58**: p. 82–115.
201. Johnson, L., et al., *NMC horizon report: 2016 higher education edition.* 2016: The New Media Consortium.
202. Uzuner, Ö., B.R. South, S. Shen, and S.L. DuVall, *2010 i2b2/VA challenge on concepts, assertions, and relations in clinical text.* Journal of the American Medical Informatics Association, 2011. **18**(5): p. 552–556.
203. Irvin, J., et al. *Chexpert: A large chest radiograph dataset with uncertainty labels and expert comparison.* in *Proceedings of the AAAI conference on artificial intelligence.* 2019.
204. Codella, N.C., et al. *Skin lesion analysis toward melanoma detection: A challenge at the 2017 international symposium on biomedical imaging (isbi), hosted by the international skin imaging collaboration (isic).* in *2018 IEEE 15th international symposium on biomedical imaging (ISBI 2018).* 2018. IEEE.
205. Jack Jr, C.R., et al., *The Alzheimer's disease neuroimaging initiative (ADNI): MRI methods.* Journal of Magnetic Resonance Imaging: An Official Journal of the International Society for Magnetic Resonance in Medicine, 2008. **27**(4): p. 685–691.
206. Johnson, A.E., et al., *MIMIC-CXR, a de-identified publicly available database of chest radiographs with free-text reports.* Scientific data, 2019. **6**(1): p. 317.
207. Rückert, J., et al., *Rocov2: Radiology objects in context version 2, an updated multimodal image dataset.* Scientific Data, 2024. **11**(1): p. 688.
208. Liang, P., et al., *Holistic evaluation of language models.* arXiv preprint arXiv:2211.09110, 2022.
209. Jin, S., et al., *Inference and analysis of cell-cell communication using CellChat.* Nature communications, 2021. **12**(1): p. 1088.
210. DeYoung, J., et al., *ERASER: A benchmark to evaluate rationalized NLP models.* arXiv preprint arXiv:1911.03429, 2019.
211. Wang, B., et al., *Pre-trained language models in biomedical domain: A systematic survey.* ACM Computing Surveys, 2023. **56**(3): p. 1–52.
212. Tiu, E., et al., *Expert-level detection of pathologies from unannotated chest X-ray images via self-supervised learning.* Nature biomedical engineering, 2022. **6**(12): p. 1399–1406.

Legal and Regulatory Considerations

Azadeh Zamanifar, Negar Moslehi Jenabian, Amirfarhad Farhadi, Mohammad Asadollahi, and Homayoun Safarpour Motealegh Mahalegi

Abbreviations

Large Language Models (LLMs)	AI models trained on vast corpora capable of producing human-like medical text.
Healthcare Regulation	Legal frameworks governing the use of AI technologies in clinical and public health contexts.
Explainability	The degree to which an AI model's decision-making process can be understood by humans.
GDPR	The European Union's General Data Protection Regulation, focusing on privacy and automated decision-making.
Accountability	Legal responsibility and oversight mechanisms for AI-driven healthcare applications.
Governance and Regulation	Systems and laws ensuring safe and lawful AI deployment in healthcare

A. Zamanifar (✉) · N. Moslehi Jenabian · M. Asadollahi
Department of Computer Engineering, SR.C, Islamic Azad University, Tehran, Iran
e-mail: azamanifar@iau.ac.ir

N. Moslehi Jenabian
e-mail: negar.moslehijenabian@iau.ir

M. Asadollahi
e-mail: mohammad.asadolahi@iau.ir

A. Farhadi
School of Computer Engineering, Iran University of Science and Technology, Tehran, Iran
e-mail: am_farhadi@mail.iust.ac.ir

H. S. M. Mahalegi
University of Limerick, Limerick, Ireland
e-mail: safarpourmotealegh.homayoun@ul.ie

A. Zamanifar et al. (eds.), *Explainable Large Language Models in Healthcare Applications*, Bio-IT and AI, https://doi.org/10.1007/978-3-032-15088-2_11

1 Introduction

Unlike traditional medical equipment that passes through a definite sequence of the approval process and rarely changes since the time it was certified, LLMs are continually learning and evolving. And as their behavior may evolve with each retraining or model update, making the NP system in effect new in the legal sense of this term, one not necessarily addressed by the regulatory framework [6]. The legal and ethical consequences of all of this are enormous. This new validation, surveillance and liability management models are necessitated by the need of health regulators to shift away down the balance sheet and away to new models of validation, surveillance and managing liability as opposed to simply being in a state of dynamism but under lifecycle control as Mello et al. [1] suggest.

The revolution in regulation is already occurring in some geographical regions. The AI Act (2024) by the EU is one of the earliest wholesome efforts to risk tier an AI system. High-risk apps in the area of healthcare need a conformity test, rightful technical documentation, and human control as outlined in the EU medical device regulation [7]. In the USA, FDA Digital Health Innovation Action Plan speaks of a more iterative process that allows learning and adjusting to real-world performance, through so-called Good Machine Learning practices (GMLPs) [8].

Meanwhile, the Global South also has a governance problem of its own: according to comparative legal studies [9], most of the low- and middle-income nations have neither the technical nor the regulatory capability to evaluate advanced AI systems. This disconnect could aggravate global inequalities in healthcare innovation and healthcare access, indicating that there is a necessity of global collaboration and the need to share knowledge in AI regulation. LLM is also black box, whose logic is not visible to the users or developers. This is troublesome in the medical practice where diagnosis and treatment being proposed can be life-threatening: failure to clarify the logic behind the proposals can hinder the user to trust the system and consequences may be legal [10]. The proponents of legal scholarship posit that explainability must be regarded as a technical end as well as a matter of law [3] because of the rights of the patients and the duty of care of clinicians.

However, the technological progress has not removed technical obstacles to the ethical application of AI, such as the privacy of the data, the bias of the algorithms, and fairness. Using the example, when a specific demographic group is underrepresented in the training data of an LLM, the latter will issue predictions or recommendations that biased it [11]. It also overlaps with other anti-discrimination and anti-human rights law.

In such a way, the use of LLMs in health care is an area where interdisciplinary legal regulation is needed, which includes the principles of data protection law, medical ethics, and AI governance. This chapter uses the advancements in the scholarly and policy literature to examine the manner these factors collide on empirical levels. Recent studies have also demonstrated the role of deep learning in specific medical subdomains; proposed an autonomous dental treatment planning system using object detection on panoramic X-rays, illustrating the capacity of AI to augment

Table 1 LLMs versus traditional medical devices

Aspect	Traditional medical devices	Large language models	Regulatory implication
Development	Fixed design, pre-certified	Dynamic, continuously updated	Requires lifecycle-based oversight
Approval process	Premarket clearance only	Needs ongoing post-market evaluation	Post-deployment audits required
Risk nature	Hardware failure	Algorithmic bias or drift	Ethical and transparency issues
Accountability	Manufacturer primarily liable	Shared among developers, clinicians, institutions	Calls for shared liability model

precision in medical decision-making [12]. By so doing, it aims at answering some of the following important questions:

Which types of legal and regulation systems are possible to evolve to ensure that a state will remain involved with an AI that is constantly evolving over time? How can we guarantee transparency and responsibility and at the same time bar innovation? Who takes the blame and responsibility: delevopers, clinicians and institutions? And what would international collaboration become so as to encourage equitable and harmless usage of LLMs in the medical context? In dealing with these concerns, this chapter offers an expansive insight on how regulation authorities, legislation, and moral codes can syntactically organize a conscientious adoption of clarifiable AI in the medical field (Table 1).

2 Governance and Regulatory Frameworks

2.1 Developing Prescriptive Medical Regulation into AI Supervision with Flexibility

Regulation has historically taken the form of pre-market approval (that is, prior to the certification of the product) of medical technologies, which may not require further updating and follow-up (unless it happens that their safety meant that they required closer monitoring). This hardware-based model originally intended to work with such hardware as pacemakers or MRI scanners obviously does not fit the concept of artificial intelligence systems that continuously evolve as more data is provided in and retrained on [13].

Healthcare—there are quite different with large Language Models (LLMs). They are also programmed to innovation themselves over time and any performance or behavior that was not initially done during the time of regulatory clearance may change [14]. The notion behind the so-called lifecycle regulation was mentioned by scholars like Mello et al. [1] and Amann et al. [10], and the latter is applied to AI tools, which are targeted at post-market monitoring (synonymous with pharmacovigilance).

2.2 Good AI Governance Principles in Healthcare

The four principles that the AI governance literature in health care highlights are interrelated and they include:

- Accountability: There should be clear accountability by each of the parties, developers, clinicians, institutions, as to the results of AI [15].
- Transparency: The users and patients should be aware of the decisions AI is making (when and how) [16].
- Equity: Systems must not be discriminative against groups of population who can disadvantage certain groups of people [11].
- Sustainability: There should be governance systems that are long term follow up and not only point assessment [17].

According to the report published by the World Economic Forum (2025), the principles should only work under the condition of coordination of regulators on the legal, technical and ethical levels in their report titled Earning Trust for AI in Health. It recommends the establishment of multi-stakeholder councils of clinicians, technologists and patient advocates that evaluate AI systems during any phase of their lifecycle [18].

Moreover, practice governance should stop being compliant to ethics-by-design and become fair, accountable and transparent, therefore the fairness, accountability and transparency should be incorporated in the system development and not added later [19].

Blockchain-enabled authentication frameworks, such as the secure and efficient SEBAKA protocol proposed [20] exemplify how decentralized trust models could reinforce accountability and data integrity in AI-driven healthcare systems.

2.3 Regional Regulatory Models

The AI Act established by EU has raised the level in the world as regards the risk-based regulatory approach. Healthcare AI is classified as a high-risk and thus there are specifications regarding the quality of data, technical documentations, human controls and explanations [7]. The regulation directly refers to the AI governance since it is consistent with the already existing laws like GDPR [3] where the rights and autonomy of patients are protected [12].

The US has a more decentralized system of regulation in this issue. Flexibility promotes innovation by the FDA Digital Health Center of Excellence and 21st

Century Cures Act. Instead of the strict certification procedure, what regulating bodies do is that they concentrate on the post-market performance monitoring and update [8, 20].

The arising economies that involve South Africa and India, introduce a different form of the regulation issues due to their AI evaluation skills and data control [9]. International partnerships are also beginning to be viewed as the sufficient tool to resolve these mismatches, such as Global Health AI Alliance [21]. Lessons from other smart-infrastructure domains, such as [22] energy-efficient communication model for smart cities, highlight the need for scalable and resource-sensitive frameworks that could inform AI governance in low-capacity healthcare environments.

2.4 *Common Accountability and Regulatory Sandboxes*

To date, there exists no single institution that can be called upon to give guarantee to AI safety. Within the field of legal studies, the concept of shared accountability is put forward to describe the collective responsibility on the side of the developers, the healthcare institutions, and regulators with respect to the results of AI [23].

Controlled environments, such as regulatory sandboxes that enable the testing of AI systems in real-life settings, provide a chance to test them under observative eyes. The medicines and healthcare products regulatory agency (MHRA) in UK and the health sciences authority (HSA) in Singapore have established sandbox models that facilitate innovation without posing any risk to the safety of patients [24].

Such efforts inform us that there is no clear boundary between governance and innovation, that the two can co-exist and that the source of such co-existence lies in guided experimentation along with continuous auditing and disclosure of the results to the community (Table 2).

Table 2 Evolution overview

Phase	Regulatory Characteristic	Example or Reference
Traditional Regulation	Static approval for medical devices; limited updates	Pre-market device certification
Lifecycle Regulation	Continuous monitoring, iterative re-approval	FDA good machine learning practices
Collaborative Governance	Shared responsibility between regulators, developers, clinicians	AI Act (EU) & WHO-AU partnership

3 Transparency and Explainability Requirements

3.1 The Legal Bases of Explainability

Transparency and explainability are at the centre of the ethical and legal discussions of AI in healthcare. Article 22 of the GDPR provides a right to object a decision made on the basis of a mere automated processing that has some legal consequences on him or her in the absence of meaningful information on the logic applied [3]. This legal principle is the most significant one, which helps to justify the obligations to explain the issues within EU system.

AI Act goes a step further as it provides that the developers of AI systems that are classified under high risk category should provide the reasons of model outputs in a clear, intelligible and comprehensible manner. Professionals, as well as laypersons, should be able to read these [7]. This has been interpreted by the judiciary and academia not just as a technical transparency, but also as a binding on oneself to communicate, and thus to make complex algorithms understandable to human beings [25].

3.2 Technical Legal Explainability Versus Legal Explainability

Accountability and comprehensibility are the focus of legal explainability, as AI researchers often define explainability based on algorithms or math [26]. As an example, a clinician should be able to express why an AI model suggested a particular diagnosis not simply that it suggested a diagnosis. Technical approaches like saliency mapping, counterfactual explanations and retrieval-augmented generation can be interpreted technically however it is recommended that implementation is done considering the legal framework so as to guarantee conformity to the law [27].

The proponents of explainability state that it is informed by the three criteria:/ on legal adequacy of explainability. The most important thing is that the description of the model is truthful and rather shows how the model actually functions rather than a simplistic account. Things that are to be discussed must obviously be so as in the case of those things that are useful in making a diagnosis.

These explanations can, therefore, be replicated in order to ensure that when similar inputs are used, consistent outputs are obtained [10].

3.3 Transparency Accountability by Institution

The developer is not the only one that needs to be transparent. Facilities using LLMs ought to have audit logs, model card, and data sheets to capture the system updates, retraining incidences, and constraints [5].

The NEJM AI DEAL Checklist [10] suggests a standardized reporting model that guarantees regulator consistency and implicated inter-institutional consistency. In this checklist, model development, data provenance, and validation, as well as bias testing, must be documented, making them possible to audit in a post-deployment manner. Also, the principles of explainability-by-design support the concept that interpretability should be incorporated into the initial phases of the model creation. In this way, organizations will not need to undertake expensive retrofits and improve the compliance with the AI Act and GDPR at the same time [19].

Similarly, Lataran et al. [28] introduced a BERT-enhanced blockchain framework for health-insurance fraud detection, demonstrating how explainable AI and immutable audit trails can coexist to promote algorithmic accountability.

3.4 Ethical and Practical Tensions

Although it is widely accepted that explainability is the appropriate thing to do, a conflict exists between interpretability and performance. LLMs and other deep neural networks are highly likely to lose transparency in pursuit of accuracy. Clinicians might favor solid black-box systems as opposed to open but inaccurate systems- the dilemma which requires both ethics and law to come up with solutions [24].

Additionally, too much disclosure is contradictory because it may jeopardize the privacy of the patients in case sensitive training information is disclosed due to the process of clarification. This is what legal scholars refer to as the transparency paradox [3] in which excessive transparency destroys confidentiality.

Therefore, regulators promote proportionate transparency, that is an accountability level of explanation that does not impair privacy or trade secrets [7]. The expectations of explainability are different around the world.

4 Case Studies

4.1 The European Union: Implementation of AI Act in Healthcare

European Union (EU) has been leading the pack with its Artificial Intelligence Act (AI Act) to adopt a legally binding framework of Artificial Intelligence. The Act treats the vast majority of AI systems, including LLMs, as high-risk in the medical sector, which restricts its application to serious transparency, explainability, and data management requirements [7].

A real-life case in point of this example can be illustrated with the European Health Data Space (EHDS) project, which aims to cross-analyze health-data across borders yet in a way that will not breach the General Data Protection Regulation (GDPR). The introduction of patient history summarization and diagnostics assistance on the basis of LLM is being introduced in hospitals such as Charite - Universitaatsmedizin Berlin and is regulated within an algorithm audit-based framework, such as bias tests, and human-in-the-middle validation [21].

The European developers would be required to adhere to the standards concerning AI Act by including technical documentation to explain the rationale of the decision-making process, record all data-processing processes and human experts must be capable of overriding automated outputs where necessary. Such governance measures are approved by ethical boards and they evaluate the performance of AI after a specified period of time [6].

4.2 The United States: Adaptive Oversight Through the Assistance of the FDA

The United States is more flexible, more innovation-based, as opposed to the thorough and written form that the EU takes. The U.S. Food and Drug Administration (FDA) gives attention to post-market monitoring and learning through the framework of Digital Health Innovation Action Plan and SaMD [8, 29].

The other interesting case is the FDA and Google health and Mayo Clinic were in the process of testing the LLM-based radiology assistants. In other words, the FDA allows the controlled changes that are within pre-approved thresholds named Predetermined Change Control Plans (PCCPs) [29].

It is a dynamic oversight that is particularly useful in rapidly changing models but has also been discussed to not provide sufficient protection of patients and transparency [8]. According to the researchers, the U.S. model allows cultivating innovation spirit yet creates a risk of irregular practice and irresponsibility in the event of AI error [19].

4.3 Global South: Regulatory Cooperation to Venture into Capacity Building

The majority of the low- and middle-income states (LMICs) lack the infrastructure and expertise to control AI at a wholesome level. The South African Health Products Regulatory Authority (SAHPRA) has selected a hybrid model that consists of ethics committees, AI audit centres at universities, and government regulation among others

Table 3 EU, US, Global South

Region	Regulatory framework	Oversight approach	Key strengths	Main challenges	Example initiatives
EU	AI Act (2024) + GDPR	Pre-market conformity + periodic audits	Strong transparency, explainability, data quality	High compliance burden, transparency versus privacy	EHDS Project, Charité Hospital
US	FDA Digital Health Innovation Plan, SaMD, PCCPs	Post-market adaptive monitoring	Flexibility, continuous learning, innovation	Liability gaps, limited transparency	FDA–Google–Mayo Clinic Radiology LLM
Global South	Hybrid ethical-regulatory models	Cooperative governance and capacity-building	Fairness, contextual ethics, WHO collaboration	Weak infrastructure, dependency on external models	SAHPRA Audit Centers, NDHM India Guidelines

[9, 21]. The guidelines of AI ethics with its insistence on fairness and explainability are also a part of the national digital health mission (NDHM) of India that does not become legally binding [9].

These case studies indicate that there must be contextual adaptation. Introduction of the western types of regulation without capacity building initiatives normally leads to bad governance. The multilateral collaboration between the WHO and the African Union on AI towards Health is important in the delivery of technical assistance in line with cultural and ethical values [9, 30] (Table 3).

5 Challenges and Limitations

5.1 Legal Uncertainties and Overlap of Jurisdictions

One of the most burning questions that are ambiguous is the legal liability issue. When it is exceptionally clear that an AI-generated recommendation harms the patient, it is not evident who is to be held accountable: a clinician, an institution, or a programmer [13, 15, 23]. The tort law and product liability as it stands today are biased towards a physical machine that is not a self learning algorithm. Scholars suggest that there may be a need to have new hybrid forms of liability that would entail professional negligence and strict product liability [4].

The other complication is as a result of overlapping legal regimes. To give an example, in the case of Europe, an AI product must comply not only with the GDPR but also with the AI Act which at times may impose conflicting demands, e.g., the requirement to be transparent and protect intellectual property and privacy [3].

5.2 The Explainability Performance Trade-Off

XAI techniques are sometimes lower in performance than black-box models. It presents a dilemma of precision against explainability to the clinicians [24]. In high-stakes areas of intensive care or oncology, performance may override the right to the explanation as it may be anticipated [26]. The developers sometimes use post-hoc explanations that appear transparent but cannot be interpreted [2]. Whether such explanations are adequate and in good faith remains an issue that regulators can wrangle in the process of compliance audits [31].

5.3 Data Quality, Inequity, and Bias

LLLM can continue to perpetuate health disparities by being trained on biased or incomplete data. As indicatively, the misdiagnosis or inconsistent recommendations of the treatment may be attributed to underrepresentation of the minority groups in the medical datasets [11, 17]. Legislation like the AI Act has attempted to limit the bias by making data quality requirements but there is still very little technical implementation of that [7]. The necessity to have demographic auditing and provenance documentation of datasets is increasingly becoming a requirement [17].

There is a perilous risk of the algorithms to undermine the ethical standards that are not related to the law—they cause people to be less trusting and contradict the principles of non-maleficence in medical ethics [11].

5.4 Transnational Data Flows and Data Privacy Threats

The AI healthcare systems in the world often require transnational data interchange. The privacy laws like the GDPR however restrict the cross-border data transfer without necessarily having adequate level of protection in the country of destination [3]. This creates barriers to global cooperation especially of the LMICs who take part in multicenter AI studies. Solutions that do not invade the privacy of users, such as Data Trusts and Federated Learning, have been proposed [15, 32].

5.5 Restrictions and Social Acceptance of Ethics

Besides legal needs, healthcare AI acceptability within the society will be subject to societal perception of fairness and transparency [9]. It has been shown that AI is more trusted by the patients when the clinicians remain a part of decision-making and the limitations of the models are familiar [30, 33]. Everything technocratic solution

Table 4 Challenges and possible responses

Challenge	Impact	Suggested Response
Legal uncertainty	Ambiguity on liability allocation	Develop hybrid liability models (shared negligence + product liability)
Explainability versus performance	Trade-off between accuracy and interpretability	Use proportionate transparency under AI Act
Data bias and inequity	Reinforces health disparities	Demographic audits, dataset documentation
Cross-border privacy	Limits data sharing for LMICs	Adopt federated learning and data trusts
Social acceptance	Low trust in AI decisions	Include participatory ethics and clinician involvement

is a risk of loss of users. The improved governance should then entail participatory ethics: where patients, clinicians and the civil society assist in the creation of AI control systems [9, 19] (Table 4).

6 Conclusion

The emergence of explainable LLMs in healthcare offers immense potential to transform. The development of healthcare AI is gaining momentum more than ever concerning the legal and regulatory environment. Not only may LLLMs prove to be disruptive to the practice of diagnostics and personalized medicine but also to administrative efficiency, they are also incompatible with the law and ethics of today.

The review of the available frameworks has shown that there are three general lessons:

1. Dynamism Oversight is required: The adaptive AI systems should not be based on pre-market approvals that are fixed. Throughout the model lifecycle, analysis and re-designed reporting is needed [1, 13].
2. Transparency and Explainability Both legal: Legal and regulatory frameworks such as the GDPR and the AI Act have made interpretability more of a right than an option with technical and communicative requirements [3, 7, 30].
3. Cooperation and Equity at the International Level: Differences in regulatory capacity can cause the development of the two-layer world of AI medical care: highly regulated and insufficiently regulated [9, 21].

In the future, there is need to cease paying attention to compliance and begin to create trust. This incorporates the multi-disciplinary activities of technologists, lawyers, ethicists and clinicians. The purpose of AI in healthcare of the future cannot be reduced to the concept of regulation use but instead serves as a general move towards transparency, justice and human dignity.

References

1. Mello, M.M. and I.G. Cohen, *Regulation of health and health care artificial intelligence.* JAMA, 2025. **333**(20): p. 1769–1770.
2. Mennella, C., et al., *Ethical and regulatory challenges of AI technologies in healthcare: A narrative review.* Heliyon, 2024. **10**(4).
3. van Kolfschooten, H.B., *A health-conformant reading of the GDPR's right not to be subject to automated decision-making.* Medical law review, 2024. **32**(3): p. 373–391.
4. Bughin, J., et al., *Artificial intelligence the next digital frontier.* 2017.
5. Mesinovic, M., P. Watkinson, and T. Zhu, *Explainability in the age of large language models for healthcare.* Communications Engineering, 2025. **4**(1): p. 128.
6. Bilal, A., D. Ebert, and B. Lin, *Llms for explainable ai: A comprehensive survey.* arXiv preprint arXiv:2504.00125, 2025.
7. Van Kolfschooten, H. and J. Van Oirschot, *The EU artificial intelligence act (2024): implications for healthcare.* Health Policy, 2024. **149**: p. 105152.
8. Loloei, A., et al., *Evolution of the FDA Regulation of Digital Health Technologies.*
9. Naidoo, T., *Overview of AI regulation in healthcare: A comparative study of the EU and South Africa.* South African Journal of Bioethics and Law, 2024. **17**(3): p. 121–126.
10. Tripathi, S., et al., *Development, Evaluation, and Assessment of Large Language Models (DEAL) Checklist: A Technical Report.* 2025, Massachusetts Medical Society. p. AIp2401106.
11. Nasir, M., K. Siddiqui, and S. Ahmed, *Ethical-legal implications of AI-powered healthcare in critical perspective.* Frontiers in Artificial Intelligence, 2025. **8**: p. 1619463.
12. Farhadi, A. and A. Taheri, *Application of GenAI in Synthetic Data Generation in the Healthcare System*, in *Application of Generative AI in Healthcare Systems*, A. Zamanifar and M. Faezipour, Editors. 2025, Springer Nature Switzerland: Cham. p. 67–89.
13. Pham, T., *Ethical and legal considerations in healthcare AI: innovation and policy for safe and fair use.* Royal Society Open Science, 2025. **12**(5): p. 241873.
14. Amann, J., et al., *Explainability for artificial intelligence in healthcare: a multidisciplinary perspective.* BMC medical informatics and decision making, 2020. **20**(1): p. 310.
15. Nguyen, S.M., M.B.M. Liyanage, and S. Wang, *Enhancing Accountability, Resilience, and Privacy of Intelligent Networks with XAI.*
16. Spagnolli, A., et al., *Transparency in Healthcare AI: Testing European Regulatory Provisions against Users' Transparency Needs.* arXiv preprint arXiv:2505.17105, 2025.
17. Bharati, S., M.R.H. Mondal, and P. Podder, *A review on explainable artificial intelligence for healthcare: Why, how, and when?* IEEE Transactions on Artificial Intelligence, 2023. **5**(4): p. 1429–1442.
18. Dwivedi, D., *Artificial Intelligence and the Future of Business.* AI in Business Management: Bridging the Gap between Technology and Strategy, 2025.
19. Gille, F., et al., *From theory to practice: viewpoint on economic indicators for Trust in Digital Health.* Journal of Medical Internet Research, 2025. **27**: p. e59111.
20. Shahparian, J., S. Erfani, and A. Zamanifar, *A secure and efficient authentication and key agreement protocol in blockchain-enabled VANETs.* Computers and Electrical Engineering, 2025. **122**: p. 109947.
21. Udechukwu, L.M., et al., *AI-Driven Adversarial Defense Framework with Generative Adversarial Network for Secure Healthcare IoT Ecosystems.* Archives of Current Research International, 2025. **25**(10): p. 148–165.
22. Zamanifar, A. and E. Nazemi, *EECASC: an energy efficient communication approach in smart cities.* Wireless Networks, 2020. **26**(2): p. 925–940.
23. Nabil, R., *Artificial Intelligence Regulatory Sandboxes.* JL Econ. & Pol'y, 2024. **19**: p. 295.
24. Giorgetti, C., G. Contissa, and G. Basile, *Healthcare AI, explainability, and the human-machine relationship: a (not so) novel practical challenge.* Frontiers in Medicine, 2025. **12**: p. 1545409.
25. van Kolfschooten, H., et al., *AI Chatbots for Promoting Healthy Habits: Legal, Ethical, and Societal Considerations.*

26. Bienefeld, N., et al., *Solving the explainable AI conundrum by bridging clinicians' needs and developers' goals.* npj digital medicine, 2023. **6**(1): p. 94.
27. Muren, P.D., et al., *Workers in transition: the impact of artificial intelligence in OECD countries.* Beyond bits and algorithms.
28. Lataran, P.N., et al., *Developing a BERT-enhanced blockchain model for HealthInsurance fraud detection.* Journal of Information Systems Research and Practice, 2024. **2**(3): p. 56–61.
29. Yang, S.-R., J.-T. Chien, and C.-Y. Lee, *Advancements in Clinical Evaluation and Regulatory Frameworks for AI-Driven Software as a Medical Device (SaMD).* IEEE Open Journal of Engineering in Medicine and Biology, 2024.
30. Aaronson, S.A., *A Missed Opportunity to Further Build Trust in AI: A Landscape Analysis of OECD. AI.* 2022.
31. Farhadi, A., A. Zamanifar, and M. Faezipour, *Application of Generative AI in Drug Discovery*, in *Application of Generative AI in Healthcare Systems*, A. Zamanifar and M. Faezipour, Editors. 2025, Springer Nature Switzerland: Cham. p. 155–174.
32. Rashidieranjbar, F., A. Farhadi, and A. Zamanifar, *Revolutionizing Healthcare with Generative Artificial Intelligence Technologies*, in *Generative Artificial Intelligence (AI) Approaches for Industrial Applications*, N.R. Vajjhala, et al., Editors. 2025, Springer Nature Switzerland: Cham. p. 189–221.
33. Mahalegi, H.S.M., et al., *Application of Generative AI in Resource Optimization in Healthcare Systems*, in *Application of Generative AI in Healthcare Systems*, A. Zamanifar and M. Faezipour, Editors. 2025, Springer Nature Switzerland: Cham. p. 175–195.

Future Directions—Human-AI Collaboration, Adaptive Explanations, and Regulatory Readiness

Farshid Babapour Mofrad and Midya Yousefzamani

1 Introduction

The recent rapid integration of large language models (LLMs) into healthcare has revealed not only transformational potential but also longstanding challenges [1, 2]. Similar to prior Human–Machine Interaction (HMI), whether focused on usability or task performance, Human-AI Collaboration (HAIC) compels a reassessment of transparency, trust, control, and ethics [3, 4]. For example, current LLMs are stochastic and opaque algorithms; therefore, healthcare professionals cannot track their reasoning or validate AI-generated recommendations. This raises pertinent clinical concerns about error detection and automation bias [5]. Trust is impaired by opacity and may, in turn, impair decision-making: too much trust can lead to blindly accepting outputs, while too little trust limits the willingness to adopt valuable insights [6]. Beyond transparency and trust, HAIC requires negotiation of control, and the collaborative process should be understood in this context. Unfortunately, most evaluation methodologies consider each agent's performance separately, rather than assessing the collaborative outcomes that result from the fusion of agents and immediate task sharing [7]..Ethical issues such as accountability, fairness, bias reduction, and data privacy are also essential considerations, particularly in high-stakes healthcare contexts. The potential for these issues to be amplified, alongside the need to achieve satisfactory outcomes, suggests that stronger frameworks are required to address ethics sensibly within healthcare settings. Taken together, these interrelated challenges demand a rethinking and revision of approaches to the design, evaluation, and governance of collaborative human-AI systems in healthcare [7, 8]. It is important to avoid siloed thinking in developing human-AI systems [7, 8]. In high-stakes situations, it is often counterproductive to optimize for a single dimension such as algorithmic accuracy, transparency, or user satisfaction [7]. Rather, there is a higher demand for holistic development

F. B. Mofrad (✉) · M. Yousefzamani
Department of Medical Radiation Engineering, SR.C., Islamic Azad University, Tehran, Iran
e-mail: babapour@iau.ac.ir

A. Zamanifar et al. (eds.), *Explainable Large Language Models in Healthcare Applications*, Bio-IT and AI, https://doi.org/10.1007/978-3-032-15088-2_12

processes, which combine the machine-oriented performance with human concerns, organizational workflows, and equity/ethical considerations [1, 2]. Recent literature across diverse domains explains how some current research from a variety of fields supports this shift. For example, mixed-methods evaluation frameworks support this shift by combining objective performance metrics and subjective human experiences, thus aligning the development of an AI system with expectations in the real world [9, 10]. In the creative domains, Arias-Rosales [9] used a well-developed combination of objective metrics around shape-generation, and designers' qualitative judgements, which illustrated that technical novelty alone is not sufficient to predict perceived aesthetic value. In finance, Chakravorti et al. [10] Demonstrated that maximized accuracy in AI prediction markets did not engender trust, thus highlighting trust in an AI system required transparency and interpretability. Together, these examples offer a broader lesson, human-AI collaboration needs more than explainable models, it needs explainable systems made within the prism of context-aware, trust-sensitive, and ethically aligned infrastructures [5, 6]. Given the rapid pace of large language model deployment in clinical environments, recent developments can enhance the creation of comprehensive evaluation and development frameworks centered on both machine-level performance and human-oriented perspectives [1, 2]. Once this comprehensive stance has been established the following sections of this chapter explore potential future pathways for collaborative, adaptive, and regulation-ready LLM integration in healthcare [7, 9]. This chapter presents a future-oriented framework that is designed to address the multi-faceted issues presented. The framework comprises three anchor points: Collaboration, Adaptation, and Regulation. First, human-AI collaboration is no longer about just automation and tasking of work, it is a dynamic, iterative process, during which humans and AI systems co-design, adapt, and create value based on mutual capabilities [11, 12]. This approach is evolving from the collaborative theory and hybrid intelligence, and as a process of joint goal striving, real-time knowledge co-production, and mutual trust [13]. For example, adaptative explanation systems that leverage and enhance the knowledge, goals, and roles of different made this real-world adoption of explainable AI systems more feasible [14]. In high-stakes healthcare contexts, we must not only be technically correct but also provide explanations that fit the cognitive and situational needs of patients, clinicians, and regulators [15]. Finally, ensuring that AI-enabled systems are ethical, lawful, and trustworthy and accountable at scale requires regulatory readiness [16, 17]. Together, these architectural three dimensions provide a means, by which future research and implementation can focus, to further develop responsible, safe, effective, and human-centred uses of LLMs in healthcare [16, 18] (Fig. 1).

In conclusion, the future of explainable AI in healthcare will not emerge from linear advances in technology, but through the intersection of collaboration, adaptation, and regulation. The present moment calls for putting these notions into practice (creating systems that individuals can engage with) that evolve alongside their users, withstand ethical challenges over time, and remain compatible with governance. The figure below visualizes these three foci and their expected outputs. Together, they provide a strong scaffold for ensuring the responsible impact of high-stakes AI.

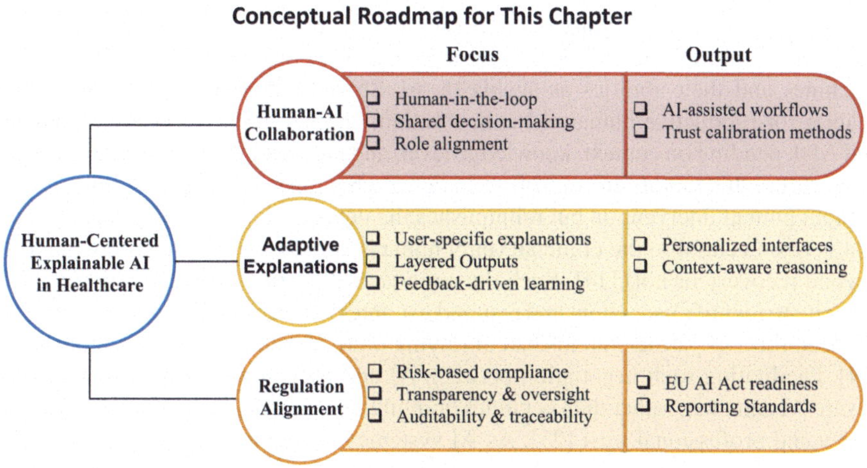

Fig. 1 Conceptual roadmap for human-centered explainable AI in healthcare

2 Human-AI Collaboration

Human-AI collaboration is not just automation or using an AI tool, it requires consideration of shared autonomy, mutual interpretation, and collective cognition [19, 20]. As AI systems are operationalized as teammates and not simply as tools, developing collaborative workflows, communication interfaces, and trust calibration will take on increasing importance [21, 22]. This section builds up the elements of what true collaboration looks like, including task splitting and interface design, as well as trust dynamics and ethical co-responsibility.

2.1 Beyond Tools and Automation: Defining True Collaboration

Traditional views often regard AI as a tool, an auxiliary engine that provides staff with the support needed to make decisions while keeping the locus of control with humans [23]. This position can severely preclude the kind of human-AI collaborations possible in high-stakes domains, such as healthcare [6]. When clinicians view AI systems as tools rather than a teammate capable of making autonomous recommendations and acting similarly to a collaboration partner, Users are more likely to simply override the recommendations or ignore the system, regardless of how well the system was programmed or trained [23]. By perceiving AI as a team member, rather than merely a technological resource, nurses and doctors could open up opportunities to collaborate, share responsibility, and develop trust [5, 23]. As noted by researchers, working in cooperation with AI means more than just incorporating AI into existing workflows; it also means reconsidering how humans see machines and their abilities as agents [5, 6]. Shared autonomy is a fundamental enabler for productive human-AI collaboration: control is shifted between human and AI depending on context, knowledge level, and task complexity [14]. In a healthcare setting, this fluidity in control could occur where the AI interprets clinical data without human intervention but relinquishes the decision-making under uncertainty that is often central to the clinician (only humans can decide whether the patient is beyond recovery or not) [14]. Such dynamic task allocation is enabled by the AI's capacity to identify user intent and respond to contextual cues in real time, fostering a nuanced form of partnership instead of relying on rigid, one-size-fits-all automation [24]. Facilitating such negotiation between AI and user, shared autonomy helps to ensure AI enhances rather than occludes or inhibits human skill that regulates human safety and professional trust [25]. As AI systems become more sophisticated, they are increasingly evolving from simple tools to augmenting collaborators augmenting humans' abilities by providing context, adaptive reasoning, and shared intent rather than just replacing their role [26]. This evolution is beginning now to take the shape of new conceptual models like human-centred human-AI collaboration (HCHAC), where AI agents assist humans to make decisions, while humans maintain leadership, responsibility, and strategic oversight [26]. Artificial intelligence should aim to reduce human cognitive workload rather than replace human decision-making. In hybrid human-AI teams, each agent should leverage its unique strengths to achieve outcomes that neither could attain alone [27]. The notion of joint cognitive systems refers to situations where humans and AI share mental models(common representations of goals, expectations, context, and task states) which help them coordinate, predict actions, and manage uncertainty in real time [19]. In recent investigations, proposed architectures exist that infer teams' situation awareness, and alignment of mental models by AIs from healthcare teams to potentially enable less friction in future collaboration and resilience when interacting [28].

2.2 *Designing Collaborative Interfaces and Workflows*

Establishing workflows that enable seamless human-AI collaboration hinges on interfaces that feel natural, are context-aware, and allow users to stay meaningfully in control. First, interfaces that use natural communication channels, such as natural language and rich visuals, allow users to convey complex intents and interpret AI responses, as Zhao et al. recently reviewed for human-AI teaming [29]. Second, proactive support and context-aware solutions offer clinicians insights and enable them to offer the greatest relevancy at the right moment; these proactive supports help a user feel supported without compromising their agency, but they need to take care not to cross the line and diminish user satisfaction [30]. Finally, effective collaboration relies on supporting human oversight with understandable feedback, lucid audit trails, and control by a user, thereby fostering trust and preventing automation bias [31]. Together, the design principles just described can be utilized synthesize collaborative AI systems that use the user as a partner, as opposed to merely receiving and observing the results of automations.

2.3 *Building Mutual Trust and Understanding*

Building mutual trust is critical for human-AI collaboration to be successful. To build mutual trust, we need more than technical correctness: we need calibrated trust, the ability to act in context, and a shared accountability between a human and an intelligent machine. An overview of these three components is provided in Table 1.

Calibrating Human Trust

In human-AI collaboration, trust has to be calibrated just right—not too much, but not too little [32]. An over-reliance on AI can lead users to accept erroneous AI recommendations with no challenge, a common outcome of automation bias [33]. Under-reliance can mean abdication of really useful recommendations, more so in high-stakes contexts like clinical care where mistakes can have deadly consequences [33]. In fact, the literature has established that trust is dynamic, developed over time, formed by the system's previous results, transparency and user-feedback mechanisms [34]. Importantly, next-gen systems are developing trust calibration interfaces, like uncertainty visualizations, confidence measures and explanation layers, to provide users with real-time cues for trust adjustment [34, 35]. The current focus is on creating mechanisms to foster critical engagement, while maintaining a healthy level of vigilance, ultimately making sure the human remains the active reflective decision-maker, not a passive operator [35].

Table 1 Key components of building mutual trust and understanding in human-AI collaboration

Component	Main idea	Examples
Calibrating human trust	Avoid over/under-reliance by adjusting user trust dynamically	Uncertainty visualizations, explanation layers, feedback loops
Understanding human intent	AI should infer implicit goals, adapt to user expertise and urgency	Context-aware prompting, user modeling, multimodal sensing
Shared goals and accountability	Clarify roles, track decisions, align team objectives to ensure collaboration and ownership	Shared dashboards, traceable recommendations, clinician-AI responsibility split

AI Understanding Human Intent and Context

For collaboration to be flexible, smooth and adaptive, AI systems need to do more than just obey commands; they require the ability to infer human intention and respond to context cues [36]. This would include an awareness of user objectives, foreseeing their needs as well as reacting to them, and modifying their behavior accordingly [37]. In a dynamic practice like healthcare, for instance, AI tools must be able to analyze whether inputs are routine or urgent, if the inputs are coming from a registered or experienced clinician, or a new trainee, and alter their prompting style accordingly [38]. User modeling, real-time dialogue tracking, and multimodal sensing (like speech tone, the user's camera screen focus, or pauses in decision making) give systems the ability to account for not just explicit commands but implicit intentions [39]. When users find that AI shows an understanding of the context, they are more likely to see the AI system as a collaborative partner. This helps to create a greater sense of fluidity, trust, and safety in a team, especially in high-stakes situations [38, 39].

Establishing Shared Goals and Accountability

To function effectively in a team(whether as humans or working as a human-AI team) there should be clear goals and a responsibility foundation [40]. In human-AI teams, confusion about the roles or responsibility can lead to ambiguity, blame, disengagement [40]. If the AI systems share reasoning, track its decision paths, and record the team, they create a clear structure, that can be reprised and transparency can be shown for accountability [20]. For instance, in healthcare, shared dashboards which identify if the recommendation came from the AI or from the clinician and approved and finalized by the clinician may establish some level of clarity and legal formality [21]. Likewise, having a shared purpose, especially in teams of AI, e.g., improve patient outcomes, or reduce diagnostic error, serves to emphasize a level of urgency towards shared goals. In this case, the AI need not serve as a technical assistant, but rather as a collaborative, and accountable, partner in the work [21, 41].

2.4 *Challenges in Collaboration*

While Human-AI Collaboration (HAIC) has enormous promise, especially in complex domains like health care, it also has its challenges. One of the obstacles is even determining what we consider "good" collaboration. It's not something you can look at and measure based on accurate responses or action time of the system. Good collaboration also considers things like, the extent of cooperation between agents, the ability to adjust to participants' capability changes, and mutual understanding of the collaborative task [7]. In creative work or language-structured work, for example, it is complicated to separate the contribution of the human in collaboration to the

AI contribution, because the value is often context-dependent and can't be factored as simply right or wrong [7]. From a technical and organizational perspective, AI systems seem to have circumstances of reliability impacted by internal factors such as biased training data, adverse downstream input distributions, or woeful validation (a sobering consideration in safety–critical applications like diagnostics). Likewise, in existing workflows and infrastructure, the potential for HAIC collaboration is limited to workflows that successfully implement compatible change from the introduction of AI, require significant investment in hardware, software, and training, and align to the end-user's and organizational intent [37]. Where AI is not comfortably integrated the space for collaboration is likely broken or misaligned. A lack of common understanding, mismatched skills, and unclear accountability structures make collaboration more challenging especially in interdisciplinary teams [7, 10]. This interrelatedness of technical, social, and evaluative barriers needs to be addressed to release the full potential of HAIC. Lastly, human factors (e.g., trust dynamics) are major barriers to successful HAIC collaboration, as users seem to either over-trust AI systems like risking automation bias or under-trust them, thus failing to act on valuable input. The balance of trust depends on transparency, explainability, and perceived competence [7, 37]. Furthermore, users often need targeted training on how they might apply AI-derived outputs as part of their real-time decision making. Figure 2 displays these inter-related issues including clear limitations of evaluation metrics, technical and integration barriers, and human-centered challenges like trust Miscalibration and on the job training.

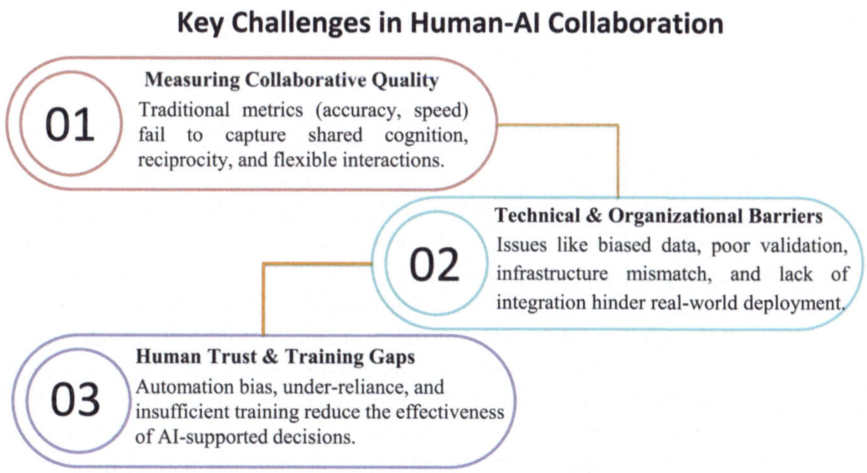

Fig. 2 Key challenges in human-AI collaboration

3 Adaptive Explanations

Especially within organizational settings, static explanations in collaborative human-AI situations are unlikely to create the psychological safety and trust necessary for effective human-AI interaction. Trust in AI systems is ultimately shaped by several contextual considerations, including perceived task fit, command structure, and user factors such as the ability to exercise autonomy and expertise within the team [42–44]. Providing explanations that are generic or fixed ignores the context and dynamics of interpersonal and organizational behaviors and processes, particularly when the AI moves toward being a teammate [45, 46]. Also noteworthy, if the AI is being used for managerial oversight, users may feel vulnerable or surveil lanced which diminishes trust in the system [47, 48]. By contrast, systems that had adaptive explanations, which are created to support psychological safety (for example, through the use of social contracts or the practice of privacy tolerances), can reduce anxiety and enable trust and adoption [49]. Taken together, these forces demonstrate that static and one-size-fits-all explanation models do not account for dynamic interpersonal and contextual nature of collaboration and bring to the forefront the importance of adaptive XAI strategies that are user- and context-sensitive [50]. These complexities highlight the disassociation of static explanation paradigms needing to adapt explanations user and context-aware explanation systems that build and maintain trust, provide cohesion for the team, and optimize collaborative tasks in real-world settings [51].

3.1 Defining Adaptivity in XAI

While language models have advanced quickly in their capabilities, explainability is still stagnant and poorly defined [52]. Most leaderboard models (e.g., MMLU) focus on the outcome performance of a model, thus they are rewarded for getting correct answers, rather than justifying answers [53]. This lack of congruency has raised the challenge of developing adaptable explanations that respond equally well to user needs and roles (i.e., user-adaptable), task types and the risk domain (i.e., context-adaptable), and the nature and function of the explanation (i.e., content-adaptable) [54]. While a generic explanation format (e.g., a list of important features) would suffice for a data scientist for justification in the reasoned action of the model, it would fail for a clinician who needs to reason through cases or counterfactuals, for example, when justified action is high-risk. Moving towards explanations that follow protocols such as ask why then probe or causal-consistency checks, from which explanation systems can only aspire, are non-existent in the fragile and illusory rationales provided by explanation (hallucinated rationales). Adaptive XAI attempts to step into this gap by producing those situationally and cognitively variable explanations in real-time and determining when transparency is functional, rather than simply formal [55].

3.2 Key Enablers of Adaptive Explanations

To design adaptive explanations effectively, we require an infrastructure that supports flexibility, personalization, and real-time responsiveness. There are a number of technological and cognitive enablers that have been identified that we would need to ensure successful delivery of this vision in practice and these four enablers, outlined in Table 2, provide the foundation for delivering adaptive and context-based AI explanations in real-world, high-stakes environments such as healthcare.

User Modeling and Profiling

An adaptive, user-centred provision of explanations must be within the full range of the user dealing with individual differences of expertise levels, trust levels, cognitive preferences, and cognitive styles. The literature shows there is a lot of variability in every user's own engagement with the AI outputs and how much they are using or aware of the explanations [56]. In order to develop individualized explanation strategies, it is necessary to accurately profile the user's characteristics (for example, their prior knowledge, mental models, and experiences with the task) by either providing explicit input or making inferences from implicit cues [57, 58].

Table 2 Key technological and cognitive enablers for adaptive explanations

Enabler	Description	Example use cases
User modeling and profiling	Captures user-specific characteristics like knowledge level, trust, cognitive style, or preferences to personalize explanations	Tailoring explanation detail based on clinician's experience level
Context sensing and interpretation	Recognizes situational factors (e.g., urgency, risk level, user role) to adapt timing, depth, and style of explanations	Providing real-time alerts during high-pressure surgeries based on phase or device data
Explanation generation and multimodality	Supports diverse types of explanations (e.g., counterfactuals, visual summaries) and media formats (text, visual, bio-signal, etc.)	Showing annotated scan images alongside textual summaries for radiologists
Interactive dialogue systems	Allows users to question, explore, and refine explanations via conversational or visual interaction	Enabling clinicians to ask "why" or "what-if" queries during diagnosis assistance

Context Sensing and Interpretation

To be optimal, systems must be able to adapt appropriately to the situation context such as task time pressure and difficulty, contextually relevant domains (e.g., health care), or even the cognitive load on the user. In high-stakes situations, high-risk environments, as is the case in clinical decision-making and emergency care, contextual cues (e.g., action taken, user role, indication of a device, or phase of procedure) should be taken into consideration of the style of description or explanation, as well as the timing of the descriptions or explanations [59]. Real-time sensing, especially in contexts that are critical, such as a surgical context, enables the system to capture and interpret data as dynamic input in order to deliver contextually dependent supporting assistance [55].

Explanation Generation Engines and Multi-Modal Outputs

Comprehensive explanation engines must be able to flexibly produce various types of explanation content, from feature-level attribution to example-based reasoning or counterfactuals, depending on user needs [56]. As LLMs are extended beyond text to incorporate biomolecular, physiological, or visual data (e.g., scans, or device outputs), and the need for such abilities will be increasingly important, especially in fields such as precision medicine or genomics. The ability to provide multi-modal adapted explanations enhances interpretability and action ability [52, 55].

Interactive Explanation and Dialogue Systems

There are situations in which static explanations (those that provide fixed, one-way information without the possibility of further interaction) are insufficient. In such cases, many users will seek clarification, justification, and the ability to make comparisons. Meaning, the capacity for interactive explanation, based on dialogue-enabled systems, gives users the opportunity to cross-question, drill-down, or verification on elements of the AI situation to engage and learning toward more cognitive understanding and a higher level of accountability and transparency [56, 58]. Again, these interactive systems do not merely provide information; rather, they function as tools for empowerment. They support the user's intention, enabling them to recognize and identify the parameters inherent in the system, discover their own mistakes, and correct both the excessive trust placed in the system and any misuse of their access [52, 56].

3.3 Challenges and Research Frontiers

Although interest in adaptive explainability has expanded, there remain challenges to work through for development and deployment. The first issue is at the center of scalability and real-time adaptation. It is trouble-some in many domains to even think about designing and constructing explanation systems that operate in real-time millisecond timeframes and are responsive and adaptable to myriad user profiles and contexts that change sometimes rapidly even in high-stakes situations such as medicine or autonomous driving. Product implementations will need a lightweight architecture and usable robust inference pipelines for explanation-generation [52, 56]. A second issue is that evaluating adaptive explanations as effective is still in an open space of research. Adaptive explanations need dimensions novel to this space, beyond fidelity and human satisfaction, dimensions that identify usefulness across levels of expertise, levels of emotional states, and levels of cognitive loads [60]. One more challenge relates to deception and honesty [61]. Adaptive explanations might particularly lend themselves to presentations that are framed for persuasion rather than to provide information. This raises ethical questions about possible fidelity to the model and whether the system would even optimize for compliance rather than follow-through, especially in regulatory or medical contexts [61]. Lastly, user profiling gives rise to complex and nuanced privacy concerns. Personalization tends to depend on implicit user modeling (following patterns of interaction, gaze, tone, fallibility, correction, etc.) and developing patterns of thought and behavior. This raises questions of consent, data-minimization, and ultimately the line on personalization in AI interfaces [52]. These challenges are illustrated in Fig. 3.

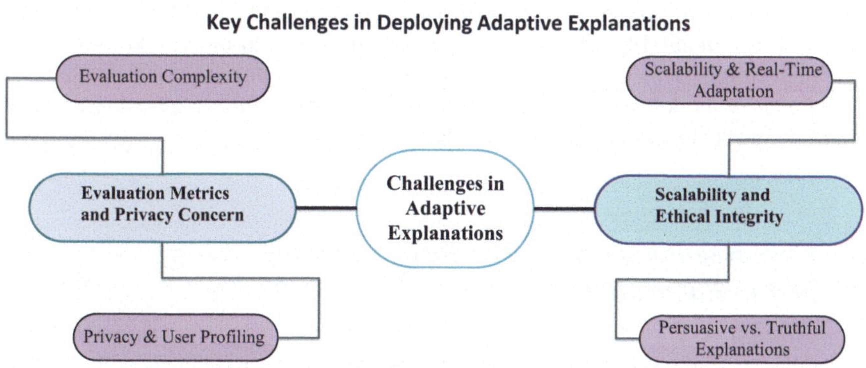

Fig. 3 Key challenges in adaptive explainability systems

4 Regularity Readiness

At present, the international regulatory environment for AI is undergoing dramatic change, with governments working to navigate various considerations and balance the tradeoff between advancing an innovative landscape and being ethically accountable [62]. Of the various potential frameworks now available, the EU AI Act represents the most extensive framework as it takes a risk-based approach to the categorization of AI and imposes significant obligations on High-risk systems such as: Biometric ID (e.g., facial recognition) and Critical Infrastructure AI systems. Other obligations include transparency requirements (e.g., the transparency obligations for chatbots and deep fakes) and further requirements for foundational models, like GPT-4, such as documentation and energy documentation [63]. In addition, a new European AI Office and enforcement team will be created and they will have the ability to impose penalties up to 7% of your global revenue [62].

The U.S. Executive Order on Artificial Intelligence (2023) promotes voluntary standards and responsible development through the NIST AI Risk Management Framework. In contrast,, Canada's AIDA (Bill C-27) creates obligations to protect that citizens from high-impact AI systems by imposing obligations around transparency and contributes enforcement obligations to the Minister of Innovation [62]. A number of countries are advancing national AI regulations. For instance, Brazil's AI Bill sets forth civil liability, user rights, bans on high-risk uses, and China's Interim Measures emphasize respecting legal frameworks when collecting and using data, alongside algorithm accountability over generative AI. On a global level, organizations such as the OECD and the UN AI Advisory Body are working on shared principles (e.g., transparency and fairness). The G7 Hiroshima Process and the Bletchley Declaration mark examples of a growing multilateral approach. However, regulation is still fragmented across the world, especially between the Global North and the Global South. This indicates a need for more coherent, risk-based, and adaptable regulatory models [62, 63]. A comparative overview of these global regulatory initiatives and their policy focus areas is presented in Fig. 4.

4.1 Core Regulatory Requirements Impacting XAI and Collaboration

As regulatory frameworks adapt to the specific obstacles associated with the use of AI systems, and LLMs, there are a few common requirements that are central and imperative to responsible deployment. These principles intend to support the design and governance of AI that leads with safety, transparency and human-oriented collaboration on a foundation of equity. As summarized in Table 3, current regulatory frameworks emphasize a layered approach to governance through risk-based classification, transparency, oversight, and accountability.

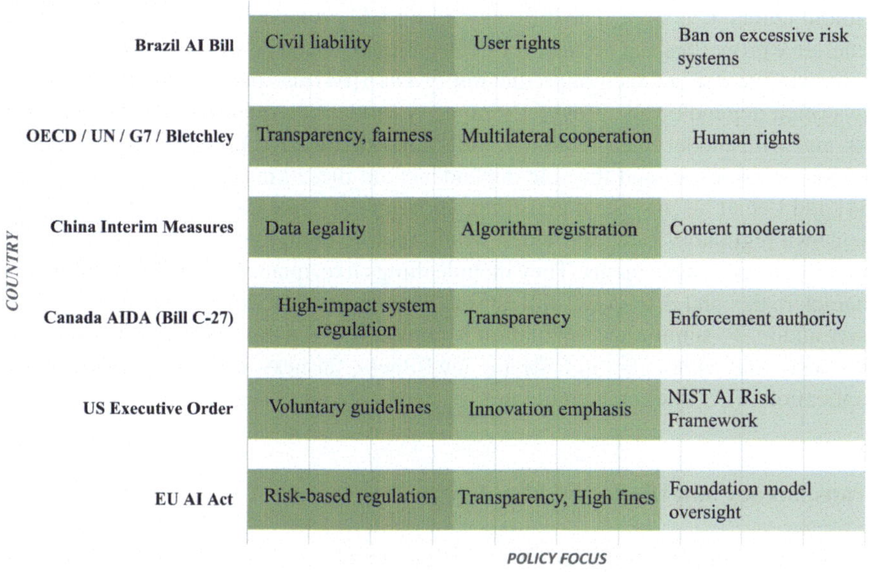

Fig. 4 Comparative overview of key global AI regulatory initiatives and their policy focus areas

Table 3 Core regulatory requirements for XAI and collaboration

Requirement	Description
Risk-based approach	Classifies AI systems by risk level; high-risk LLMs need stricter controls
Transparency and disclosure	Requires clear information on model use, limitations, and data origin
Human oversight	Ensures users can supervise, intervene, and override AI decisions if needed
Accountability and auditing	Supports traceability of decisions and clarifies responsibility across stakeholders

Risk-Based Approaches and High-Risk AI Categories

Regulatory frameworks, such as the EU AI Act and Canada's AIDA, are increasing a focus on a risk-based approach to reduce the risk associated with the usage of large language models (LLMs). Each framework classifies AI systems into distinct risk categories, with high-risk systems are AI systems that could have significant impacts on safety, fundamental rights or trust in public services [64]. Under the EU AI Act, certain instances of LLMs may fit into the high-risk tiers based on intended use (e.g., medical diagnosis, employment decisions, or public service and critical infrastructure). These systems require a pre-market conformity assessment, more transparency for users, and monitoring after use [65]. Some of the major risks for LLMs would be opacity of the system, potential for mischief and autonomous decision making without human oversight. The difficulty with LLMs is their risk opacity due to their generative properties and because they can hallucinate and be subject to jailbreaks for malicious purpose and their unpredictability. Diab [64] also point out that the existing laws often depend on the provider's capacity to identify, quantify, and mitigate the risk and with model interpretability and LLMs that evolve in their behavior raises serious questions around the sole dependence on these kinds of frameworks. Diab [64] and Lee et al. [65] State that foundation models, like GPT-4, would constitute their own regulatory category which would require continued risk assessments and not only ex-ante assessments. They include things like, quantification of training data characteristics and disclosure obligations, documented model measures, and third-party audits. To summarize, regulation is not viewed as a discretionary measure, but as an essential component in fostering lawfulness, fairness, and transparency in the application of AI system technologies.

Transparency and Disclosure Mandates

Transparency and disclosure are core principles in AI governance frameworks, particularly for the deployment of LLMs. The EU AI Act outlines transparency obligations for high-risk and general-purpose AI systems (which includes LLMs), requiring the developer to provide details to comply with obligations after deployment of the system, including documentation of capabilities, technical limitations and intended use [66]. The EU AI Act also holds developers of foundation (or base) models (LLMs, or GPT-type systems) to transparency obligations at the pre-deployment stage, requiring disclosure about the characteristics of the training data, energy usage information, and provide proper documentation to make an external audit feasible [66]. Moreover, the General Data Protection Regulation (GDPR) is vital in reading disclosure obligations for LLMs, specifically in regard to personal data. Scholars have argued that meeting the transparency obligations will be challenging for complex models, given that the pathways for input to output are frequently opaque, creating opportunities for both technical and legal concerns regarding proportionality [67]. Research further shows that LLMs often provide neither faithful explanations and/or easily traceable reasoning paths, thereby creating (at a minimum) legal compliance issues in relation to the rights of data subjects, which are established by the GDPR [67]. Finally, transparency mandates are not only legal obligations, but ethical ones. There are risks to user trust and user agency by not disclosing AI interactions (e.g., if the user is engaging with a chatbots versus a human). As a result, there are current efforts by regulatory bodies around the world—including the OECD AI Principles and US Executive Order on AI—that stress explainability and meaningful disclosure to ensure responsible use of AI [68].

Preparing for Compliance: Technical and Organizational Measures

In the implementation of large language models (LLMs), human supervision and control is critical, especially in relation to domains of high consequence including healthcare, finance or legal decision-making [69]. Fortunately, LLMs should not be allowed to operate as fully autonomous agents; rather they should be emphasized to allow meaningful human intervention at fundamental decision points [70].

Supervision in this instance should not be a passive act, but representatives must be provided with the information, context and resources, that help them to understand, assess, and override outputs from the model if desired [69, 70]. This supervision also consists of explainability methods that can allow humans to understand the route, logic, or reasoning, undertaken by the AI system. Decisions that have been designed to offer human in the loop control ensures responsibility and accountability lie with the human stakeholders and are not forfeited to algorithmic systems [71]. Providing transparency and interfaces that enable users to exert control in an LLM system means building user's trust in the system that they do not over rely upon its outcomes [72]. Therefore, human supervision should not just be a protective measure, but a regulatory and ethical requirement in the responsible development and implementation of LLM systems [70].

Accountability and Auditability

In healthcare, where AI systems' decision-making may have a direct influence on patient outcomes, establishing accountability and auditability is regulation expectation, and moral obligation. For accountability to occur, It is necessary to identify who is responsible for diagnostic or treatment decisions related to LLMs; at present, responsibility is allocated to individual human stakeholders (e.g., developer, deployers, clinician) [73]. Auditability on the other hand, is how to make it clear what happened when an LLM system performed an action based on some input (e.g. electronic health record or radiology report) and produced an output from a specified prompt. In order to support auditability users would need to record specific documentation, such as logging how the prompts were produced, which version of the model used, which version of software used, and documenting any human actions to allow for tracing [69]. For example, as applied to LLMs in clinical decision support systems (CDSSs), audit log would provide the source of the diagnosis suggestion whether it was generated by automatically developing from the model, from a suggestion of a human-in-the-loop, or provided to the model system prompt [72]. Without having an accountability and auditability framework, we cannot hold any actor accountable for malpractice, and thus cannot engage in error analysis to improve models. Ultimately, achieving accountability and auditability for healthcare-related LLMs requires not only robust technical components such as explainability tools and audit trails—but also clear governance frameworks that define who will be responsible for validation, monitoring, updating, and maintaining system components [70]. Together, these elements form the foundation for ensuring patient safety, regulatory compliance, and ethical accountability in AI-supported care [70]. Thus, establishing patient safety, regulatory compliance, and ethical accountability in the provision of AI-supported care.

4.2 Technical and Organizational Foundations for Responsible and Innovative XAI

In order to meet changing AI regulations such as the EU AI Act, organizations that develop LLMs must implement both technical and organizational measures to mitigate risks in operational LLMs. This encompasses undertaking measures to guarantee that audit trails and documentation are made available to log system behavior, and alterations made in the system, which is especially imperative in higher risk settings, such as health care, in which accountability, and traceability is vital [74]. LLM systems need to develop ways of providing human oversight for human experts to intercede and identify dangers associated with relying on automation output. Transparent and interpretable interfaces can also assist in providing sufficient control to provide human stakeholders autonomy but most importantly ensure there is no loss of governance of LLMs and that human stakeholders are ultimately responsible [69]. In addition to the preceding guidelines, developing regularized reporting formats for XAI will facilitate productive discussions with regulators and align expectations across various sectors. Regulation does not have to be viewed as a barrier to producing trustworthy AI, when trust and the regulations governing development are built against frameworks like Technology Readiness Level (TRL), which help illustrates the readiness to deploy LLMs and meet compliance [74]. Table 4 has Technology Readiness Levels (TRLs) to Stages of AI Adoption in Healthcare [17].

Table 4 Mapping of technology readiness levels (TRLs) to stages of AI adoption in healthcare

TRL	Description	Healthcare example
TRL 1	Basic principles observe	Literature-based concept for AI-based triage
TRL 3	Experimental proof of concept	Initial clinical testing with synthetic data
TRL 5	Technology validated in relevant environment	AI tested in simulated emergency room
TRL 7	Prototype demonstration in real-world setting	AI deployed in live hospital workflow
TRL 9	System proven in operational setting	Fully regulatory-approved AI tool in daily use

4.3 Challenges in Regulatory Implementation

Continued structural challenges remain, at a global level, to effective regulatory implementation, despite growing regulatory efforts worldwide. One fundamental challenge is to respond to technological change. Large Language Models (LLMs) and other AI, especially in their more interactive forms, change faster than the legislative process allows; by the time the new implemented rules come into effect, it is possible the current technological context have fundamentally changed [62]. It is also difficult to maintain a balance between flexibility and prescriptiveness. Detailed regulations can provide clarity, but sometimes rigid requirements can stymie innovation, or, incredibly quickly become outdated. There is certainly an argument for a more adaptive, principles-based approach, but that approach may not be enforceable in the complex, high-risk areas such as health care [74]. The fragmentation of international regulatory efforts is another significant challenge. Different competencies and frameworks between jurisdictions have been illustrated in the differences of approaches between the EU's AI Act which operates on a risk-based procedure and the more voluntary guidelines developed by the US. These differences create challenges for compliance for a global AI System, and impede horizontal governance across jurisdictions [62]. One more current challenge relates to the paucity of rigorous verification and enforcement mechanisms. Many regulatory regimes lack both technical tools, as well as institutional capacity, to conduct audits and facilitate compliance (and do this codified with repeatable scale) especially with opaque model approaches such as LLMs [69]. It seems that without a means to ever-scale, transparently, oversight is always going to difficult, and this will likely reduce any prospects brand well-intentioned-regulations will be realized in practice.

5 Collaborative XAI: Synergies, Challenges, and Open Questions

As artificial intelligence (AI) systems are adopted in high-stakes contexts such as healthcare, the concepts of explainability, human oversight and regulatory requirements play an important role in trust, accountability and safety. Explainable AI (XAI) techniques can assist users to better understand the logic behind an automated decision, which are necessary for effective human intervention and responsible utilization—especially in high risk sectors including those involving diagnostics, and treatment planning [75]. There also exist regulatory frameworks, such as the EU AI Act, which requires human oversight of high-risk systems, and thereby confirms a legal need for interpretable models so that individuals are able to assess, as well as override model outputs. This intersection causes a kind of feedback loop, where regulatory pressures push [76]. The utilization of XAI, which subsequently increases the ability for oversight and compliance. In a healthcare context, and particularly when discussing medical AI, interpretability becomes not just a technical design goal, it also a normative obligation in orders of patient safety, and legal accountability [76].

5.1 The Feedback Loop: From User Needs to Regulation

In high-risk AI systems, human collaboration is an essential safeguard. When a system (for instance, in recruitment, healthcare or the legal domain) is determined to be high-risk, human beings should remain as an oversight mechanism. As Busari illustrates, human-in-the-loop (HITL), or human and AI collaboration systems, relieve some of the most problematic aspects of algorithmic bias while, simultaneously, creating ethical accountability for human overseers. Importantly, HITL systems allow for humans to intervene at important decision points associated with the actions of the machine that may not be subject to ethical consideration [77]. The EU AI Act incorporates human oversight in a legal framework. For instance, systems considered high-risk require human supervision to be active, informed and technically able [78]. In addition, regulatory frameworks require manifestation of human oversight systems that permit a human to challenge or override machine-generated decisions only through collaborative human engagement with AI systems [78]. As well, from an explicability perspective, explainability tools including XAI are designed to not only attach clearly defined mechanisms to the AI system's logic, but to help the human collaborators identify when an error occurs or when a deliberate dispassion between intentions and machine- generated output may occur [77]. This feedback also allows human collaborators to take a more responsible approach to oversight while also ensuring that the AI systems are acting in the service of goals desired by human beings and not simply acting instead of humans. To summarize, regulation does not see oversight as discretionary, but rather as ingredients in the growth of lawfulness, fairness, and transparency relative to the application of AI systems technologies.

5.2 Key Challenges in Collaborative XAI Systems

Collaborative, explainable artificial intelligence (XAI) systems will face numerous challenges as the dual goals of enabling effective human-AI interaction and adhering to ethical and regulatory expectations are inherently opposed. One of the key challenges is the varying needed explanations that align with the different user groups or individuals, such as clinicians and regulators, similarly individuals in a non-user role, developers, all of whom require different levels of detail and different types of detail, leaving no opportunity for standardization [69]. The second challenge is related to the complexity of a model and its interpretability. As large language models and other increasingly complex systems are seemingly improving in overall performance, they are simultaneously growing in complexity which makes their internal decision

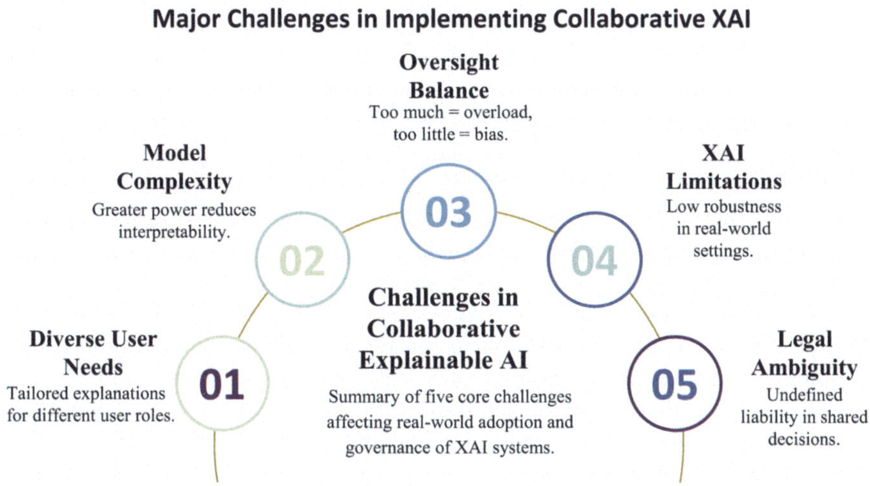

Fig. 5 Key challenges in collaborative explainable AI systems

guidelines even harder to interpret and limits the potential transparent AI applications [79]. Moreover, providing human oversight is not straightforward. Insufficient human control can create automation bias while excessive human control could overload the humans thus reducing demonstrated accountability [69]. This is especially critical in high-stakes domains, such as in healthcare where errors can lead to severe consequences for individuals or for individual outcomes. As well, XAI methods often lack robustness, adaptability, and domain-specificity, and, therefore, can be difficult to scale and may not serve the intended purpose in real-world, high-stake domains [69, 80]. Finally, legal liability regarding collaborative systems is somewhat undefined, which generates questions regarding liability for AI supported decision-making gone awry, mainly in shared decision-making contexts [74]. These intertwined challenges are illustrated in Fig. 5 and provide an understanding of how collaborative XAI systems coexist in complexity on technical, human, and legal dimensions.

5.3 Legal and Ethical Open Questions

Collaborative XAI systems face some legal and ethical complications that are not well understood as they are used in autonomous and collaborative ways, and in high-stakes settings, such as healthcare, law, and public service. For instance, one of the most prevalent legal issues pertains to uncertainty about who might be liable for harm. When humans and AI systems collaborate to co-decide and, as a consequence, harm occurs, it is not clear where liability would potentially fall (would the liability reside with the developer, deployers, or human operator of the decision support system) [69, 81]. In addition, ethical transparency is an additional open issue. While XAI approaches try to make AI-based decision-making clear, they do not often manage to genuinely inform the end-user of the consequences of AI supported decisions (especially in domains that require informed consent) pose challenges [69]. Large Language Models make this worse, as their probabilistic and data-driven approaches prevent any honest mapping of the decision-making logic in a manner that would allow transparent auditing [75]. Possibly the most pressing ethical dilemma involves bias and fairness. Adaptive explanation systems can personalize explanations that help perpetuate existing inequalities, thereby reinforcing filter bubbles and, ultimately, manipulation [81]. Lastly, the absence of regulatory clarity on how to classify adaptive and collaborative AI systems within existing instruments (e.g., GDPR, EU AI Act) continues to pose an increasing barrier to accountability and trust. Taken together, these open questions convey the need to have clearly defined legal terms, meaningful ethical propositions, the development of effective regulations that specifically aim to resolve the complex ethical and whole-systems issues posed by collaborative and adaptive XAI systems [63, 68].

6 Envisioning the Future: Towards Responsible and Beneficial AI

As human-AI collaboration advances, new ways of working in knowledge work and complex decision making become possible. One common benefit of human-AI collaboration is better creativity and problem solving, as AI systems can provide new angles or suggestions that can be used in tandem with human intuition. This is particularly useful in creative contexts, including scientific research, design, and healthcare. Clearly, automation of routine tasks effectively creates efficiencies and increased productivity, allowing human agents to engage in more complex cognitive tasks [2]. Furthermore, AI tools (especially explainable and collaborative ones) promise democratized access to expertise that is normally available only to specialists. They allow non-experts to use simple interfaces and adaptive explanations to access knowledge that would not otherwise be attainable. For example, decision support opportunities may now be available clinicians in remote locations and educators in contexts that may be considered deprived compared to elite academic institutions. Collectively, these tools will empower societies to deal with complex and multi-layered issues and challenges, including climate resilience, global health emergencies, and policy assessment, where no one actor or model has the answers [3, 15, 18]. Realizing these and other benefits will depend on the coordinated actions of many people and organizations across relevant stakeholder groups [14]. Researchers need to build interpretable, human-compatible models. Developers need to design systems that engage the user, establish trust, and offer control [5]. Regulators need to create enforceable regulations that give public value while encouraging innovation. End-users (including professionals and the public) need to critically engage with the outputs of AI systems [16]. Society also has a role in establishing expectations, ethical norms, and other dimensions of collective action, ensuring that collaborative AI is structured to be accountable, equitable, and positive [12, 56].

To take advantage of human-AI collaboration opportunities and properly mitigate potential risks requires bold action: the design and development of AI systems will need to be human-centered, explainable, and governable. Explainability is essential because users must understand system outputs and therefore have a reasonable trust in those outputs, especially in high-stakes fields like health care and finance [14, 15]. Governability, which is the ability to monitor, audit, and intervene over a system is equally important to trustworthiness; stakeholders expect organizations to act ethically, legally, and professionally through their engagement [69, 73]. This collective framework calls for engagement in interdisciplinary collaboration: technologists need to partner with knowledgable experts, those working in ethics and/ or policy, to create systems that are both novel and accountable. The future of AI is not the erasure of human capability, but the reconceptualization of what we can do, together with machines. A sustainable AI future is one which enables collaboration, not solely automation. When AI operates within approaches to explainability, accountability, and shared control. As Topol [18] and Makarius et al. [3] Articulate, AI should be a facilitator for empowerment, not a displacer. This is a vision of a sustainable AI future where we continue to work together vigilantly, adapting regulation and policy, and commitment to inclusivity to ensure that we stay at the center of AI (as well as human dignity, agency, and well-being). Creating such a future is not up for debate; it is necessitated.

7 Conclusion

To summarize, the introduction of large language models into healthcare is both promising and challenging. This chapter has shown that advancing explainable AI is not just a technological issue, but a challenge situated at the intersection of collaboration, adaptation, and governance. Human-AI collaboration must be understood in a broader sense than automation alone, involving the redefinition of responsibility between humans and AI and the trust that arises from shared decision-making. Adaptive explanations are needed to ensure that users can understand how algorithms shape performance, whether through transparency or accountability in collaborative workflows. Regulation, in turn, should be seen not as a constraint but as a framework that guarantees lawful, equitable, and ethical accountability. Taken together, these insights emphasize that explainable AI in healthcare is not a final destination but an evolving practice that must continually align technology, human actors, and governance systems. This chapter has brought these dimensions into focus, offering a foundation for thinking about how explainable AI can be responsibly integrated into high-stakes healthcare settings.

References

1. Lai, Y., A. Kankanhalli, and D. Ong, *Human-AI collaboration in healthcare: A review and research agenda.* 2021.
2. Tschandl, P., et al., *Human–computer collaboration for skin cancer recognition.* Nature medicine, 2020. **26**(8): p. 1229–1234.
3. Makarius, E.E., et al., *Rising with the machines: A sociotechnical framework for bringing artificial intelligence into the organization.* Journal of business research, 2020. **120**: p. 262–273.
4. Parasuraman, R., T.B. Sheridan, and C.D. Wickens, *A model for types and levels of human interaction with automation.* IEEE Transactions on systems, man, and cybernetics-Part A: Systems and Humans, 2000. **30**(3): p. 286–297.
5. Okamura, K. and S. Yamada, *Adaptive trust calibration for human-AI collaboration.* Plos one, 2020. **15**(2): p. e0229132.
6. Yue, B. and H. Li, *The impact of human-AI collaboration types on consumer evaluation and usage intention: a perspective of responsibility attribution.* Frontiers in psychology, 2023. **14**: p. 1277861.
7. Woelfle, T., et al., *Benchmarking Human–AI collaboration for common evidence appraisal tools.* Journal of Clinical Epidemiology, 2024. **175**: p. 111533.
8. Rouse, W.B., N.D. Geddes, and R.E. Curry, *An architecture for intelligent interfaces: Outline of an approach to supporting operators of complex systems.* Human-computer interaction, 1987. **3**(2): p. 87–122.
9. Arias-Rosales, A., *The perceived value of human-AI collaboration in early shape exploration: An exploratory assessment.* PloS one, 2022. **17**(9): p. e0274496.
10. Chakravorti, T., et al. *Artificial prediction markets present a novel opportunity for human-AI collaboration.* in *Proceedings of the 22nd International Conference on Autonomous Agents and MultiAgent Systems (AAMAS 2023).* 2023. London, United Kingdom: International Foundation for Autonomous Agents and Multiagent Systems (IFAAMAS).
11. Cao, S., et al., *From man vs. machine to man+ machine: The art and AI of stock analyses.* Journal of Financial Economics, 2024. **160**: p. 103910.
12. Wilson, H.J. and P.R. Daugherty, *Collaborative intelligence: Humans and AI are joining forces.* Harvard business review, 2018. **96**(4): p. 114–123.
13. Noble, S.M., et al., *The Fifth Industrial Revolution: How harmonious human–machine collaboration is triggering a retail and service [r] evolution.* Journal of Retailing, 2022. **98**(2): p. 199–208.
14. Amann, J., et al., *Explainability for artificial intelligence in healthcare: a multidisciplinary perspective.* BMC medical informatics and decision making, 2020. **20**(1): p. 310.
15. Holzinger, A., et al., *Causability and explainability of artificial intelligence in medicine.* Wiley Interdisciplinary Reviews: Data Mining and Knowledge Discovery, 2019. **9**(4): p. e1312.
16. Guidance, W., *Ethics and governance of artificial intelligence for health.* World Health Organization, 2021.
17. AI, N., *Artificial intelligence risk management framework (AI RMF 1.0).* https://nvlpubs.nist.gov/nistpubs/ai/nist.ai, 2023: p. 100–1.
18. Topol, E., *Deep medicine: how artificial intelligence can make healthcare human again.* 2019: Hachette UK.
19. Senoner, J.S., S.; Kratzwald, B.; Feuerriegel, S.; Netland, T, *Explainable AI improves task performance in human–AI collaboration.* Scientific Reports, 2024. **14**(1): p. 31150.
20. McGrath, M.J., P.S. Cooper, and A. Duenser, *Users do not trust recommendations from a large language model more than AI-sourced snippets.* Frontiers in Computer Science, 2024. **6**: p. 1456098.
21. Chen, Z., Y. Luo, and M. Sra, *Engaging with AI: How Interface Design Shapes Human-AI Collaboration in High-Stakes Decision-Making.* arXiv preprint arXiv:2501.16627, 2025.
22. Ashktorab, Z., et al. *Trust and Reliance in Evolving Human-AI Workflows (TREW).* in *CHI '24: CHI Conference on Human Factors in Computing Systems.* 2024. Honolulu HI USA: Association for Computing Machinery, New York, NY, United States.

23. Wynne, K.T. and J.B. Lyons, *An integrative model of autonomous agent teammate-likeness.* Theoretical Issues in Ergonomics Science, 2018. **19**(3): p. 353–374.
24. Holzinger, A. *From machine learning to explainable AI.* in *2018 world symposium on digital intelligence for systems and machines (DISA).* 2018. IEEE.
25. 26.Shin, D., *The effects of explainability and causability on perception, trust, and acceptance: Implications for explainable AI.* International journal of human-computer studies, 2021. **146**: p. 102551.
26. Xu, W. and Z. Gao, *Applying HCAI in developing effective human-AI teaming: A perspective from human-AI joint cognitive systems.* interactions, 2024. **31**(1): p. 32–37.
27. Chiriatti, M., et al., *System 0: Transforming Artificial Intelligence into a Cognitive Extension.* Cyberpsychology, Behavior, and Social Networking, 2025.
28. Seo, S., et al. *Towards an AI coach to infer team mental model alignment in healthcare.* in *2021 IEEE Conference on Cognitive and Computational Aspects of Situation Management (CogSIMA).* 2021. IEEE.
29. Liu, J. and Y. Zhao. *Role-oriented task allocation in human-machine collaboration system.* in *2021 IEEE 4th International Conference on Information Systems and Computer Aided Education (ICISCAE).* 2021. IEEE.
30. Kong, X., et al., *Explainable interface for human-autonomy teaming: a survey.* arXiv preprint arXiv:2405.02583, 2024.
31. Diebel, C., et al., *When AI-Based Agents Are Proactive: Implications for Competence and System Satisfaction in Human–AI Collaboration.* Business and Information Systems Engineering, 2025: p. 1–20.
32. 33.Miller, T., *Explanation in artificial intelligence: Insights from the social sciences.* Artificial Intelligence, 2019. **267**: p. 1–38.
33. Romeo, G. and D. Conti, *Exploring automation bias in human–AI collaboration: a review and implications for explainable AI.* AI & SOCIETY, 2025: p. 1–20.
34. 35.Marusich, L.R., et al. *Trust Calibration for Joint Human/AI Decision-Making in Dynamic and Uncertain Contexts.* in *International Conference on Human-Computer Interaction.* 2025. Springer.
35. 36.Reyes, J., A.U. Batmaz, and M. Kersten-Oertel, *Trusting AI: does uncertainty visualization affect decision-making?* Frontiers in Computer Science, 2025. **7**: p. 1464348.
36. 37.Belardinelli, A., *Gaze-based intention estimation: principles, methodologies, and applications in HRI.* ACM Computing Surveys, 2024. **13**(3): p. 1–30.
37. Chauhan11, A., *Human-AI Collaboration: Exploring Synergies and Future Directions.* 2024.
38. Li, M.M., et al., *One Patient, Many Contexts: Scaling Medical AI Through Contextual Intelligence.* arXiv preprint arXiv:2506.10157, 2025.
39. 40.Belardinelli, A., *Gaze-based intention estimation: principles, methodologies, and applications in HRI.* ACM Transactions on Human-Robot Interaction, 2024. **13**(3): p. 1–30.
40. 41.Procter, R., P. Tolmie, and M. Rouncefield, *Holding AI to account: challenges for the delivery of trustworthy AI in healthcare.* ACM Transactions on Computer-Human Interaction, 2023. **30**(2): p. 1–34.
41. Lee, W., J. Park, and C.H. Park. *Acceptability of tele-assistive robotic nurse for human-robot collaboration in medical environment.* in *HRI '18: ACM/IEEE International Conference on Human-Robot Interaction.* 2018. Chicago IL USA: Association for Computing Machinery, New York, NY, United States.
42. Bao, Y., et al., *Investigating the relationship between AI and trust in human-AI collaboration.* 2021.
43. 44.Jain, R., N. Garg, and S.N. Khera, *Effective human–AI work design for collaborative decision-making.* Kybernetes, 2022. **52**(11): p. 5017–5040.
44. 45.Chhibber, N., J. Goh, and E. Law, *Teachable conversational agents for crowdwork: Effects on performance and trust.* Proceedings of the ACM on Human-Computer Interaction, 2022. **6**(CSCW2): p. 1–21.
45. Schelble, B.G., et al., *Let's think together! Assessing shared mental models, performance, and trust in human-agent teams.* Proceedings of the ACM on Human-Computer Interaction, 2022. **6**(GROUP): p. 1–29.

46. 47.Seeber, I., et al., *Machines as teammates: A research agenda on AI in team collaboration.* Information & management, 2020. **57**(2): p. 103174.
47. 48.Hancock, J.T., M. Naaman, and K. Levy, *AI-mediated communication: Definition, research agenda, and ethical considerations.* Journal of Computer-Mediated Communication, 2020. **25**(1): p. 89–100.
48. Benke, I., *Towards design principles for trustworthy affective chatbots in virtual teams.* 2020.
49. 50.Cardon, P.W., H. Ma, and C. Fleischmann, *Recorded business meetings and AI algorithmic tools: Negotiating privacy concerns, psychological safety, and control.* International Journal of Business Communication, 2023. **60**(4): p. 1095-1122.
50. Fleischmann, C., P. Cardon, and J. Aritz, *Acceptance of speech-to-text technology: Exploring language proficiency and psychological safety in global virtual teams.* 2021.
51. De Silva, C., T. Halloluwa, and D. Vyas, *A Multi-Layered Research Framework for Human-Centered AI: Defining the Path to Explainability and Trust.* arXiv preprint arXiv:2504.13926, 2025.
52. 53.Zhao, H., et al., *Explainability for large language models: A survey.* ACM Transactions on Intelligent Systems and Technology, 2024. **15**(2): p. 1–38.
53. Xu, C., et al., *Benchmark data contamination of large language models: A survey.* arXiv preprint arXiv:2406.04244, 2024.
54. Wu, X., et al., *Usable XAI: 10 strategies towards exploiting explainability in the LLM era.* arXiv preprint arXiv:2403.08946, 2024.
55. Zhu, Z., et al. *Explanation in the era of large language models.* in *Proceedings of the 2024 Conference of the North American Chapter of the Association for Computational Linguistics: Human Language Technologies (Volume 5: Tutorial Abstracts).* 2024. Mexico City, Mexico: Association for Computational Linguistics.
56. 57.Buçinca, Z., M.B. Malaya, and K.Z. Gajos, *To trust or to think: cognitive forcing functions can reduce overreliance on AI in AI-assisted decision-making.* Proceedings of the ACM on Human-computer Interaction, 2021. **5**(CSCW1): p. 1–21.
57. 58.Küper, A., et al., *Psychological factors influencing appropriate reliance on ai-enabled clinical decision support systems: experimental web-based study among dermatologists.* Journal of Medical Internet Research, 2025. **27**: p. e58660.
58. 59.Nourani, M., et al., *On the importance of user backgrounds and impressions: Lessons learned from interactive AI applications.* ACM Transactions on Interactive Intelligent Systems, 2022. **12**(4): p. 1–29.
59. 60.Naiseh, M., et al., *How the different explanation classes impact trust calibration: The case of clinical decision support systems.* International Journal of Human-Computer Studies, 2023. **169**: p. 102941.
60. 61.Mersha, M.A., M.G. Yigezu, and J. Kalita, *Evaluating the Effectiveness of XAI Techniques for Encoder-Based Language Models.* Knowledge-Based Systems, 2025. **310**: p. 113042.
61. Jacovi, A. and Y. Goldberg, *Towards faithfully interpretable NLP systems: How should we define and evaluate faithfulness?* arXiv preprint arXiv:2004.03685, 2020.
62. Adedokun, A., *Global ai regulatory landscape challenges, trends, and future outlook.* Trends, and Future Outlook (February 29, 2024), 2024.
63. Musch, S., M. Borrelli, and C. Kerrigan, *The EU AI Act: A comprehensive regulatory framework for ethical AI development.* Available at SSRN 4549248, 2023.
64. Diab, R., *Too dangerous to deploy? the challenge language models pose to regulating ai in canada and the eu.* University of British Columbia Law Review, Forthcoming, 2024.
65. Lee, D., C. Todorova, and A. Dehghani. *Ethical Risks and Future Direction in Building Trust for Large Language Models Application under the EU AI Act.* in *Proceedings of the 2024 Conference on Human Centred Artificial Intelligence-Education and Practice.* 2024. Naples Italy Association for Computing Machinery, New York, NY, United States.
66. Veale, M. and F. Zuiderveen Borgesius, *Demystifying the Draft EU Artificial Intelligence Act—Analysing the good, the bad, and the unclear elements of the proposed approach.* Computer Law Review International, 2021. **22**(4): p. 97–112.

67. Voigt, P. and A. Von dem Bussche, *The eu general data protection regulation (gdpr)*. A practical guide, 1st ed., Cham: Springer International Publishing, 2017. **10**(3152676): p. 10–5555.
68. 69.Feretzakis, G., et al., *GDPR and Large Language Models: Technical and Legal Obstacles*. Future Internet, 2025. **17**(4): p. 151.
69. 70.Cavalcante Siebert, L., et al., *Meaningful human control: actionable properties for AI system development*. AI and Ethics, 2023. **3**(1): p. 241–255.
70. Santoni de Sio, F. and J. Van den Hoven, *Meaningful human control over autonomous systems: A philosophical account*. Frontiers in Robotics and AI, 2018. **5**: p. 323836.
71. 72.Wallach, W., *Implementing moral decision making faculties in computers and robots*. Ai & Society, 2008. **22**(4): p. 463–475.
72. 73.Robbins, S., *A misdirected principle with a catch: explicability for AI*. Minds and Machines, 2019. **29**(4): p. 495–514.
73. 74.Reddy, S., et al., *A governance model for the application of AI in health care*. Journal of the American medical informatics association, 2020. **27**(3): p. 491–497.
74. 75.Kalodanis, K., P. Rizomiliotis, and D. Anagnostopoulos, *European artificial intelligence act: an AI security approach*. Information & Computer Security, 2024. **32**(3): p. 265–281.
75. 76.Jowarder, R.A., *The Ethics of AI Decision-Making: Balancing Automation, Explainable AI, and Human Oversight*. International Journal of Science and Research Archive, 2025. **14**(3): p. 435–443.
76. Holzinger, A., K. Zatloukal, and H. Müller, *Is human oversight to AI systems still possible?* 2025, Elsevier. p. 59–62.
77. Capasso, M., et al., *On the Right to Work in the Age of Artificial Intelligence: Ethical Safeguards in Algorithmic Human-Resource Management*. 2024.
78. 79.Torkamaan, H., et al., *The role of human-centered ai in user modeling, adaptation, and personalization—models, frameworks, and paradigms*, in *A Human-Centered Perspective of Intelligent Personalized Environments and Systems*. 2024, Springer. p. 43–84.
79. Feretzakis, G. and V.S. Verykios, *Trustworthy AI: Securing sensitive data in large language models*. AI, 2024. **5**(4): p. 2773–2800.
80. 81.Kasneci, E., et al., *ChatGPT for good? On opportunities and challenges of large language models for education*. Learning and individual differences, 2023. **103**: p. 102274.
81. Hulsen, T., *Explainable artificial intelligence (XAI): concepts and challenges in healthcare*. Ai, 2023. **4**(3): p. 652–666.

Correction to: Role of Explainable AI

Amirfarhad Farhadi, Atoosa Hosseinian, Azadeh Zamanifar, and Amir Taherkordi

Correction to:
Chapter 2 in: A. Zamanifar et al. (eds.), *Explainable Large Language Models in Healthcare Applications*, Bio-IT and AI, https://doi.org/10.1007/978-3-032-15088-2_2

The book was inadvertently published with chapter author's incorrect Given name. This information has been updated from "Elahe Hosseinian" to "Atoosa Hosseinian" in the initially published version of chapter "2".

The updated version of this chapter can be found at
https://doi.org/10.1007/978-3-032-15088-2_2

A. Zamanifar et al. (eds.), *Explainable Large Language Models in Healthcare Applications*, Bio-IT and AI, https://doi.org/10.1007/978-3-032-15088-2_13

The manufacturer's authorised representative in the EU is Springer
Nature Customer Service Centre GmbH, Europaplatz 3, 69115 Heidelberg,
Germany. If you have any concerns regarding our products, please
contact ProductSafety@springernature.com

Printed and bound by CPI Group (UK) Ltd, Croydon, CR0 4YY
17/09/2026
02218456-0012